Exploring Logical Dynamics

Studies in Logic, Language and Information

Studies in Logic, Language and Information is the official book series of the European Association for Logic, Language and Information (FoLLI).

The scope of the series is the logical and computational foundations of natural, formal, and programming languages, as well as the different forms of human and mechanized inference and information processing. It covers the logical, linguistic, psychological and information-theoretic parts of the cognitive sciences as well as mathematical tools for them. The emphasis is on the theoretical and interdisciplinary aspects of these areas.

The series aims at the rapid dissemination of research monographs, lecture notes and edited volumes at an affordable price.

Exploring Logical Dynamics

Johan van Benthem

CSLI Publications
Center for the Study of Language and Information
Stanford, California
&
FoLLI
The European Association for
Logic, Language and Information

Center for the Study of Language and Information
Leland Stanford Junior University
Printed in the United States
00 99 98 97 96   5 4 3 2 1

Library of Congress Cataloging-in-Publication Data

Benthem, J. F. A. K.van, 1949–
Exploring logical dynamics / Johan van Benthem.
p. cm. - (Studies in logic, language and information)
Includes bibiliographical references.
ISBN 1-57586-059-7(alk. paper). — ISBN 1-57586-058-9 (pbk. : alk. paper)
1. Logic. I. Title. II. Series.
BC71.BC76 1996
121—dc20
96-23338
CIP

Contents

Introduction

In recent years, there has been a growing interest in the logical structure of cognitive actions underlying human reasoning or natural language understanding. Traditional logic and philosophy have been mainly concerned with the products of these actions, such as thoughts, proofs, propositions. But in various disciplines, including philosophy, computer science and linguistics, the mechanisms of information flow themselves are becoming primary objects of study. This interest reflects a broader cultural influence of computational paradigms, emphasizing not the hardware of Turing Machines, but the epistemic software: what are the 'programs for the mind'? The broad purpose of this Book is to convey this perspective, to develop its logical structure, and to demonstrate the many surprising repercussions of the dynamic stance — even for apparently settled standard parts of logic. The result may be called a dynamic logic of information flow. Even so, we shall only scratch the surface, mostly ignoring such aspects as many-person information exchange, the interplay between cognitive and physical actions, or the alternative action structure in logical proofs or games.

Successful paradigms have many parents, while failures tend to have few. The dynamic turn in logical semantics is foreshadowed in the work of philosophical logicians, linguists and cognitive scientists in the 70s and 80s, it is currently flourishing in the 'Dutch School' of philosophers of language, logicians and computer scientists, and it is still finding new adherents in new fields. The list can be extended, and we strive for due credits throughout. But the new age of one part of science may be the received wisdom of another. Dynamic systems navigating through state spaces are the heart of physics. Update actions on belief functions underlie the foundations of probability — and also, the days are not long past when the activity of language games dominated the philosophy of language. Still, there are many new twists to our proposed paradigm, partly influenced by 'dynamic logic' (broadly conceived) as found in computer science.

This book documents some five years of research into these matters. It presents a technical framework which includes spaces of dynamic operators, varieties of dynamic inference, different abstraction levels for processes, as well as expressive and deductive complexity for dynamic architectures. This framework is developed by techniques from modal logic which return throughout. These include labeled transition systems (i.e., possible worlds models), bisimulations between these, frame correspondences for dynamic axioms, translations into standard logics, and fine-structure hierarchies of formalisms measuring expressive power. We prove many mathematical results, such as theorems on semantic preservation, functional completeness and axiomatization. Our inspiration comes from the semantics of natural language, and from universal algebra, mathematical linguistics, logic programming and process theories in computer science. Our narrower aim is threefold. We try to sensitize the reader to the new dynamic way of viewing things. We try to convey the sweep and elegance of its logical theory. And we show how all this affects a variety of disciplines, from computer science to philosophy. Even so, this is not an elementary textbook. We presuppose familiarity with first-order model theory, as well as fundamentals of modal logic and universal algebra. (A crash course for Platonic learners is provided in Chapter 3.) Moreover, we merely emphasize main issues, techniques and results — while providing references to a fast-growing literature (often in the form of recent dissertations) for elaboration and applications.

During the preparation of this book, I have been fortunate in living in a stimulating international environment, both at the *Institute for Logic, Language and Computation* (Amsterdam) and the *Center for the Study of Language and Information* (Stanford). In particular, I have profited from many fruitful and pleasant collaborations on talks, papers, books and dissertations. Therefore, I wish to dedicate this book in gratitude to all my co-authors and Ph.D. students who have been directly involved in its genesis: *Natasha Alechina*, *Hajnal Andréka*, *Jon Barwise*, *Martin van den Berg*, *Jan Bergstra*, *Giovanna Cepparello*, *Alla Frolova*, *Willem Groeneveld*, *Marco Hollenberg*, *Jan Jaspars*, *Marianne Kalsbeek*, *Megumi Kameyama*, *Makoto Kanazawa*, *Natasha Kurtonina*, *Maarten Marx*, *Wilfried Meyer Viol*, *Szabolcs Mikulás*, *Reinhard Muskens*, *István Németi*, *Maarten de Rijke*, *Edith Spaan*, *Vera Stebletsova*, *Yde Venema*, *Albert Visser*. In particular, several joint results from an Amsterdam-Budapest cooperation with Andréka and Németi are acknowledged separately in Chapter 4, and the same holds for results from an Amsterdam-Amsterdam cooperation with Bergstra in Chapter 10. In addition, Maarten de Rijke and Dikran Karagueuzian have provided guidance and editorial assistance far beyond the call of duty, with the help of Maureen Burke and Tony Gee.

All people listed have given me solid support and bona fide inspiration. The subsequent irresponsible and esoterical twists in this book I must claim as all my own...

Amsterdam/Stanford, March 1996
Johan van Benthem

Part I

The Dynamic Turn

The general perspective of this book has emerged out of several recent developments in logical semantics. Their shared concern is to put cognitive actions and procedures at centre stage, rather than static representation structures. We are going to present a logical account of this paradigm, which has arisen partly from reflection on existing systems in the literature, and partly from a priori speculation about information flow. This exemplifies the usual interplay between logical theory and descriptive practice. Moreover, there is a matter of appropriate strategic depth. The present level of abstraction high-lights similarities across many different fields (linguistics, computer science, philosophy) that might not be so apparent otherwise. In this first part, we set out our philosophical motivations, and explore some consequences of the dynamic stance (Chapter 1). Next comes a technical tour along a number of current dynamic systems updating such diverse things as information states, variable assignments or preferences, which modify standard extensional and intensional logics (Chapter 2). Finally, we review the relevant mathematical tools, coming from first-order logic, modal logic, relational algebra and dynamic logic (Chapter 3). These will be essential in constructing the general theory of dynamics found in Part II of this book.

1

Cognitive Actions

In this first chapter, we convey the flavour of the dynamic viewpoint on cognitive information flow, we identify its immediate repercussions for the agenda of logic, and we give a sense of the main issues it engenders. In the course of this discussion, there emerge the key topics included (and those omitted) in the rest of this book.

1.1 Recognizing Cognitive Processes

1.1.1 From Products to Processes

Many cognitive notions have a dual character. For instance, *judgment* stands for both an intellectual action and the content of that action, and likewise *reasoning* denotes both an intellectual process and its products. This interplay between static contents and dynamic actions also occurs in our ordinary use of the term 'natural language' — which can be either a static mathematical structure of words and rules, or a dynamic social activity with systematic conventions not necessarily encoded in explicit syntax. In mainstream logic and linguistics, static aspects have been predominant, witness the emphasis on 'truth conditions' for declarative sentences, describing some situation. Here, Boolean propositional structure is paramount, with logical operators creating complex forms of description, such as negation, conjunction or disjunction. Thus, the emphasis lies on "that" or "whether" statements are true about a situation, not so much on "how" they are seen to be true. To some, this declarative stance, as opposed to procedural views, is a hall-mark of logical analysis as such. But human cognitive competence consists in procedural facility, as much as communion with eternal truths. And the logical structure of cognitive actions shows many analogies with other kinds of activity, including computation or physical games. This new perspective does not invalidate the old. It says that we need a logical account of cognitive "knowledge how" in addition to "knowledge that". But the two aspects of cognition occur intertwined. This is clear with many activities. Computation involves both instructions to a machine and periodic tests whether static statements are true or false about the current state. Examples are instructions like "if P then A else B" or "while P do A" in computer programs. Likewise, physical games involve both possible movements (such as 'kicking a ball') and tests whether some standard statement is true ("the ball is outside the lines"), which again determine which further actions are appropriate. Thus, our aim must be to explore the nature and consequences of this duality in an appropriate mathematical framework, thereby enriching the traditional agenda of Logic.

1.1.2 Dynamic Procedures in Natural Language

Many concrete phenomena in natural language turn out essentially dynamic in nature. For instance, linguistic *anaphora* involves changing references for pronouns ("he", "she", "it") in discourse, with reassignment of relevant objects (mostly introduced by quantified noun phrases) occurring throughout. This process is somewhat like the reassignment of values to local variables in execution of programs, and its dynamic structure has been brought out by a line of authors, using both explicit and implicit computational analogies. Another example is *temporal reference*. In narrative discourse, the focus of attention moves through a temporal space, indicated by various linguistic devices. This movement is triggered systematically by past, present and future verb tenses, temporal adverbs like "now", "then", or temporal connectives like "before", "since". At a somewhat higher discourse level, another long-standing phenomenon is that of linguistic *presupposition*, such as the unicity conveyed by definite descriptions, or the factual presuppositions of verbs. A descriptively adequate account of changing presuppositions across discourse involves the systematic computation of changing 'contexts of assertion'. Also, understanding the role of *conditional statements* "if A, then B" naturally involves an account of moves like jumping to a hypothetical A-world, or in more recent default theories, changing one's preferences among worlds confirming or refuting this regularity. Finally, dynamic flow of information is evident with speech acts like *questions*, whose purpose it is precisely to invite an update of the interlocutor's information state. The result is an appealing picture of *natural language as a programming language for cognition*. Understanding discourse amounts to navigation across a space of rich information states, either for single listeners or readers, or (the reality) combined states including mutual knowledge for groups of cognitive agents.

1.1.3 Computational Influences

The computational metaphor is not accidental. It reflects a growing influence of ideas and techniques from computer science in adjoining fields. In particular, several strands may be mentioned that are quite congenial to the above movement. First, there is the well-known Dynamic Logic of programs, which emerged from the Hoare correctness calculus of imperative programs. Dynamic logic describes the execution of programs across a state space, while allowing statements about the states traversed in the process. It has been applied in broader settings by many authors (cf. Chapters 3, 6). Then, there are recent theories of informational update, contraction and revision, which have become quite influential in computer science and artificial intelligence. These view propositions as dynamic transformations of information states (data bases, theories, sets of possible worlds), which enrich the current information, or modify it in other ways. Also in computational linguistics, data base updates are a standard tool for natural language processing. More generally, the dual cognitive perspective advocated here reflects the computational interaction between *representation* and *computation*, as being on an equal footing. There are still further sources for dynamic accounts of propositions. For instance, the philosophical literature on *language games* and *speech acts* has always been action-oriented, describing the intended affects of utterances. Also, much work in the foundations of probability has been oriented towards learning as computing *updates of probability functions* serving as information states.

1.2 Mathematical Process Models

1.2.1 Available Formal Paradigms

A movement like the above needs a broad mathematical framework, highlighting significant questions and analogies, which can serve as a laboratory for designing further applied systems. Such a role is played for intensional logic by possible worlds semantics, and for linguistic semantics by mathematical type theories. It is not yet clear what the eventual framework for dynamic semantics will be like. Competitors include mathematical category theory or universal algebra — and many people in the area even feel at the dawn of a new age, where the whole world is made anew, and hence some mysterious new form of mathematics will emerge. To some extent, this choice depends on the guiding *metaphor* for cognitive actions. There are several broad options for this purpose. One powerful metaphor is that of *games*, and hence of mathematical game theory. Games are a highly structured activity, with powerful concrete intuitions concerning moves and strategies. This view has been around as an undercurrent in logic for quite a while, emerging every so often, and it would be of interest to exploit it further. Another option is the paradigm of *proofs*. Proof is a cognitive activity by which we establish the truth of statements — and proof theory has developed a subtle account of proofs-as-procedures in the Curry-Howard-deBruyn isomorphism with typed lambda calculi. Yet another dynamic viewpoint arises with central algorithms for logical validity.

Notably, *semantic tableaus* provide an attractive account of cognitive actions of model-building and model-checking. Nevertheless, in this book, we opt for a different paradigm, being that of *programs*. We shall think of actions in some (computational) state space, triggered by texts viewed as imperative programs. This allows us to use a standard model of computer science, viz. 'labeled transition systems'. When all is said and done, these are just the possible worlds models of the philosophical tradition, now exploited to a higher degree.

1.2.2 Abstract Procedures over Labeled Transition Systems

Our model of dynamics is simply this: a system moves through a space of possibilities. Thus, there is a set S of relevant *states* (cognitive, physical, etcetera) and some family $\{R_a \mid a \in A\}$ of *binary transition relations* among them, corresponding to an 'atomic repertoire' of actions that can be performed to change from one state to another. Even such a simple model with labeled transitions between states displays a number of fundamental features. For instance, cognitive actions may be deterministic (producing unique output states) or indeterministic (one input — several outputs), total (always defined) or partial (not always defined). Moreover, special cases may be of independent interest. For instance, a fixed point where an action 'loops' is a state where the update does not yield new information: intuitively, the state already verified the propositional content of the update. Here are some pictures for these situations.

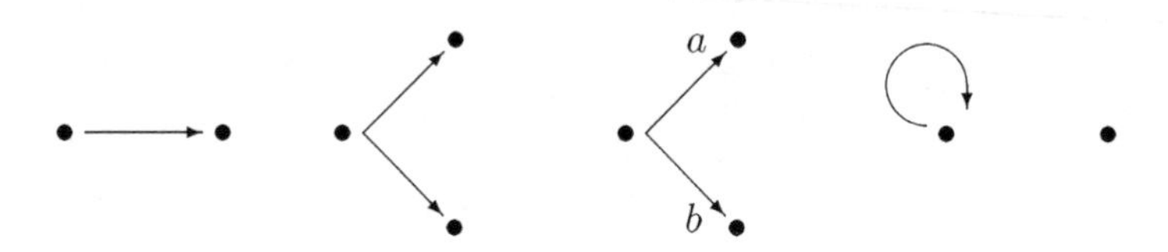

Procedural aspects of action or cognition concern the ways in which atomic transitions can be combined to obtain further desired state changes (the *procedural repertoire*). Think of a complex instruction manual for installing a smoke detector, or a computer program guiding some computation, or a linguistic text as it modifies the information state of its reader. Classical descriptive truth can be encoded by a valuation function V for standard propositional atoms over states, just as in possible worlds semantics. 'Logic' now enters as a study of complex procedures on binary relational models and their general effects. This may be contrasted with the standard approach, which merely aims at description of *unary* propositional *properties* of states, which can be tested for current truth or falsity. The difference is reflected in the greater number of natural operators for constructing dynamic procedures, over and above the Boolean Algebra of sets of states. Even so, classical logic retains its value too, both as the useful limit case of test actions which do not change information states, and in a more abstract sense of being able to *reduce* the dynamic approach by technical 'translation' if the need arises.

1.2.3 Concrete Dynamic Systems

Specific dynamic systems will usually specify a more detailed set of states, supporting more concrete *basic actions*. The relevant atomic repertoires can have diverse forms, ranging from physical to cognitive ones:

real action	put the block on the table
playing games	move the queen
computation	assign some value to a register
information flow	update the current epistemic state

Relevant states can be concrete physical situations, or abstract procedural ones, or mixtures of both. In computational practice, constructive information states arise, allowing syntactic manipulation of their contents via deductive calculi or update algorithms. Such states include data bases, syntactic theories, semantic tableaus, or other concrete symbolic objects. Not only different types of activity induce different kinds of state. The same activity may be studied with different degrees of detail. For instance, in transition models, one can shift from individual states to *finite sequences* of states, being 'traces' of computational processes or physical activities. Single states may also have many components, recording different aspects of a process, as is the reality of natural language understanding.

1.3 Standard Logics Dynamified

A rich source of dynamic systems is reanalysis of standard logics. Semantic procedures often underlie their *motivation*, and they occasionally show up in didactic presentations. We mention three examples, to be considered at greater length in Chapter 2 below.

1.3.1 Dynamic Propositional Logic

The folklore account of information states takes them to be sets of alternative possible worlds (valuations, models). Such sets provide a range of uncertainty as to the actual world (which lies somewhere inside), and incoming propositions increase knowledge by reducing uncertainty via elimination of possibilities. A vivid illustration is the game "Master Mind", where a player must guess the exact position of a number of colored pegs. In each round, she is allowed one guess — which is evaluated with black marks for correct positions and white marks for correct colors placed at incorrect positions. In each round, all possibilities are eliminated which are incompatible with the additional information received. The aim is to arrive at one single situation.

guess	*answer*	*open options*
START		24
red, orange, white	• ◦	6
white, orange, blue	• ◦	2
blue, orange, red	◦ ◦	1

Example 1.1 (Master Mind) Consider a game with 4 colors and 3 pegs, where no colors may be repeated. Let the actual color arrangement be ⟨red, white, blue⟩, with a total color set also including orange. Here is a game table, with numbers of open possibilities indicated.

In actual games, players make more focused deductions. Thus, after the first round, we know that the color blue must occur. And after the second round, we know that red must occur, and that the white peg cannot have been in a correct position on the right. Such assertions represent more constructive information, which gets modified via an interplay of incoming external information and internal *deduction*. To represent this, other notions of information state are needed, such as deductive data bases or semantic networks. In the above example, the internal structure of propositions is still classical. In Chapter 2, we shall consider a more radical 'dynamification' of propositional logic.

1.3.2 Dynamic Modal Logic

Another point of departure from standard logic are the intuitionistic possible worlds models proposed by Kripke. Here, worlds are information states ordered by inclusion, and the intuitive picture is that of a cognitive agent traversing such states in the quest for knowledge. Intuitionistic logic refers to transitions in this information pattern:

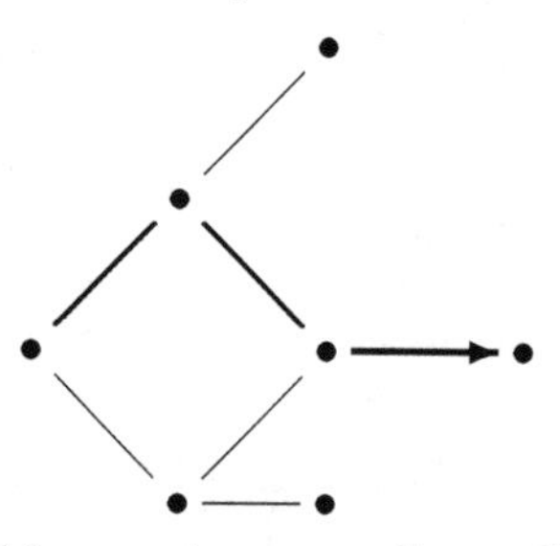

E.g., to see that $\neg\phi$ holds, one inspects all possible extensions of the current state for absence of ϕ. Again, this implicit dynamics may be made explicit by adding a system of cognitive transitions, such as informational 'updates' taking us to some minimal extension where a certain proposition has become true. Standard intuitionistic logic is a forward-looking system, of Dutch mathematicians who never forget and never err, whereas ordinary mortals will display a zigzagging travel pattern of cognitive advances — and also cognitive retreats, including 'downdates' and zigzagging 'revisions'. (Indeed, the ability to forget is a precondition for civilised human life.) Providing such an explicit dynamic system takes the original 'constructivism' to its logical conclusion.

1.3.3 Dynamic Predicate Logic

Consider Tarski's well-known truth definition for a formula ϕ in a first-order model $\mathbf{M} = (D, I)$ under some variable assignment s. Its atomic clauses involve static tests whether some fact obtains. But intuitively, there is also a clear procedural flavour. The clause for an existential quantifier $\exists x$ tells us to shift assignment values for x, searching until some verifying object d is found:

$$\mathbf{M}, s \models \exists x\, \phi \text{ iff there exists some } d \in D \text{ with } \mathbf{M}, s(x|d) \models \phi.$$

Thus, intuitively, the existential quantifier is tied up with a computational process of 'random assignment'. We can make the latter process explicit, by assigning to each formula ϕ a binary relation $s_{\mathrm{in}}[\![\phi]\!]s_{\mathrm{out}}$ between assignments encoding those transitions that result in successful verification of ϕ. We shall present a concrete system to this effect in Chapter 2, inspired by analogies with Hoare semantics for imperative programs in terms of input-output relations on assignments to their program variables. The first-order truth definition has even further dynamic aspects. 'Tarskian Variations' of domains D, interpretation functions I and formulas ϕ are found in Chapters 2, 12.

1.4 Propositions as Procedures

The preceding state-transition model is extremely poor. Nevertheless, even at this high level of abstraction, one can already highlight many intriguing questions. Let us consider what happens to the central concerns of logic when we take the dynamic turn.

1.4.1 From Truth Conditions to Update Conditions

When propositions ϕ become cognitive procedures that change states, the standard semantic truth format $\mathbf{M}, s \models \phi$ no longer works, and standard truth conditions for logical operators and other expressions get out of focus. We need *update conditions* instead, defining the processes (binary transition relations between states) denoted by complex propositions, telling us how their components contribute to this effect. It is even unclear what unary 'truth' at a state s would mean for a cognitive procedure P — although we may get by with the earlier-mentioned states that are *fixed points* for P. In this connection, a natural conceptual concern is the following. What are the cognitive processes that we want to study? Can we delimit them any more closely than as just some family of binary transition relations? For this, we need some notion of *process equivalence*, providing a criterion of identity that tells us when two different transition models represent one and the same process. Such notions are largely absent from the literature on dynamic semantics — but we shall take our cue from computer science, viz. the theory of bisimulation (see Chapters 4, 5). To a first approximation, we shall study only those process operations whose update conditions are 'safe' for bisimulation. But there are other broad computational issues involved. Are cognitive processes sequential or parallel, or both? We will concentrate mainly on the former — although we feel that actual cognitive processes can be partly parallel, too.

1.4.2 Logical Constants

What do the familiar logical constants mean when applied to procedures? For some of them, this seems obvious. A *conjunction* $P \wedge Q$ is naturally read as the sequential composition of the two actions P and Q. Of course, its behaviour will change then, as composition is no longer commutative. A *disjunction* $P \vee Q$ is already somewhat more complicated, although the computational analogy with programs might suggest an action of indeterministic choice. But, e.g., a procedural interpretation for a *negation* $\neg P$ is much less obvious (refraining?, avoiding?) — and we shall see different technical proposals for its proper treatment later on. In any case, this way of stating things is overly conservative. There is no need to assume that classical logical constants have one unique meaning in the richer dynamic perspective. For instance, computationally, there are different natural 'conjunctions'. In addition to sequential composition, there are forms of *parallel merge* $P \parallel Q$ for actions without any order dependence. In classical logic, such finer distinctions were invisible: all conjunctions collapsed into the Boolean one. Moreover, there may be natural dynamic operators that lack a classical counterpart altogether. One example is *reversal* $P^{\smile}$: running procedure P backwards. Thus, we find a new space of logical constants for constructing complex procedures. One of our main interests will be to chart this terra incognita, and its most attractive inhabitants. The eventual aim here is a stable vocabulary of logical 'control operators' for dynamic procedures comparable in power and elegance to the classical core.

All examples so far are propositional in nature. But there are also dynamic counterparts to standard quantifiers. With the latter, however, a further subtlety emerges. From a dynamic stance, quantifiers are different in character from logical connectives. Indeed, they need not be thought of as logical constants at all — but rather as atomic instructions for performing some specific action over states. In Section 2.1, an existential quantifier $\exists x$ is read as a 'random assignment' ($x :=?$), and in Section 2.4, a modality $\Diamond$ by itself denotes a 'random upward jump' in an inclusion ordering over information states. In this way, quantifiers and modalities become more like 'active counterparts' for propositional atoms, whose corresponding actions are static tests on unchanging states.

1.4.3 Valid Inference

The dynamic turn has striking consequences for the archetypal inferential setting:

$$\frac{P_1 \ldots P_n}{C} \quad \text{conclusion } C \text{ follows from premises } P_1, \ldots, P_n.$$

The standard account of valid inference says that the conclusion must be true in all models where the premises are true ('transmission of truth'). But what is the sense of this explication when all propositions involved are dynamic procedures changing information states? When is such a proposition 'true' — and does truth remain the central notion anyway? Here, various options arise. As we said, one may call a proposition P true at a state s if P has a *fixed point* at s: sPs. (That is, updating with P does not have any effect on the information in s.) Then, the classical account may be maintained. But this still ignores the dynamic nature of actual arguments. The premises of an argument invite us to update our initial information state, and then, the resulting transition has to be checked to see whether it 'warrants' the conclusion (in some suitable sense). Two prominent candidates for such an account are as follows. The premise transition might be required to be a correct update for the conclusion also ('update-update consequence') — or, the resulting state might have to be a fixed point for the conclusion ('update-test consequence'). Whatever precise explication is chosen, the procedural viewpoint on validity will be quite unlike its classical counterpart. If a premise sequence is a complex instruction for achieving some cognitive effect, then its *presentation* will be crucial. The sequential order of premises matters, the multiplicity of their occurrence matters, and each premise move has to be relevant. And this will clash with the basic *structural rules* of standard logic (allowing us to disregard such aspects in classical reasoning). Think of meeting a date, where one has all the right moves available: flowers, tickets, sweet talking, kisses, and imagine the various ways in which successful seduction might fail by Permutation of actions, Contraction of identical actions, or Monotonic Insertion of arbitrary additional actions. Of course, in certain settings, deviations from classical reasoning will be slight — for instance, when all premise actions correspond to tests, or steady updates (upward in the information ordering on states). But in general cognition, our information may be more complex, brought about also by retractions ("no, forget about A after all") or qualifications ("unless B, that is"). And then, a more delicate dynamic logic becomes imperative.

1.4.4 Dynamic Architecture

Declarative statements and dynamic procedures both have strong motivations. There is no need to favor one over the other — say, through some a priori 'reduction'. Actual reasoning is a mixture of more dynamic sequential short-term processes and more static long-term ones, presumably over different representations. Thus, both kinds of system must co-exist. This requires, at least, the two-level logical architecture pictured here:

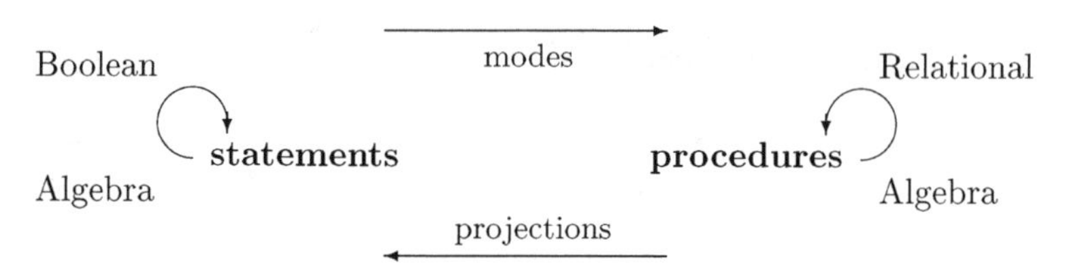

Here, connections between the two levels are essential components in their own right. *Modes* take standard propositions P to procedures having P for their informational content, such as 'updating' with P or 'testing' for P. Conversely, *projections* assign to each procedure R a standard proposition recording some essential feature of its action, such as the above fixed point operator describing those states where R is already satisfied, or the set-theoretic image of R, describing the possible outcomes of the latter action. Thus, dynamic architecture generates new kinds of logical operator, which may be analyzed in tandem with those combining statements or procedures. In particular, the use of projections is known from the computational literature. Intuitively, dynamic evaluation of ϕ moves to states where some static version of ϕ holds, while new values for its program variables have been set. For this 'static tracing', the following technique may be used. Imperative programs are often explained in terms of 'preconditions' and 'postconditions'. Starting from some set of states defined by an input predicate A, execution of a program π moves us to a new set of states (viz. the set-theoretic π-image of A), defined by the *strongest postcondition* of A under π ($\mathrm{SP}(A, \pi)$), which may often be computed recursively (Chapters 2, 6). More generally, if logical architecture becomes important, with many calculi working at the same time, then issues arise of 'logical management'. In particular, what are the mechanisms managing the exchange of information between components?

There is also another sense to the term 'architecture'. The above schema provides an abstract logical model for dynamic phenomena. But what is the actual modus operandi, both as to software and hardware, of human cognition? The present study does not tell. But at least, it provides an exact setting for raising more articulate questions concerning natural reasoning or language processing.

1.4.5 Problems of Complexity

Finally, there is another concern in the present perspective. The dynamic account of cognition seems closer to actual human performance than a static classical description. But then, a mathematical mismatch threatens. The richer universe of logical operators, inference and architecture outlined above comes at a price. Its mathematical paradigm is a relational algebra of procedures, which is known to be more complex than its Boolean counterpart (even becoming undecidable in some cases). There are various responses to this paradox. One can accept the inefficiency of human behaviour as an empirical fact. Or, dynamic logics might allow simpler *representation* of the data structures involved in cognition — making the total package more efficient than that of classical logic. In this book, we shall explore a different route — scrutinizing the innocuous 'standard tools' that go into the usual set-theoretic presentation of labeled transition systems. What we find then are weaker decidable 'logical cores' for relational algebra and even first-order predicate logic, which might serve just as well, and which will increase our sensitivity for (sometimes debatable) dynamic presuppositions in the standard tools of our trade. We shall look into these matters successively. Chapter 2 provides concrete illustrations from current systems of dynamic semantics. Chapter 3 reviews the necessary technical tools. And Part II contains the makings of a general dynamic logic in our broad sense. Finally, Part III elaborates a number of these issues for specific application areas.

This concludes our first tour of the dynamic paradigm as understood in this book.

1.5 Notes

Dynamic logic and semantics are 'in the air'. We already hinted at the fact that there are other sources than the ones we discussed, other mathematical paradigms than the one we employed, and richer research agendas than the one we have chosen to pursue. We mention a few of these issues here, to delimit the scope of our investigation, pointing at some relevant literature. More generally, each chapter of this book has a final section listing related research which goes into greater depth and width.

Historical Precursors. Sources for a dynamic cognitive paradigm abound. Even mathematicians and physicists have sought to transfer their models to language — although we do not follow up on that particular lead. The following suggestive quote is from Thom 1973:

"... We propose ... a geometric interpretation of language's function, and of traditional grammatical categories: noun, adjective, verb. While the noun is described by a potential well in the dynamics of mental activities, the verb is ... an oscillator in the unfolding space of a spatial catastrophe. ... Such a model may describe the emission and reception of meanings, as well as some facts in linguistic theory ... and ... it offers a dynamical interpretation of the tree of generative grammar."

Philosophical Motivation. More elaborate discussion of the dynamic paradigm is found in the General Dynamics program of van Benthem 1991a, 1991b, 1994b. A poster description for the resulting logic of information flow is van Benthem 1993b. Earlier scattered sources exist: cf. the suggestive account of information updates in Belnap 1977. A Fregean motivation for dynamic denotations in the foundations of mathematics is given in Moschovakis 1991. Much richer agendas exist in the philosophy of action and intentionality: cf. Harman 1985, Bratman 1987, 1991.

Linguistic Sources. Speech act theories highlight actions performed by language users. Notably, a question invites the hearer to update the current information state of the speaker. From the start, motivations for the logic of questions have been dynamic: cf. Harrah 1984. For this purpose, one can use information states partitioning the space of possibilities, treating true answers as updates refining these: cf. Groenendijk and Stokhof 1996. Dynamics of queries also occurs in the computational literature on databases (cf. the survey Kanellakis 1990 in the *Handbook of Theoretical Computer Science*). In linguistic pragmatics, dynamic phenomena have always been at centre stage: cf. Stalnaker 1972. But also in syntax, grammars based on rewrite rules and their corresponding automata have a dynamic character, as opposed, e.g., to static descriptions of their admissible grammatical analysis trees (cf. Chapter 12). A broad survey of the strands that fed into current systems of dynamic semantics may be found in the recent chapter Muskens et al. 1996 in the *Handbook of Logic and Language*.

Computational Sources. Dynamic logic in the computational tradition is surveyed in Harel 1984. See also the textbooks Goldblatt 1987, Harel et al. 1995. The basic source for the so-called AGM theory of update and revision is Gärdenfors 1988, an excellent up-to-date survey including connections with 'knowledge maintenance' in computer science and artificial intelligence is the chapter Gärdenfors and Rott 1995 in the *Handbook of Logic in Artificial Intelligence and Logic Programming*.

Psychological Connections. Our talk of cognitive actions suggests connections with cognitive psychology. These have not been explored yet — but we refer the reader to Clark 1996 for a comprehensive interdisciplinary treatment of language use.

Mathematical Alternatives. Logical *game theories* of a dynamic slant exist since the 50s. Hintikka 1973 has games for interpretation, Lorenzen and Lorenz 1979 for argumentation, and van Benthem 1988a presents a general survey (including also e.g., the Ehrenfeucht-Fraïssé model comparison games of Chapter 3). *Category-theoretic approaches* to cognitive dynamics are proposed in Gärdenfors 1988, MacNamara and Reyes 1994, Pratt 1994b ('Chu Spaces' as a bridge between relational algebra and linear logic) and Visser 1994a, 1994b (the 'Utrecht Program' in Muskens et al. 1996). A general plausibilistic approach to belief revision is Friedman 1996. A broad presentation of dynamic paradigms and a comparison of their peculiarities is in van Benthem 1993b.

Logical Relatives. *Epistemic logic* in the tradition of Hintikka 1962 is clearly related to the work presented here, even though its emphasis has been on indifference relations between epistemic states rather than cognitive updates. Connections are even stronger with its contemporary manifestations in the computational literature: cf. Fagin et al. 1995. Another related enterprise is the philosophical tradition on action logics, of which Belnap 1989 is a fair sample. As for newer logical accounts of information flow, we mention the *Channel Theory* of Barwise and Seligman 1996.

Proofs, Types and Labels. There is another broad algorithmic tradition in the logical literature, using the Curry-Howard-deBruyn isomorphism to develop assertions in tandem with (verification) actions. Van Benthem 1991b proposes a lambda calculus-based hierarchy of semantic procedures for natural language, Girard 1987 provides a new proof-theoretic account of computation. Recent richer proof formats of Gabbay 1992, Van Lambalgen 1991, Meyer Viol 1995 manipulate additional dynamic resources (labels, dependencies, epsilon term management) with computational, mathematical and linguistic applications. The dynamics of proofs shows interesting new cognitive actions — such as management of changing levels of assumption, dependency and commitment. (Cf. the connections between logic and argumentation in van Benthem 1996b.) One would like to understand the exact relation between the two paradigms, which meet in intuitionistic logic — an issue which we shall discuss briefly in Chapter 12.

Constructive Information States. Our state-transition model is fully abstract. But there are also more constructive information states allowing syntactic manipulation of their contents, via deductive calculi or update algorithms. The latter states include such diverse items as data bases (cf. Allen 1983), syntactic theories (cf. Ryan 1992, van Eijck 1993), semantic tableaus (Beth and Piaget 1966 discuss their psychological reality), or yet other concrete symbolic objects. Compare also the 'data semantics' of Veltman 1984, Landman 1986, as well as Gerbrandy 1994.

Cognitive Action and Real Action. Actual intelligent behaviour takes place in a changing physical environment. Thus, cognitive dynamics occurs intertwined with physical action (observation, movement, etc.), within real-time constraints. This double perspective of bounded rationality for human agents is studied by Halpern and Friedman 1994a, 1994b, Huang et al. 1996, Ramanujan 1996. Cf. also the connections between epistemic logic and non-monotonic logics, inspired by various strands from the philosophical tradition, explored by the Utrecht computer science group, documented in van der Hoek et al. 1994, 1995. Amongst other things, the latter publications deal with norms, actions, propositional attitudes and capabilities.

Many-Agent Settings. Nature is not the only other player on the cognitive stage. Cognition usually takes place in a larger context, with speakers (authors) and listeners (readers), or still larger group settings. (Our earlier dating example is also a typical two-person activity, with a more complex mathematics than was suggested.) Thus, information states are social constructs, and they may engender collective epistemic phenomena which need not make sense at an individual level — such as 'common knowledge'. We refer to an extensive literature, including Lewis 1969, Barwise 1988, Halpern and Moses 1985. Recent attempts to transpose these things into our brand of dynamic logic include Jaspars 1994, Borghuis 1994, Huang 1994 and Groeneveld 1995. These dissertations model both private and public information states, studying diverse attitudes (knowledge, belief, desire) as well as cognitive interactions ('believe on the authority of', 'learn from'). Even so, no generally accepted notion of collective update has been devised so far. The pioneering book Fagin et al. 1995 has many suggestive pointers — solving, e.g., the puzzle of the Muddy Children via simple elimination of collective states in multi-S5 Kripke models. (Here, the indifference relation for a child holds between those distributions of mud and cleanliness which it cannot visually distinguish.) But the simplicity of this example is not representative for updates with more complex epistemic assertions. Gerbrandy 1995 takes these issues further, linking them up with dynamic update semantics (cf. Chapter 2), using tools from modal logic and non-well-founded set theory. Despite all this exciting work, it seems fair to say that there is still an unsolved 'many-body-problem' in the logic of cognition.

2

Dynamification

In logic, new paradigms usually present themselves in the form of new systems, which compete with the old ones. But the dynamic perspective can also be seen as a mentality, suggesting a fresh look at existing logical systems, rather than their speedy overthrow. In particular, the motivation for standard logics often contains procedural elements visible in text book presentations — and one can make this implicit dynamics explicit. In that process, we also arrive at simple and concrete examples of dynamic systems. We shall present a number of these, and discuss some of their features — noting the broader issues that emerge for the agenda of our general dynamic logic in later chapters. These include the format of update conditions, the choice of proper logical constants, dynamic notions of validity, and useful interfaces with standard logics. Examples run through propositional logic, predicate logic, modal logic, as well as some less standard options. This survey also reveals recurrent patterns that motivate our eventual mathematical paradigm, based on relational algebra and modal logic.

2.1 Updating Information in Propositional Logic

Processing of informational content is an intuition even behind propositional logic, which may be brought out in several ways. In Chapter 1, we illustrated this with the propositional game Master Mind, which eliminates successive options from the player's range of uncertainty. Similar intuitions drive possible worlds semantics for intuitionistic or epistemic logic, which will return later on. Here we continue with a propositional system demonstrating the eliminative perspective carried through compositionally.

2.1.1 Eliminative Propositional Update Conditions

Consider the fine-structure of eliminative updates. What dynamic role is played, inductively, by propositional connectives? Here is a system from Veltman 1991, which is concerned with the 'deterministic' case where information states are sets of propositional valuations, and transition relations for formulas ϕ are functions $[\![\phi]\!]$ sending sets of propositional valuations to new sets of valuations. The following clauses are plausible dynamic counterparts for the classical truth conditions:

$$\begin{array}{rcll} [\![q]\!](X) & = & X \cap Q & \text{where } Q \text{ is the set of valuations verifying } q \\ [\![\phi \wedge \psi]\!](X) & = & [\![\psi]\!]([\![\phi]\!](X)) & \text{function composition} \\ [\![\phi \vee \psi]\!](X) & = & [\![\phi]\!](X) \cup [\![\psi]\!](X) & \text{Boolean union} \\ [\![\neg\phi]\!](X) & = & X - [\![\phi]\!](X) & \text{relative complement.} \end{array}$$

All these operations move an agent 'forward', to ever smaller sets of possibilities, representing increased information. Backwards movement to larger sets of valuations would correspond to retraction or revision of information. Note that there are other options for update operators. For instance, another conjunction would use intersection:

$$[\![\phi \cap \psi]\!](X) = [\![\phi]\!](X) \cap [\![\psi]\!](X).$$

This is still a purely descriptive system. The following observation is easy to prove by induction on the above formulas:

Fact 2.1 $[\![\phi]\!]_{\text{dyn}}(X) = X \cap [\![\phi]\!]_{\text{class}}$, *where* $[\![\phi]\!]_{\text{class}}$ *is the set of valuations verifying* ϕ *in the standard sense.*

This will change once we exploit further resources of this dynamic framework. But let us first compare this single-valued functional approach with the relational one over transition models from our previous chapter.

2.1.2 Relations and Functions over State Spaces

There is no conflict between relational and functional approaches. Functions are *deterministic total* relations. And conversely, every binary relation R on S induces a function $R^{\#}$ from pow(S) to pow(S), by setting

$$R^{\#}(X) =_{\text{def}} R[X] \quad (= \{y \in S \mid \exists x \in X\, Rxy\}).$$

Nothing is lost in this larger setting, as these 'lifted' functions can be uniquely retrieved via the well-known mathematical property of *continuity*, stating that a function commutes with arbitrary unions of its arguments:

Fact 2.2 *A function* $F : \text{pow}(S) \to \text{pow}(S)$ *is of the form* $R^{\#}$ *for some binary relation* R *on* S *if and only if* F *is continuous.*

Proof. All functions of the form $R^{\#}$ are continuous. Conversely, continuous maps F can be computed locally from their values at singleton arguments. I.e., they satisfy $F(X) = \bigcup_{x \in X} F(\{x\})$, for all sets X. Now, define a relation

$$R_F s_1 s_2 \text{ iff } s_2 \in F(\{s_1\}).$$

It is easy to show that $F = R_{F^{\#}}$. $\dashv$

Lifts and lowerings between various levels of set-theoretic representation can be studied systematically. E.g., here is a similar reduction of maps on sets of states to propositions in the standard style, being subsets of S (van Benthem 1986, Chapter 3):

Fact 2.3 *A unary operation F on* pow(S) *is both continuous and* intro-spective *(i.e., $F(X) \subseteq X$ for all $X \subseteq S$) if and only if it represents some unary property P via the rule $F(X) = X \cap P$.*

2.1.3 New Logical Operators: Epistemic Modalities

In addition to *updates* driving the current state forward, one can have various epistemic *tests* checking what is still possible. Here is a dynamic reading for epistemic possibility:

$$[\![\Diamond\phi]\!](X) = \begin{cases} X & \text{if } [\![\phi]\!](X) \neq \emptyset \\ \emptyset & \text{otherwise.} \end{cases}$$

Now, the system diverges from the classical case. Here is a linguistic example illustrating this. Consider, e.g., the following two discourses:

(i) "Maybe it's snowing. ... It is not snowing. ..."

(ii) $*$"It is not snowing. ... Maybe it's snowing. ..."

The former seems acceptable, whereas the latter is odd. But their two standard modal renderings (i)′ $\Diamond s \wedge \neg s$ and (ii)′ $\neg p \wedge \Diamond p$ would be equivalent. The difference will show up as one of sequential processing, however. Given initial options $\{s, \neg s\}$, the instructions (i) will produce successive states $\{s, \neg s\}$ (successful test), $\{s\}$ (successful update), whereas the instructions (ii) will produce $\{s\}$ (successful update), $\emptyset$ (failed test). The difference with the earlier purely descriptive updates F can be brought out as follows. The latter are all continuous and introspective, and hence admit of representation via their corresponding standard proposition. To be precise, the latter corresponds with the set of states $\{s \in S \mid F(\{s\}) = \{s\}\}$. Epistemic tests, on the other hand, are not continuous. For instance,

$$\begin{aligned} [\![\Diamond s]\!](\{s, \neg s\}) &= \{s, \neg s\} \quad \text{but} \\ [\![\Diamond s]\!](\{s\}) \cup [\![\Diamond s]\!](\{s\}) &= \{s\} \cup \emptyset = \{s\}. \end{aligned}$$

2.1.4 Dynamic Inference

There is a substantive issue now of defining appropriate dynamic notions of inference. Before doing so, we recall the classical logical definition of valid consequence $X \Rightarrow C$ for a conclusion C from a sequence of premises X. This requires that the conclusion C be true in all situations where all the premises X are true. Classical consequence satisfies the following *structural rules*, which regulate combination of inferences:

$/C \Rightarrow C$	Reflexivity
$X \Rightarrow D$ and $Y, D, Z \Rightarrow C \ / \ Y, X, Z \Rightarrow C$	Cut Rule
$X, P_1, P_2, Y \Rightarrow C \ / \ X, P_2, P_1, Y \Rightarrow C$	Permutation
$X, P, Y, P, Z \Rightarrow C \ / \ X, P, Y, Z \Rightarrow C$	Contraction
$X, P, Y, P, Z \Rightarrow C \ / \ X, Y, P, Z \Rightarrow C$	Contraction
$X, Y \Rightarrow C \ / \ X, P, Y \Rightarrow C$	Monotonicity.

Together, these say that only the bare set of premises matters to a classical conclusion: details of presentation, such as ordering or multiplicity are irrelevant. Moreover, Monotonicity says 'the more the better': new information cannot invalidate conclusions. As is shown in Chapter 7, these rules completely characterize classical consequence. Thus, they must register changes in behaviour for every deviant notion of inference. Here is a plausible candidate for dynamic consequence. One processes the successive premises, thereby absorbing their informational content into the initial information state. At the end, one checks if the resulting state is rich enough to satisfy the conclusion:

Definition 2.4 (Update-to-Test Consequence) $\phi_1, \ldots, \phi_n \models_{\text{upd}} \psi$ iff for every set X,

$$[\![\psi]\!]([\![\phi_n]\!](\ldots([\![\phi_1]\!](X))\ldots)) = [\![\phi_n]\!](\ldots([\![\phi_1]\!](X))\ldots).$$

As expected, this notion of inference differs from classical consequence in its structural rules. For instance, Reflexivity, Monotonicity, Cut and Contraction fail, even for formulas of our modal propositional logic. (The easy check of the following assertions will make the reader familiar with the workings of this kind of update system.)

not $\Diamond p \wedge \neg p \models_{\text{upd}} \Diamond p \wedge \neg p$

$\Diamond p \models_{\text{upd}} \Diamond p$ but not $\Diamond p \wedge \neg p \models_{\text{upd}} \Diamond p$

$p, \Diamond q \models_{\text{upd}} \Diamond p$ and $\neg q, \Diamond p \models_{\text{upd}} \Diamond p$ but not $\neg q, p, \Diamond q \models_{\text{upd}} \Diamond p$

$\Diamond p \wedge \neg p \wedge \Diamond p \models_{\text{upd}} p \wedge \neg p$ but not $\Diamond p \wedge \neg p \models_{\text{upd}} p \wedge \neg p$

What does hold, though, are certain *variants* of the above structural rules, which are sensitive to the processing order, such as

Left Monotonicity if $X \models_{\text{upd}} \psi$, then $\phi, X \models_{\text{upd}} \psi$

Left Cut if $X \models_{\text{upd}} \phi$ and $X, \phi, Y \models_{\text{upd}} y$, then $X, Y \models_{\text{upd}} \psi$.

We shall study Update-to-Test consequence more closely in Chapter 7.

One can also investigate this whole update system via more standard logical methods. For instance, it supports an *algebraic calculus* of equivalent expressions, denoting the same update function everywhere. Here are some algebraic peculiarities, again showing how dynamic inference diverges from its classical counterpart. First, consider the classical law of

Non-Contradiction. This comes in two sequential forms, both of which express non-trivial (and different) properties:

- $\phi \wedge \neg\phi = 0$ holds only for those propositions F which satisfy, for all states X, $F(X) - FF(X) = \emptyset$, i.e., $FF(X) \supseteq F(X)$.

Adding Introspection makes these update functions F even idempotent. A formula without the preceding property is $\phi = \Diamond p \wedge \neg p$. (Take the state $X = \{p, \neg p\}$.)

- $\neg\phi \wedge \phi = 0$ holds only for those propositions F which satisfy, for all X, $F(X - F(X)) = \emptyset$.

A formula without this property is $\phi = \neg\neg(\neg\Diamond p \wedge p)$. Next, the classical principle of *Distribution* also comes in two forms. Interestingly, these show a divergence:

- $\phi \wedge (\psi_1 \vee \psi_2) = (\phi \wedge \psi_1) \vee (\phi \wedge \psi_2)$ is valid,
- $(\phi_1 \vee \phi_2) \wedge \psi = (\phi_1 \wedge \psi) \vee (\phi_2 \wedge \psi)$ is invalid.

2.1.5 Standard Reductions

Despite the previous differences, our dynamic calculus is close to propositional logic, or to the modal logic S5. This proximity can be brought out by various *translations*. E.g., let P describe an arbitrary set of states, and, for any formula ϕ, let $T(P, \phi)$ denote the updated state $[\![\phi]\!](P)$. We can think of these sets of states as models for the modal logic S5 in an obvious way. Then the following recursion clauses hold for on-line computation of classical content in the modal logic S5.

Fact 2.5

$$\begin{aligned} T(P, q) &= P \wedge q \\ T(P, \phi \wedge \psi) &= T(T(P, \phi), \psi) \\ T(P, \phi \vee \psi) &= T(P, \phi) \vee T(P, \psi) \\ T(P, \neg\phi) &= P \wedge \neg T(P, \phi) \\ T(P, \Diamond\phi) &= P \wedge \Diamond T(P, \phi). \end{aligned}$$

Proof. These clauses reflect the earlier update conditions in a straightforward manner. For the last one, either $T(P, \phi)$ is non-empty, and $\Diamond T(P, \phi)$ holds everywhere in P, or $T(P, \phi)$ is empty, and $\Diamond T(P, \phi)$ is false. This is the clause for the S5 modality. ⊣

An alternative formulation takes epistemic update logic into monadic predicate logic:

$$\begin{aligned} T^*(P, q) &= Px \wedge Qx \\ T^*(P, \phi \wedge \psi) &= T^*(T^*(P, \phi), \psi) \\ T^*(P, \phi \vee \psi) &= T^*(P, \phi) \vee T^*(P, \psi) \end{aligned}$$

$$\begin{aligned} T^*(P, \neg\phi) &= Px \wedge \neg T^*(P, \phi) \\ T^*(P, \Diamond\phi) &= Px \wedge \exists x\, T^*(P, \phi). \end{aligned}$$

As a consequence, we have an embedding into a simple standard system:

Fact 2.6 $\phi_1, \ldots, \phi_n \models_{\text{upd}} \psi$ *iff*

$$\models_{\text{class}} T^*(T * (P, \phi_1 \wedge \ldots \wedge \phi_n), \psi) \leftrightarrow T^*(P, \phi_1 \wedge \ldots \wedge \phi_n)).$$

Hence, dynamic consequence is decidable, and it has the same pleasant meta-properties as monadic first-order consequence.

2.1.6 Further Connectives: Cognitive Retreat

All operators so far take sets 'forward' to subsets, whether they are tests or updates. But backward movement is also possible. An example is the qualifier "*unless* ϕ":

$$[\![unless\ \phi]\!](X) = X \cup [\![\phi]\!](S).$$

This adds the whole range of ϕ back to the current information state — computed over the initial state S of minimal information. Such additional operators of 'epistemic retreat' fit easily into the above update framework. For instance, the revision operator "unless" satisfies ordinary logical principles, such as the distribution law

$$\text{"}unless\text{"}\ \phi \vee \psi = \text{"}unless\text{"}\ \phi \vee \text{"}unless\text{"}\ \psi.$$

In actual reasoning, revisions are common. Eventually, their dynamic logic will require a more complex notion of state, keeping track of previous moves (Segerberg 1995).

2.2 Changing Perspectives in Predicate Logic

Dynamic changes do not always concern informational content. They may also affect 'semantic perspective', computing transient values for linguistic expressions. The latter phenomenon shows clearly with another logical core system. As noted in Chapter 1, Tarski's truth definition suggests changing variable assignments. Its usual format defines truth of a formula ϕ in a first-order model $\mathbf{M} = (D, I)$ under a variable assignment s. We now make the dynamic transitions visible, making ϕ the binary relation of all transitions between assignments that result in its successful verification.

2.2.1 Dynamic Predicate Logic: Core System

The main issue on this point of view is how to interpret the standard logical constants. The update conditions that follow are a well-known proposal from Groenendijk and Stokhof 1991 (the approach as such was first proposed in Barwise 1987):

Definition 2.7 (Update Conditions for Dynamic Predicate Logic)
atoms are tests

$$s_1[\![Pt]\!]^{\mathbf{M}}s_2 \text{ iff } s_1 = s_2 \text{ and } \mathbf{M}, s_1 \models Pt$$

conjunctions are relational compositions

$$s_1[\![\phi \wedge \psi]\!]^{\mathbf{M}}s_2 \text{ iff for some } s_3,\ s_1[\![\phi]\!]^{\mathbf{M}}s_3 \text{ and } s_3[\![\psi]\!]^{\mathbf{M}}s_2$$

existential quantifiers are random assignments

$$s_1[\![\exists x\, \phi]\!]^{\mathbf{M}}s_2 \text{ iff for some } s_3,\ s_1 =_x s_3 \text{ and } s_3[\![\phi]\!]^{\mathbf{M}}s_2$$

negations are strong failure tests

$$s_1[\![\sim\phi]\!]^{\mathbf{M}}s_2 \text{ iff } s_1 = s_2 \text{ and for no } s_3,\ s_1[\![\phi]\!]^{\mathbf{M}}s_3.$$

Example 2.8 (Dynamic Evaluation) Simple 'flow diagrams' may be used to visualize the transitions involved in evaluation. Consider the formula

$$Bx \wedge \exists x\, Ax \wedge \sim\exists y\, Byxz \wedge \exists x \sim Axy.$$

We may omit brackets for the conjunctions because, as we shall see later, relational composition is associative. Successful evaluation involves the following sequence of states: (i) an initial one s_1, where Bx has to hold, (ii) a next one s_2 with a reset value for x satisfying Ax, (iii) a next one s_3 corresponding to a successful test for $\sim\exists y\, Byxz$ (hence $s_3 = s_2$), (iv) a next one s_4, again with a reset value for x, satisfying $\sim Axy$. In all, then

dynamically, the formula $Bx \wedge \exists x\, Ax \wedge \sim\exists y\, Byxz \wedge \exists x \sim Axy$ denotes the following set of transitions between variable assignments:
$\{(s_1, s_4) \mid I(B)(s_1(x))$ and for some $s_2 =_x s_1$, $I(A)(s_2(x))$ while for no $d \in D$, $I(B)(d, s_2(x), s_1(z))$, and $s_4 =_x s_2$, not $I(B)(s_4(x), s_1(y))\}$

We see an interplay of two processes at work here. 'Tests' do not change the current state, while 'assignments' can. Formulas in general will involve an interplay of both. Clearly, scopes and quantifier binding no longer work as in standard predicate logic. E.g., existential quantifiers scope dynamically toward the right. This is precisely the point of the new system. Its computations of shifting assignments and static conditions will be familiar to students of the Hoare Calculus for imperative computer programs.

Dynamic predicate logic treats *conjunction* as composition, like propositional update logic. Its *negation* is a strong failure test $\sim$ (which we distinguish notationally from classical $\neg$), comparable to intuitionistic negation as absence of future verification. Even more radical is the treatment of the *existential quantifier*. Despite appearances, quantifiers are not really logical connectives here, but rather atomic actions. One can read the update condition for $\exists x\, \phi$ as a composition of two separate procedures for $\exists x$ and for ϕ. Therefore, it makes more sense to interpret quantifiers *independently*:

random assignment

$$s_1[\![\exists x]\!]^{\mathbf{M}} s_2 \text{ iff } s_1 =_x s_2.$$

This is what we shall do henceforth. Thus, we can meaningfully interpret more of first-order syntax than the usual well-formed formulas, including expressions like $Px \wedge \exists x$. Next, 'random assignments' $x :=?$ naturally suggest 'definite assignments' $x := t$. These would be the natural denotation for free-floating syntactic substitution operators:

controlled assignment

$$s_1[\![\,[t/x]\,]\!]^{\mathbf{M}} s_2 \text{ iff } s_2 = s_1{}^{x}_{value(\mathbf{M},s_1,t)}.$$

In terms of these operators, other central logical notions may be defined. In particular, a *universal quantifier* $\forall x\,\phi$ may be defined as $\sim\exists x \sim\phi$, which gives the following test:

$$s_1[\![\forall x\,\phi]\!]^{\mathbf{M}} s_2 \quad \text{iff} \quad s_1 = s_2 \text{ and for all objects } d, \text{ there exists some state } s_3 \text{ with } s_1(x|d)[\![\phi]\!]^{\mathbf{M}} s_3.$$

Notice the difference with the above clause for the existential quantifier. Intuitively, checking a universal statement does not leave any preferred witness in the x-register. Likewise, dynamic implication $\phi \to \psi$ may be read via one of its classical definitions as $\sim(\phi \wedge \sim\psi)$. This works out to the following test in "for all ... there exists" form, familiar from semantic accounts of conditional statements in constructive logics:

$$s_1[\![\phi \to \psi]\!]^{\mathbf{M}} s_2 \quad \text{iff} \quad s_1 = s_2 \text{ and for every } s_3 \text{ with } s_1[\![\phi]\!]^{\mathbf{M}} s_3, \text{ there exists some } s_4 \text{ with } s_3[\![\psi]\!]^{\mathbf{M}} s_4.$$

Henceforth, we shall often write '$\mathbf{M}, s_1, s_2 \models \phi$' for '$s_1[\![\phi]\!]^{\mathbf{M}} s_2$' — to emphasize the analogy with standard Tarskian evaluation, now with one additional state parameter.

2.2.2 Linguistic Motivation

The original motivation for the preceding system lies in linguistics. This illustrates an interesting consideration. The design of a logical semantics may interact with the preferred representational use for its formalism. We present a toy example. (For further details of the natural language connection, cf. Chapter 12.) Consider the following observations about possible and impossible coreference in natural language. (Proposed identifications are underlined; while an asterisk indicates their infelicity.) To their right stand the obvious prima facie predicate-logical forms of these phrases. Intended identifications are marked by individual variables:

(i) "A man came in. He whistled." $\exists x\, Mx \wedge Wx$

(ii)* "He whistled. A man came in." $Wx \wedge \exists x\, Mx$

(iii)* "No man came in. He whistled." $\sim\exists x\, Mx \wedge Wx$

(iv)* "He whistled. No man came in." $Wx \wedge \sim\exists x\, Mx$
(v) "If a man comes in, he whistles." $\exists x\, Mx \rightarrow Wx$
(vi)* "If no man comes in, he whistles." $\sim\exists x\, Mx \rightarrow Wx$

Though natural, these formulas get various things wrong (in the standard semantics). In (i), no coreference is possible, as the scope of the initial quantifier does not extend over the second sentence. In (5), coreference is impossible, because the existential quantifier in the antecedent does not scope over the consequent. Now, the translation lore which every beginner in logic learns, makes automatic compensation for this, changing scopes and reversing quantifiers wherever appropriate. Thus, (i) becomes $\exists x\,(Mx \wedge Wx)$ and (v) $\forall x\,(Mx \rightarrow Wx)$. But dynamic predicate logic makes the original forms work without further ado, under their proper dynamic interpretation. Thus, successful evaluation of $\exists x\, Mx \wedge Wx$ results in assignment shifts (s_1, s_2) where $s_1 =_x s_2$ and s_2 satisfies both Mx and Wx. On the other hand, successful evaluation of $Wx \wedge \exists x\, Mx$ results in assignment shifts (s_1, s_2) where s_1 satisfies Wx, $s_1 =_x s_2$ while s_2 satisfies Wx. No coreference is enforced here for the first and third occurrence of x. On the other hand, in (iii) no coreference occurs, as the test for $\sim\exists x\, Mx$ does not interact with the test for Wx in the initial assignment. Finally, consider the pair (v), (vi). The former works out to a test whether all assignments setting x to some man test positive for that man to be whistling. But (vi) still contains independent tests for absence of incoming males and whistling of the object assigned to the variable x. This style of linguistic analysis has been applied to a host of other linguistic phenomena than anaphora, including temporal expressions, presuppositions and generalized quantifiers. Once again, this digression illustrates the delicacy of arguing the merits of a dynamic logic. To bring actual natural language within its scope, one has to 'decorate' linguistic forms, at least, with variables. And decisions taken then will influence the effects of the logic. (For instance, Chapter 12 discusses how regimented first-order fragments influence dynamic inference.)

2.2.3 Dynamic Inference

Next, we must look at predicate-logical reasoning. On the above view, inferences are pieces of discourse allowing anaphoric reference from premises to conclusions. E.g., if we are told that "A woman was dancing" , we may conclude that "She was moving." To make this deliver reasonable results, scopes must extend across conclusion markers ("so", "hence"). Here is what Groenendijk and Stokhof 1991 propose for this purpose. First process all premises consecutively, and then see if the conclusion can be executed. In computational jargon, the composed premises must 'enable' the conclusion:

$$\phi_1, \ldots, \phi_n \models^{\text{dyn}} \psi \quad \text{iff} \quad \text{in all models } \mathbf{M}, \text{ if } s_1[\![\phi_1 \wedge \ldots \wedge \phi_n]\!]^{\mathbf{M}} s_2, \text{ then there exists some } s_3 \text{ with } s_2[\![\psi]\!]^{\mathbf{M}} s_3.$$

This gives a Deduction Theorem for the above implication. E.g., ϕ_1, $\phi_2 \models^{\mathrm{dyn}} \psi$ holds iff $\phi_1 \models^{\mathrm{dyn}} \phi_2 \to \psi$. On the other hand, even the simple classical law $\phi \models \phi$ fails. A counter-example is $\phi = {\sim}Px \wedge \exists x\, Px$. Indeed, again this style of dynamic inference loses many of the *structural rules* of classical logic. It does not allow Monotonicity, Permutation, Contraction or Cut, witness the following counter-examples. (These are excellent exercises in grasping the peculiarities of dynamic predicate logic.)

$Px \models^{\mathrm{dyn}} Px$ but not $Px, \exists x\,{\sim}Px \models^{\mathrm{dyn}} Px$
$\exists x\,{\sim}Px, Px \models^{\mathrm{dyn}} Px$ but not $Px, \exists x\,{\sim}Px \models^{\mathrm{dyn}} Px$
$Px, \exists x\,{\sim}Px, Px \models^{\mathrm{dyn}} Px$ but not $Px, \exists x\,{\sim}Px \models^{\mathrm{dyn}} Px$
$\exists x\,{\sim}Px, Px \models^{\mathrm{dyn}} \exists x\,{\sim}Px$ and $\exists x\,{\sim}Px \models^{\mathrm{dyn}} {\sim}Px$
but not $\exists x\,{\sim}Px, Px \models^{\mathrm{dyn}} {\sim}Px$

The only structural rule which remains valid is a variant of Monotonicity, namely, 'Left-Monotonicity' prefixing arbitrary premises to a valid inference. Additional valid structural principles only emerge with the explicit introduction of special connectives. An illustration is the following valid variant of Cut using an explicit conjunction:

$$X \models^{\mathrm{dyn}} \phi \text{ and } \phi \models^{\mathrm{dyn}} \psi \text{ imply } X \models^{\mathrm{dyn}} \phi \wedge \psi.$$

2.2.4 Relational Algebra

There are also standard perspectives on validity in dynamic predicate logic. Formulas now denote binary relations, generating a *relational algebra* of equivalent expressions. This algebra explains many earlier observations. E.g., rightward scope extension of the existential quantifier in $\exists x\, \phi \wedge \psi$ exemplifies a well-known law for composition:

Associativity $((\epsilon \wedge (\phi \wedge \psi)) = ((\epsilon \wedge \phi) \wedge \psi))$.

Or, consider the calculation for the above update condition for $\forall x\, (\phi \to \psi)$:

$$\forall x\, (\phi \to \psi) \leftrightarrow {\sim}\exists x\, {\sim}{\sim}(\phi \wedge {\sim}\psi) \leftrightarrow {\sim}\exists x\, (\phi \wedge {\sim}\psi).$$

This involves the following valid relation-algebraic law (with $\epsilon = \exists x$, $\alpha = (\phi \wedge {\sim}\psi)$):

$${\sim}(\epsilon \wedge {\sim}{\sim}\alpha) = {\sim}(\epsilon \wedge \alpha).$$

More generally, the above dynamic operations of composition and negation will obey all laws that hold generally for these operations over binary relations, such as

Triple Negation ${\sim}{\sim}{\sim}\phi = {\sim}\phi$
Negations Commute ${\sim}\phi \wedge {\sim}\psi = {\sim}\psi \wedge {\sim}\phi$.

By contrast, the law of Double Negation is invalid: the test ${\sim}{\sim}\exists x\, \phi$ is almost never the same relation as $\exists x\, \phi$. At this algebraic level, further interesting divergences show up with standard predicate logic. Consider again classical Non-Contradiction, now read dynamically. This law has two sequential versions (with '0' for Boolean 'false'):

- $\sim\phi \wedge \phi = 0$
- $\phi \wedge \sim\phi = 0$.

Of these, the first is valid, but not the second. (The formula sequence $(Px \wedge \exists x \sim Px) \wedge \sim(Px \wedge \exists x \sim Px)$ is consistent.) This outcome differs from propositional update logic. The same holds for Distribution: unlike propositional update functions, relation compositions distribute over unions in *both* arguments. Here is a final example. By definition, the two dynamic quantifiers satisfied $\forall x\, \phi \leftrightarrow \sim\exists x \sim\phi$. But the dual $\exists x\, \phi \leftrightarrow \sim\forall x \sim\phi$ fails. $\sim\forall x \sim\phi$ is equivalent to $\sim\sim\exists x \sim\sim\phi$, which is equivalent only to $\sim\sim\exists x\, \phi$. (Compare intuitionistic quantifiers: cf. Troelstra and van Dalen 1988.)

Connections with Relational Algebra will be elaborated in Chapters 4, 12. One natural conjecture stated in an earlier draft of this book was that schematic validity in DPL (in the above sense) validates *only* the algebraic identities of relational set algebra. This has been proved in Visser 1995: "DPL is complete for RA". Moreover, unlike full relational algebra, the equational theory of its $\{\sim, \wedge\}$-fragment underlying DPL has a simple finite axiomatization (Hollenberg 1995a). Over this algebraic superstructure, concrete valid principles of dynamic predicate logic (without formula variables) express special relational properties of atomic actions and their underlying states. Thus,

- $\exists x \exists x \leftrightarrow \exists x$ expresses transitivity and density of the shift relation $=_x$
- $\exists x \exists y \leftrightarrow \exists y \exists x$ says that shifts may be carried out in any order.

The latter apparently innocuous statement requires the existence of 'enough' states in dynamic models, in order to provide all possible paths. Existential principles like this cause the well-known *undecidability* of predicate logic, which is inherited by its dynamic version. They will come under closer scrutiny in Chapter 9, which questions the rationale of imposing a priori existence axioms on our computational structures.

- $\sim\sim Px \leftrightarrow Px$ says that atomic tests do not change states
- $\sim\sim\exists x \leftrightarrow x = x$ says that the shift relation $=_x$ is reflexive

More generally, such special principles define specific kinds of action. For instance, here is an easy characterization of *test actions*, which never change the current state:

Fact 2.9 *The following two conditions are equivalent for any binary relation R:*

(i) R is contained in the identity relation on states

(ii) $\sim\sim(R) = R$ ('R encodes its own domain').

Both atomic actions (tests and reassignments) are *equivalence relations* over states — but complex relations DPL-definable from this atomic repertoire need no longer be.

2.2.5 Comparisons with Standard Predicate Logic

The dynamic version of predicate logic differs from the standard system in many ways. For instance, scopes of quantifiers have become open-ended towards the right, until the boundary of some test operator is encountered. On the other hand, various reductions exist connecting the two logics. First, for any classical formula ϕ, the following recursion defines a dynamic test formula $\tau(\phi)$ for its truth:

$$\begin{array}{rcl} \tau(Px) & = & Px \quad \text{for all atoms } Px \\ \tau(\phi \wedge \psi) & = & \tau(\phi) \wedge \tau(\psi) \\ \tau(\phi) & = & \sim\tau(\phi) \\ \tau(\exists x\, \phi) & = & \sim\sim\exists x\, \tau(\phi). \end{array}$$

(We do not treat $\exists x$ as a separate action here, since it does not support a useful test.) Notice that all translations $\tau(\phi)$ become test actions as above, whose domain is included in the diagonal. This observation underlies the following argument.

Fact 2.10 $\mathbf{M}, s \models_{\text{classical}} \phi$ *if and only if* $\mathbf{M}, s, s \models_{\text{dynamic}} \tau(\phi)$.

Proof. Induction on ϕ. The key facts are as follows. (i) A composition of tests for two statements is equivalent to a test for their conjunction. (ii) The strong negation of a test for a statement is a test for its Boolean negation. (Here is where we use the preceding observation. Note also that ordinary Boolean relational complement would not do the job: this is precisely where $\sim$ shows its use.) (iii) The operator $\sim\sim(R)$ defines a test for the domain of the relation R. (Thus, $\sim\sim\exists x\, \tau(\phi)$ behaves like an existential modality $\langle \exists x\, ; \phi \rangle \top$ or $\langle \exists x \rangle \langle \phi \rangle \top$ — with '$\top$' for 'true'.) ⊣

Conversely, there are also reductions from dynamic to static evaluation. Intuitively, dynamic evaluation of ϕ moves to states where some static version of ϕ holds, while some values for its variables have been set. For this 'static tracing', a technique from computer science may be used (cf. Chapter 1). Imperative programs are often explained in terms of 'postconditions'. Starting from some set of states described by an input predicate A, execution of a program π takes us to a new set of states (that is, the set-theoretic π-image of A), defined by the so-called *strongest postcondition* of A under π: $\mathrm{SP}(A, \pi)$. For the simple programs of our dynamic predicate logic, postconditions may even be computed recursively. (Explicit clauses for this purpose are found in Chapter 6 below.) We display the outcome of a sample computation on a sequence of dynamic formulas, with trace points indicated in bold-face subscripts:

$$_{\mathbf{0}}\ \exists x\ _{\mathbf{1}}\ (Ax\ _{\mathbf{2}}\ \wedge Bx\ _{\mathbf{3}})\ .\ \exists x\ _{\mathbf{4}}\ Cx\ _{\mathbf{5}}\ .\ (\exists x\, Dx)\ _{\mathbf{6}}$$

SP
0 $\top$
1 $\exists x \top = \top$
2 Ax
3 $Ax \wedge Bx$
4 $\exists x\,(Ax \wedge Bx)\ *$
5 $\exists x\,(Ax \wedge Bx) \wedge Cx$
6 $\exists x\,(Ax \wedge Bx) \wedge Cx \wedge \neg\exists x\, Dx$, etcetera

At the trace point $*$, the initial $\exists x$ becomes an ordinary quantifier after all.

This example shows how dynamic evaluation of a predicate-logical text works, via an alternation of actions and static assertions providing 'snapshots' of intermediate states. Eventually, this interplay suggests a logical system *combining* both dynamic and static assertions — which is indeed the point of the two-level architecture in Chapters 1, 6.

More direct reduction techniques exist as well. Dynamic evaluation of a formula ϕ whose variables are among $x_1, \ldots, x_n$ can be described completely by a static 'transition predicate' with $2n$ variables:

$$\mathit{Trans}(\phi)(x_1^{\text{in}}, \ldots, x_n^{\text{in}}, x_1^{\text{out}}, \ldots, x_n^{\text{out}}).$$

The latter merely transcribes the above recursive update conditions:

$$\begin{aligned}
\mathit{Trans}(Qx_ix_j) &= \bigwedge_{1\le k\le n} x_k^{\text{in}} = x_k^{\text{out}} \wedge Qx_i^{\text{out}}x_j^{\text{out}} \\
\mathit{Trans}(\exists x_i) &= \bigwedge_{1\le k\le n,\ k\ne i} x_k^{\text{in}} = x_k^{\text{out}} \\
\mathit{Trans}(\phi_1 \wedge \phi_2) &= \exists z_1 \ldots z_n\, (\mathit{Trans}(\phi_1)(x_1^{\text{in}}, \ldots, x_n^{\text{in}}, z_1, \ldots, z_n) \\
&\qquad \wedge \mathit{Trans}(\phi_2)(z_1, \ldots, z_n, x_1^{\text{out}}, \ldots, x_n^{\text{out}}) \\
\mathit{Trans}(\sim\phi) &= \bigwedge_{1\le k\le n} x_k^{\text{in}} = x_k^{\text{out}} \wedge \\
&\qquad \neg\exists z_1 \ldots z_n\, \mathit{Trans}(\phi)(x_1^{\text{in}}, \ldots, x_n^{\text{in}}, z_1, \ldots, z_n).
\end{aligned}$$

Fact 2.11 *Let s_1, s_2 be assignments to the variables $x_1, \ldots, x_n$. Let a new assignment s_{12} map duplicated variables x_k^{in} to $s_1(x_k)$ and x_k^{out} to $s_2(x_k)$ $(1 \le k \le n)$. Then we have, in any model* $\mathbf{M}$,

$$s_1[\![\phi]\!]^{\mathbf{M}}s_2 \text{ iff } \mathbf{M}, s_{12} \models \mathit{Trans}(\phi).$$

This translation implies that dynamic predicate logic inherits many meta-properties from standard predicate logic — including effective axiomatizability of its universal validities, model-theoretic Löwenheim-Skolem properties, etcetera. The translational format has also been used to describe dynamic phenomena in natural language, while staying close to the classical type theory of Montague Grammar (Muskens 1991).

2.2.6 Further Logical Constants

Dynamic predicate logic arose by interpreting the standard operators of predicate logic. There are also useful defined notions, including implication, universal quantification, or *relational domain* $\sim\sim\phi$. Another example is the *disjunction* $\phi \vee \psi$ of Groenendijk and Stokhof 1991, defined as $\sim(\sim\phi \wedge \sim\psi)$. This satisfies some classical key principles. E.g., the De Morgan Law becomes the valid $\sim\sim(\sim\phi\wedge\sim\psi) \leftrightarrow (\sim\phi\wedge\sim\psi)$. On the other hand, natural dynamic variants for classical connectives are beyond the framework so far. This holds for the Boolean false (the above '0'). Also undefinable is Boolean intersection, which expresses a *parallel composition* of relations:

$$s_1[\![\phi \cap \psi]\!]^{\mathbf{M}} s_2 \text{ iff } s_1[\![\phi]\!]^{\mathbf{M}} s_2 \text{ and } s_1[\![\psi]\!]^{\mathbf{M}} s_2.$$

'Almost' DPL-definable is Boolean union (modeling indeterministic choice in program semantics) which is a dynamic alternative for disjunction:

$$s_1[\![\phi \cup \psi]\!]^{\mathbf{M}} s_2 \text{ iff } s_1[\![\phi]\!]^{\mathbf{M}} s_2 \text{ or } s_1[\![\psi]\!]^{\mathbf{M}} s_2.$$

Under negations, one has $\sim(\phi \cup \psi) \leftrightarrow \sim\phi \wedge \sim\phi$. Also, we have distributive laws:

$$((\phi \cup \psi) \wedge \chi) \leftrightarrow ((\phi \wedge \chi) \cup (\psi \wedge \chi))$$

and

$$(\phi \wedge (\psi \cup \chi)) \leftrightarrow ((\phi \wedge \psi) \cup (\phi \wedge \chi)).$$

Applying these equivalences gives a *normal form* where unions occur only on the outside. Finally, in addition to composition and Booleans, standard Relational Algebra has one more useful operator (cf. Chapter 3), namely *conversion* of binary relations:

$$s_1[\![\phi^{\smile}]\!]^{\mathbf{M}} s_2 \text{ iff } s_2[\![\phi]\!]^{\mathbf{M}} s_1.$$

Again, a whole bunch of easy logical equivalences holds, some of a general algebraic nature, some reflecting the symmetric behaviour of our standard operations. As a result, dynamic predicate logic implicitly defines its own relational converses:

Fact 2.12 (Validities for Converse)

$$\begin{array}{rclrcl} (Pt)^{\smile} & = & Pt & (\exists x)^{\smile} & = & \exists x \\ (\phi \wedge \psi)^{\smile} & = & (\psi^{\smile} \wedge \phi^{\smile}) & (\sim\phi)^{\smile} & = & \sim\phi \\ (\phi \cup \psi)^{\smile} & = & (\phi^{\smile} \cup \psi^{\smile}) & \phi^{\smile\smile} & = & \phi \end{array}$$

Nevertheless, no explicit definition is available for $^{\smile}$ in terms of $\{\wedge, \sim\}$.

2.2.7 Analogies with Hoare Calculus

Semantics in terms of assignment change have occurred since the sixties in 'operational semantics' for imperative programs, which supports the so-called Hoare Calculus of correctness statements. Consider a toy programming language with assignments $x := t$ as its atomic instructions, and the familiar programming constructions of

sequencing $\pi_1 ; \pi_2$
conditional choice $\texttt{if } \phi \texttt{ then } \pi_1 \texttt{ else } \pi_2$
guarded iteration $\texttt{while } \phi \texttt{ do } \pi$.

Its semantics starts from standard first-order models $\mathbf{M}$ interpreting the relevant first-order test statements ϕ, and then reads programs π as binary transition relations between assignments ('the successful executions for π'):

$$\mathbf{M}, s_1, s_2 \models x := t \quad \text{iff} \quad s_2 = s_1(x|s_1(t))$$

$$\mathbf{M}, s_1, s_2 \models \pi_1 ; \pi_2 \quad \text{iff} \quad \text{there exists } s_3 \text{ with } \mathbf{M}, s_1, s_3 \models \pi_1 \text{ and } \mathbf{M}, s_3, s_2 \models \pi_2$$

$$\mathbf{M}, s_1, s_2 \models \texttt{if } \phi \texttt{ then } \pi_1 \texttt{ else } \pi_2 \quad \text{iff} \quad \mathbf{M}, s_1 \models \phi \text{ and } \mathbf{M}, s_1, s_2 \models \pi_1, \text{ or } \mathbf{M}, s_1 \not\models \phi \text{ and } \mathbf{M}, s_1, s_2 \models \pi_2$$

$$\mathbf{M}, s_1, s_2 \models \texttt{while } \phi \texttt{ do } \pi \quad \text{iff} \quad \text{there exists a finite sequence } s_1 = t_1, \ldots, t_k = s_2 \text{ with } \mathbf{M}, t_i, t_{i+1} \models \pi, \mathbf{M}, t_i \models \phi \ (1 \leq i < k) \text{ and } \mathbf{M}, t_k \not\models \phi.$$

The well-known correctness statements $\{\phi\}\pi\{\psi\}$ are then read as follows:

> whenever state s_1 verifies the 'precondition' ϕ in $\mathbf{M}$, and $\mathbf{M}, s_1, s_2 \models \pi$, then the 'postcondition' ψ holds at s_2 in $\mathbf{M}$.

Equivalently, 'the strongest postcondition $\mathrm{SP}(\phi, \pi)$ $\mathbf{M}$-implies ψ' . This semantics validates the usual principles of the Hoare Calculus, a logic for conditionals indexed by actions forming a fragment of Propositional Dynamic Logic (Chapter 3). For instance,

$$\{\phi\}\pi_1\{\psi\} \text{ and } \{\psi\}\pi_2\{\chi\} \text{ imply } \{\phi\}\pi_1 ; \pi_2\{\chi\}.$$

This shows a division of labour between static first-order formulas encoding unary properties of states and dynamic programs denoting binary relations between states. (By contrast, in the above DPL system, the first-order formulas *themselves* have acquired program behaviour.) This dual view is exploited in van Eijck and de Vries 1991 to study dynamic predicate logic through a Hoare Calculus of correctness statements. A sample validity is $\{\phi\}\, \exists x\, \{\exists x\, \phi\}$, with quantifiers read statically in the pre- and post-conditions, but dynamically in the program slot. A question suggested by this analogy is whether the *infinitary constructions* of computer programming (with arbitrary finite 'activity', such as `while`) have natural uses in linguistics and cognition.

2.3 Dynamic Modal Logic

2.3.1 Traversing Kripke Models

Dynamification carries over from extensional to intensional logic. For a clear example, consider the well-known possible worlds models for intuitionistic or relevant logic:

$$\mathbf{M} = (S, \subseteq, V).$$

Intuitively, worlds in S are information states, the relation $\subseteq$ is possible extension, and the valuation V records where propositional atoms p are true. This V satisfies a constraint of 'Heredity', stating that truth of atoms is upward preserved along $\subseteq$. (Thus, an intuitionistic reasoner only accumulates statements, that are never revised.) Over these models, an intuitionistic propositional language is interpreted as follows:

$$\begin{array}{rll} \mathbf{M}, s \models p & \text{iff} & V(s,p) = 1 \\ \mathbf{M}, s \models \phi \wedge \psi & \text{iff} & \mathbf{M}, s \models \phi \text{ and } \mathbf{M}, s \models \psi \\ \mathbf{M}, s \models \phi \vee \psi & \text{iff} & \mathbf{M}, s \models \phi \text{ or } \mathbf{M}, s \models \psi \\ \mathbf{M}, s \models \neg\phi & \text{iff} & \text{for no } s' \supseteq s, \mathbf{M}, s' \models \phi \\ \mathbf{M}, s \models \phi \rightarrow \psi & \text{iff} & \text{for all } s' \supseteq s, \text{if } \mathbf{M}, s' \models \phi \text{ then } \mathbf{M}, s' \models \psi \end{array}$$

This logic is knowledge-oriented: $\mathbf{M}, s \models \phi$ says that, at stage s, I know that ϕ. The resulting set of validities deviates from classical logic. Notably, Excluded Middle $\phi \vee \neg\phi$ is no longer valid, as it would express that we always know ϕ or its negation. (Also, this richer setting allows for more logical operators than the classical repertoire. Thus, disjunction is no longer definable in terms of the other logical operators — and one can even introduce entirely new modal operators.) Clearly, the intuitive picture behind this semantics is dynamic: with a cognitive agent traversing information states. Thus, intuitionistic formulas refer to transitions in this information pattern. E.g., the truth condition for $\neg\phi$ requires inspection of all possible extensions of the current state for absence of ϕ. As with our earlier systems, this implicit logical dynamics may be made *explicit* by adding cognitive transitions, such as informational updates upd(P) to some minimal extension where the proposition P is true. Intuitionistic logic is a forward-looking system of cognitive advance — but it is easy to design a dual version allowing cognitive retreat with 'downdates' and 'revisions'.

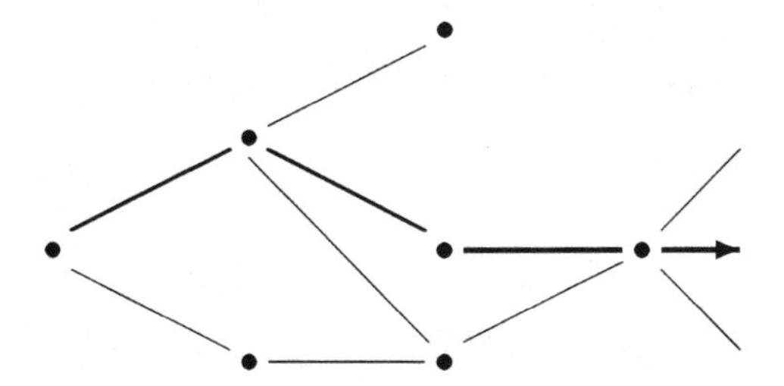

Once we take this view, dynamic aspects emerge in just about every intensional logic. This is not surprising, since most of these presuppose a form of 'multiple reference' shifting from one world to another (van Benthem 1988c). For instance, we can do the same with *relevant logics*. These use further structure over information states, such as addition or merge + of states, witness the clause for relevant implication (Dunn 1985):

$$\mathbf{M}, s \models \phi \to \psi \text{ iff for all } t \text{ with } \mathbf{M}, t \models \phi, \mathbf{M}, s + t \models \psi.$$

Other dynamic parameters occur in *temporal logic*. In temporal discourse, the center of attention moves through time (Kameyama 1994). This dynamics of perspective may be studied by dynamification of standard Prior-style temporal semantics. Further examples are found in Groenendijk et al. 1994, 1996b, Cepparello 1995, which show how dynamified versions of modal predicate logic throw new light on old issues of individuation and identity in philosophical logic and the theory of meaning.

2.3.2 Dynamic Modal Logic: Core System

The abstract transition models of Chapter 1 were merely sets of states S supporting a family of atomic actions R_p. A modest form of information structure, as in the above intuitionistic or relevant models, endows these states with a distinguished partial order $\subseteq$ of inclusion or 'possible development' — resulting in the following kind of model:

$$\mathbf{M} = (S, \subseteq, \{R_p \mid p \in P\}, V).$$

One option here is, as suggested above, to re-interpret intuitionistic or modal formulas as procedures over such spaces. But a more powerful set-up is available, merging such actions with the original static logic into one dynamic system (van Benthem 1989c). Recall the two-level architecture sketched in Chapter 1:

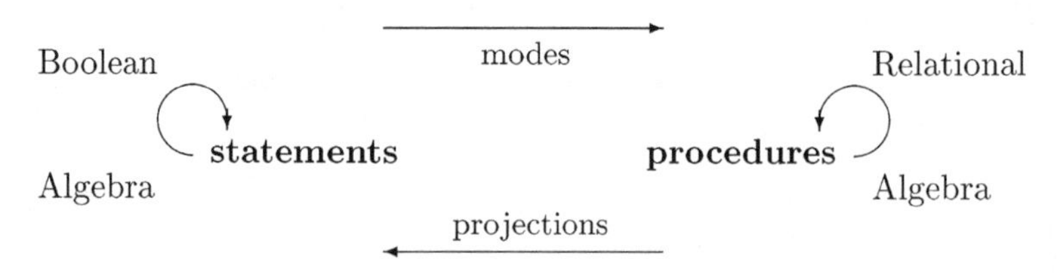

This combines a standard modal logic over information models with a dynamic algebra of actions that change states. New notions emerge in all components of this schema. In particular, new *modes* may be defined that create dynamic procedures out of static standard propositions. Notably, these include processes of cognitive advance that are essentially the earlier updates, but now with their static content indicated:

'loose updating' $\lambda P.\, \lambda xy.\, x \subseteq y \wedge Py$

'strict updating' $\lambda P\, \lambda xy.\, x \subseteq y \wedge Py \wedge \neg\exists z\, (x \subseteq z \subset y \wedge Pz)$.

Going through the information pattern in the opposite direction, these have two obvious counterparts for processes of cognitive retreat:

'loose downdating' $\lambda P.\, \lambda xy.\, y \subseteq x \wedge \neg Py$

'strict downdating' $\lambda P\, \lambda xy.\, y \subseteq x \wedge \neg Py \wedge \neg\exists z\, (y \subset z \subseteq x \wedge \neg Pz)$.

Informational downdating is no converse of updating, but we have:

$$\mathit{downdate}(P, \subseteq) = \mathit{update}(\neg P, \subseteq).$$

The resulting *dynamic modal logic* is able to perform all cognitive tasks from the well-known epistemic dynamics in Gärdenfors 1988. In particular, the latter's procedure of 'revision' may be described by combination of (first) downdate and (then) update. Here is another important example. *Conditional statements* $A \rightarrow B$ are often explained action-wise via the so-called Ramsey Test. "Given your current belief state, try to assume the antecedent A . If this leads to an inconsistent state, then make a minimal adjustment in your beliefs which restores consistency. Now, see if the result implies the consequent." In dynamic modal logic, this procedure reads as follows. "Downdate strictly with respect to $\Diamond A$. Then update strictly with respect to A. Finally, check if B holds." This connection is explored more thoroughly in de Rijke 1994, Segerberg 1995, which analyze Gärdenfors' AGM-postulates in a dynamic modal setting.

These motivations still do not fix a precise design for dynamic modal logic. For this purpose, we follow the computational tradition (cf. Chapter 3 below). That is, the static propositional component will have standard propositional and modal operators

$$\{\neg, \wedge, \vee, \Box_{\mathrm{up}}, \Box_{\mathrm{down}}\},$$

while the dynamic program component has a usual relational repertoire

$$\{-, \cap, \cup, \circ, \check{}, \Delta\}.$$

In addition, the preceding considerations suggest the following modes and projections:

- { test, loose and strict updates, as well as downdates }
- { fixed point, domain, range }.

Connections between these operators will be listed in Chapter 6 below. The universal validities of dynamic modal logic encode a general theory of cognitive processes, extending previous ones in expressive power. E.g., this modal theory describes the following interactions between actions of updating and downdating:

- $\text{upd}(P) \circ \text{upd}(P) = \text{upd}(P)$ ('idempotence')
- and the same holds for strict updates, and for downdates
- $\text{upd}(P) \circ \text{downd}(P)$ need not be the identity, witness a situation

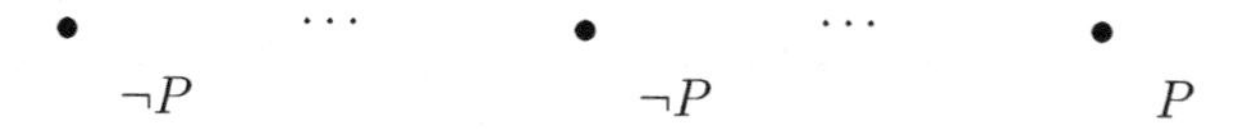

The system also handles earlier notions of dynamic inference (Chapter 7 has details). E.g., inference in dynamic predicate logic $P_1, \ldots, P_n \models^{\text{dyn}} C$ amounts to validity of the universal-existential program modality

$$[P_1 \circ \cdots \circ P_n]\langle C \rangle \top.$$

2.3.3 Dynamic Modal Logic: Theory

Dynamic modal logic raises some obvious technical questions. De Rijke 1992 presents a complete axiomatization, sketched in Chapter 6 below. Next, there is the issue whether dynamic modal logic is *decidable*. Its static relative is the the modal logic S4, which has the latter property, but its dynamic relative of relational set algebra lacks it. As it happens, undecidability holds for the full formalism (cf. again Chapter 6). Less ambitiously, we shall see in Chapter 3 that the DML fragment with only the two DPL operators $\{\wedge, \sim\}$ is decidable. But there are also reasons for increasing expressive power — as there is still an 'asymmetry' in our design. The static component of dynamic modal logic is not 'in harmony' with the dynamic one. Consider a reduction from dynamic procedures to static propositions, via strongest postconditions such as:

$$\begin{aligned} \text{SP}((P)?, A) &= P \wedge A \\ \text{SP}(\text{upd}(P), A) &= P \wedge \Diamond_{\text{down}} A \\ \text{SP}(\text{downd}(P), A) &= \neg P \wedge \Diamond_{\text{up}} A. \end{aligned}$$

In order to deal with static effects of strict update and downdate in this format, a more complex modal language over information patterns is needed, employing two well-known 'temporal' operators that go beyond Prior's basic tense logic:

$$\begin{aligned}\text{SP}(\text{strict upd}(P), A) &= P \wedge \mathit{Since}(A, \neg P)\\ \text{SP}(\text{strict downd}(P), A) &= \neg P \wedge \mathit{Until}(A, P)\end{aligned}$$

Clearly then, dynamic modal logics lie on a ladder of ascending expressive strength. In the limit, this ladder ascends to an even broader formalism over information models:

2.3.4 Translation into Standard Logic

As in standard modal logic (cf. Chapter 3 for details), there is a straightforward *translation* from our dynamic propositions and procedures to unary and binary formulas in a full first-order predicate language referring to partial orders of states decorated with unary predicates. The resulting equivalences look like this:

$$\Diamond_{\text{up}}\text{domain}(\text{downd}(Q)\,;\text{strict upd}(P))$$

translates into

$$\lambda x.\, \exists y\, (x \subseteq y \wedge \exists u \exists z\, (z \subseteq y \wedge \neg Qz \wedge z \subseteq u \wedge Pu \wedge \neg \exists s\, (z \subseteq s \subset u \wedge Ps))).$$

More systematically, there are two parts to this translation, one taking propositions to formulas $\phi(x)$ with one free variable, and another from programs to formulas $\phi(x, y)$ with two free variables. These call each other recursively via modes and projections. Details will be provided in Chapter 6. The outcome of this analysis is the following.

Proposition 2.13 *There exists a correct translation from dynamic modal logic into first-order predicate logic over state transition models.*

Corollary 2.14 *Universal validity in dynamic modal logic is recursively axiomatizable.*

This raises a general question. Modal logics, 'static' or 'dynamic', are fragments of a first-order logic over information patterns. What fragments are natural for dynamic purposes? This issue will be studied from various angles in Chapters 3 and 4 below.

2.4 Changing Preferences

Information processing has many further parameters. So far, we considered changing denotations for variables (representing local 'perspectives' on some data structure) in DPL, and updates of information states in propositional logic or DML. But in principle, any aspect of a semantic model may be dynamified. An important new kind of example stems from *conditional logic* and default reasoning. That conditionals involve dynamic instructions showed already in the earlier Ramsey Test.

2.4.1 Reasoning over Preference Models

One broad unifying mechanism which has emerged in linguistics and computer science is 'preferential reasoning', which relaxes the classical requirement that conclusions of arguments be true in all models of the premises (Shoham 1988):

The conclusion holds in all *most preferred* models of the premises.

Here, preferences may arise from probabilistic or qualitative heuristic considerations. This is a natural notion in artificial intelligence. Semantic preferences have also been invoked to model 'desires' of discourse participants who prefer being in some (information) states rather than others. Preferential reasoning is quite different from classical inference. In particular, it is defeasible: its conclusions may be overruled by new evidence (telling us that we are in less preferred, unexpected situations after all). And again, it diverges from classical consequence in its *structural rules*. For instance, Monotonicity fails. Likewise, 'transmission of validity' is no longer plausible — which shows in failures of the Cut rule. Modifications of these principles do go through, viz.

'Cautious Cut' from $A_1, \ldots, A_n \models B$ and $A_1, \ldots, A_n, B \models C$, infer $A_1, \ldots, A_n \models C$

'Cautious Monotonicity' from $A_1, \ldots, A_n \models B$ and $A_1, \ldots, A_n \models C$, infer $A_1, \ldots, A_n, B \models C$.

Proposition 2.15 *The structural properties of preferential consequence are axiomatized by the rules of Permutation, Contraction, Expansion and 'Strong Reflexivity' (i.e., $X \models B$ whenever $B \in X$), as well as Cautious Cut and Monotonicity.*

Semantically, preferential models are families of classical models, with a preference relation $<$ which is a strict (possibly well-founded) ordering. Typically, a conditional $A \to B$ will then be true if B is true in all $<$-maximal worlds where A is true.

2.4.2 Using and Modifying Preferences

Preference patterns enter dynamic logic in several ways. First, they increase options for information processing. E.g., in eliminative propositional update semantics, a classical descriptive proposition Q reduces the current state X to the intersection $X \cap Q$. But with a preference relation over valuations, we can also define an atomic update with the following 'heuristic' surplus of information:

$[\![q]\!](X)$ consists of only the most preferred states in $X \cap Q$.

This is a non-continuous update, whose formal properties have been described completely in van Benthem 1989c. (Compare the earlier account of classical updates.)

Proposition 2.16 *A preferential eliminative update operator F is determined completely by the following three properties:*

Introversion	$F(X) \subseteq X$
Splitting	$F(\bigcup_i X_i) \subseteq \bigcup_i F(X_i)$
'De Morgan'	$\bigcap_i F(X_i) \subseteq F(\bigcup_i X_i)$.

The above reference also provides an embedding into basic modal logic, which explains these properties and their consequences in more detail. With preferential updates, order of presentation becomes important, as in the earlier update systems. For instance, there will be different dynamic effects for conjunctions $p \wedge q$, $q \wedge p$ or $p \cap q$. One can also add an explicit action μ of passing to the most preferred worlds in the current state. This allows for an explicit distinction between ordinary and preferential updates, as happens in more sophisticated working systems of circumscription (Sandewal 1994).

But, a more radical dynamification is possible over preference models, which affects the preference relation itself. Some kinds of expression may *change* our expectations (cf. Gärdenfors 1988, Spohn 1988), resulting in new preferences (with or possibly without accompanying changes in factual information). Thus, a default rule "if A, then B" need not be read as excluding any worlds from the current state where $A \wedge \neg B$ holds. It only makes these worlds less relevant to deliberation. This is a semantic process of *upgrading*, of similar status to the earlier updating. Here is an upgrading instruction:

> a conditional "if A, then B' instructs us to *remove* all links preferring an $A \wedge \neg B$ world to an $A \wedge B$ one.

There are other options, too, such as adding preference links for $A \wedge B$ worlds — or giving an additional 'point' to $A \wedge B$ worlds in some numerical ranking of worlds. Veltman 1991 proposes a crisp logical system for these purposes. In addition to update and upgrade conditions, it also has static operators. Linguistic adverbs like *presumably* or *normally* are tests on the current state, whereas a default conditional triggers a change in expectation. To simplify matters, assume that ϕ, ψ are classical formulas. States C now consist of a set of worlds plus a preference order $\leq$ over them, forming an *expectation pattern*. Maximally preferred worlds in such patterns are called *normal*. Incoming propositions may either change the former 'factual' component, or the latter (or both). For instance, given C and ϕ we define the *upgrade* C_ϕ as that expectation pattern which has the same factual component as C, but whose preference relation consists of the old

$\leq$ with all those pairs (w, v) taken out in which we have $v \models \phi$ without $w \models \phi$. Here are some typical semantic upgrade-test clauses in this set-up:

$$[\![normally\ \phi]\!](C) = \begin{cases} C_\phi & \text{if } \phi \text{ is consistent with some normal world} \\ \emptyset & \text{otherwise} \end{cases}$$

$$[\![presumably\ \phi]\!](C) = \begin{cases} C & \text{if } \phi \text{ holds in all maximally preferred} \\ & \text{situations in } C \\ \emptyset & \text{otherwise.} \end{cases}$$

A much more complicated clause takes care of the binary conditional *if ϕ, then ψ*. Veltman 1991 checks the resulting predictions with intuitions about default reasoning.

2.5 General Conclusions

2.5.1 Further Degrees of Freedom

By now, it will have become clear that any standard logical system can be dynamified. Moreover, the resulting dynamic systems manipulate many different kinds of state: information states, variable assignments, preferences, as well as combinations of these. The reader should then be able to (re-)analyze other logical systems in the same spirit. Thus, we have not even exhausted the dynamic potential of Tarski's truth definition

$$D, I, s \models \phi.$$

Not just *variable assignments* s, but also its other parameters admit of dynamization. The literature has many scattered examples of this. Shifting interpretation functions I are involved in ambiguous discourse, or in answering questions, where we learn about denotations of predicates (cf. van Deemter 1991). An answer "Mary" to the question "Who is crying?" may serve as an instruction for updating the denotation of crying:

$$D, I, s[\![\text{ANSWER}(\text{“Who is crying?”}, \text{“Mary”})]\!]D', I', s' \text{ iff}$$
$$D = D' \text{ and } s = s' \text{ and } I =_{\text{cry}} I' \text{ and } I'(\text{“cry”}) = I(\text{“cry”}) \cup \{Mary\}.$$

More sophisticated dynamic views of questions exist. Harrah 1984 describes them as updates of the questioner's information state by suitably structured answers from a respondent having access to an information source. Next, shifting *individual domains* D make sense, too. For instance, there are dynamic readings of an existential quantifier which go beyond DPL in that they introduce a *new object* satisfying some description. Generally, interpretation of natural language involves shifting ranges for generalized quantifiers across sentences (cf. the 'context sets' of Westerståhl 1984). This dynamics underlies a well-known property from the earlier semantic literature. 'Conservativity' of a linguistic determiner ("all", "some", "ten", "most" ...) says that the corresponding generalized quantifier Q has a left noun argument restricting its predicate argument:

$$Q\,AB \text{ iff } Q\,A(B \cap A).$$

This suggests a dynamic refinement of the usual semantics for generalized quantifiers. Processing an expression $Qx.(\phi, \psi)$ involves initial evaluation of the left-hand argument to get a range of relevant values for the variable x, which then serves as input for evaluation of the right-hand argument. Various flavours of this idea occur (van Eijck and de Vries 1991, Chierchia 1995, Kanazawa 1993b, van der Berg 1996). Finally, and surprisingly, *the formula ϕ itself* in Tarski's truth definition is a dynamic parameter. Recent studies of 'context change' in linguistics and artificial intelligence (cf. Perry 1993, McCarthy 1993) satisfy a principle of 'minimal representation' for syntactic code, leaving as much as possible understood in the context of evaluation. This principle will induce changed assertions, as language users move between richer and poorer contexts of evaluation. We shall return to these issues in Chapter 12.

2.5.2 Mathematical Patterns

Our budget of dynamic systems also exhibited some recurrent mathematical patterns. The structure of relevant procedures is often a form of relational algebra, whose architecture is best embedded in some form of modal or dynamic logic. This identifies the main tools that we shall use — of which Chapter 3 provides a short technical survey. Also, certain broad concerns emerged repeatedly. These define the agenda for the theory to be developed in Chapters 4 through 9. In particular, we shall be dealing with (1) choice of natural atomic and procedural repertoires, (2) connections with standard systems: possibilities for reduction via translations, (3) varieties of dynamic inference, (4) architecture of many-component dynamic systems, (5) dynamic changes in the semantics of traditional systems, with a view towards decreasing their complexity. With the results and insights obtained in this way, we shall also have further things to say about the systems of this chapter. For instance, dynamic predicate logic will return in Chapter 12, where we reconsider its complexity vis-a-vis modal and dynamic logics.

2.6 Notes

Propositional Updates. Our update system for propositional modal logic is from Veltman 1991, with Heim 1982 as an ancestor. One can also implement these ideas via partial models (cf. Jaspars 1994). Other update systems use more sophisticated states for propositional update calculi, such as the information systems of Scott 1982, Belnap 1977, or the models of situation theory described in Devlin 1991. Van Eijck and de Vries 1995 analyze propositional update logic in a two-level dynamic logic.

Dynamic Predicate Logic. The DPL system comes from Groenendijk and Stokhof 1991. Assignment change semantics for anaphora was first proposed in Barwise 1987. Earlier sources are the 'discourse semantics' of Seuren 1985 (dating back to 1975), 'file change semantics' (Heim 1982) and discourse representation theory (Kamp 1984). This style of linguistic analysis has been applied to many linguistic phenomena besides anaphora (Dekker 1993, van der Berg 1996, Groenendijk et al. 1996b).

States as Stacks. DPL records only extensional input-output pairs of assignments. Finer-grained dynamic semantics have emerged in various dialects in The Netherlands. For instance, in Visser 1994a, Vermeulen 1994, states are assignments of finite stacks of objects to variables. The procedural super-structure remains the same (viz. $\{\wedge, \sim\}$), but there is a new analysis of atomic actions (where an existential formula $\exists x\, \phi$ is now decomposed into a version with labeled brackets $[_x\ \phi\ _x]$) :

atoms $Pxy\ldots$ test the objects at the top of the stacks for x, y, ...
left brackets $[_x$ add an arbitrary object on top of the x-stack
right brackets $_x]$ remove the top object from the x-stack, if possible.

This semantics will interpret formerly meaningless parts of first-order syntax as free-standing instructions for manipulating register contents. For instance, brackets now denote pops and pushes. Some logical results are in Hollenberg and Vermeulen 1994. An even more sophisticated notion of state drives the calculus of 'referent systems' of Vermeulen 1995, i.e., transducers partially coupling input variables to output variables.

Partiality. A natural modification of states uses *partial* assignments. Formulas/programs only manipulate registers mentioned by them. This introduces finer distinctions into the *atomic repertoire*. There is both 'new assignment' of a value to a new variable, and 're-assignment' of a value to a variable already in use. (Linguistic and recursion-theoretic arguments occur in Fernando 1992, van der Berg 1996). One can also partialize the *procedural repertoire*. Intuitively, test actions have three, not two values: "true", "false" occur at identity transitions, "undefined" at all others. Variable re-assignments seem either "true" or "undefined", never "false". The proper logic employs three-valued relations (marking ordered pairs as 'positive', 'negative' or 'undefined') with suitable algebraic operators. This requires a three-valued partial version of Relational Algebra, with generalized composition and test negation, plus new operators, such as the polarity reversing negation of three-valued logic (cf. Muskens 1995, Thijsse 1992, Jaspars 1994).

Combining Dynamic Systems. Combinations of dynamic logics with different design and aims (such as informational and perspectival update) are studied in van Eijck and Cepparello 1994, Dekker 1993, Groenendijk et al. 1996b, 1994.

Dynamic Modal Logic. Dynamic modal logic was first proposed in van Benthem 1989c (cf. van Benthem 1996d) — under the uninspired name of '$S4^2$' — as a theory of update and revision which might serve as a classical alternative to Gärdenfors 1988. Elaborations and applications may be found in de Rijke 1992, 1994. Many people have used propositional dynamic logics for analyzing belief update and revision, sometimes in combination with other factors, such as deontic norms or real physical change. Some examples from a vast literature are Meyer 1988, Fuhrmann 1990, van der Hoek et al. 1995, Halpern and Friedman 1994a, 1994b, Huang et al. 1996, Reiter 1994.

Stacks and Texts. DML admits of other models. Zeinstra 1990 has stacks of $\subseteq$-successive worlds as states, and lets right/left brackets refer to pops/pushes. Intuitionistic formulas denote relations between stacks. Here is an update condition:

$$s[\![\phi \to \psi]\!]t \text{ iff } \exists r\,(s[\![\phi]\!]r \wedge r[\![\psi]\!]t) \wedge \forall r\,(s[\![\phi]\!]r \Rightarrow \exists u\, r[\![\psi]\!]u).$$

One novel feature is the assimilation of conditionals inside sentences with discourse operators "suppose" in texts. Thus, logical phenomena at different levels can be studied uniformly (cf. Visser 1994a, Vermeulen 1994 on this broader theme). Interestingly, this dynamic system was inspired by a passage in the popular book Hofstadter 1979.

Conditionals as Instructions. Conditionals are a prime dynamic locus. The Ramsey Test is an early example, which has dominated research in conditional logic (Lewis 1972, Veltman 1985). Conditionals were explicit upgrade conditions above. Yet another dynamic viewpoint is this. Truth-functional implications $A \to B$ are static, but they become dynamic by adding a third argument: $A \to_\pi B$ — as in Hoare-style correctness assertions. These express that doing π can bring about B starting from A. Dynamic logic then encodes an update system for action conditionals.

Belief Revision. The Ramsey Test involves belief revision, a topic of independent importance in computer science and AI. An influential account is due to Alchourron, Gärdenfors and Makinson; cf. Gärdenfors 1988, Gärdenfors and Rott 1995. For a deductively closed theory K, and a new insight ϕ, $K * \phi$ is the *revision* of K by ϕ. Its AGM-postulates are, with $K + \phi = \{\psi \mid K, \phi \vdash \psi\}$ (the *expansion* of K by ϕ):

(∗1) $K * \phi$ is deductively closed

(∗2) $\phi \in K * \phi$

(∗3) $K * \phi \subseteq K + \phi$

(∗4) If $K * \phi$ is consistent then $K + \phi \subseteq K * \phi$

(∗5) $K * \phi$ is consistent if $\{\phi\}$ is consistent

(∗6) If ϕ is equivalent with ψ, then $K * \phi = K * \psi$
(∗7) $K * \phi \wedge \psi \subseteq (K * \phi) + \psi$
(∗8) If $(K * \phi) + \psi$ is consistent, then $(K * \phi) + \psi \subseteq K * \phi \wedge \psi$.

The result of giving up a belief is $K \dot{-} \phi$, the *contraction* of K w.r.t. ϕ:

($\dot{-}$1) $K \dot{-} \phi$ is deductively closed
($\dot{-}$2) $K \dot{-} \phi \subseteq K$
($\dot{-}$3) If $\phi \notin K$, then $K \dot{-} \phi = K$
($\dot{-}$4) If $\phi \in K \dot{-} \phi$, then $\vdash \phi$
($\dot{-}$5) $K \subseteq (K \dot{-} \phi) + \phi$
($\dot{-}$6) If ϕ is equivalent with ψ, then $K \dot{-} \phi = K \dot{-} \psi$
($\dot{-}$7) $(K \dot{-} \phi) \cap (K \dot{-} \psi) \subseteq K \dot{-} (\phi \wedge \psi)$
($\dot{-}$8) If $\phi \notin K \dot{-} (\phi \wedge \psi)$, then $K \dot{-} (\phi \wedge \psi) \subseteq K \dot{-} \phi$

The operations $*$ and $\dot{-}$ are related:

L $K * \phi := (K \dot{-} \neg\phi) + \phi$ Levi Identity

H $K \dot{-} \phi := K \cap K * \phi$ Harper Identity

The AGM-postulates have several models. Let $K \perp \phi$ be the set of all maximal subsets of K which do not imply ϕ. Let γ be a map with $\gamma(K \perp \phi) \neq \emptyset$, $\gamma(K \perp \phi) \subseteq K \perp \phi$ if $K \perp \phi \neq \emptyset$, $\gamma(K \perp \phi) = \{K\}$ otherwise. The *partial meet contraction* $K \dot{-} \phi$ is $\bigcap \gamma(K \perp \phi)$.

Theorem 2.17 *Partial meet contraction satisfies* ($\dot{-}$1)–($\dot{-}$6)*. Conversely, any operation that satisfies postulates* ($\dot{-}$1)–($\dot{-}$6) *is a partial meet contraction operation.*

Another construction uses *epistemic entrenchment*. Giving up some beliefs has more drastic consequences as giving up others. The resulting preference relation $\phi \leq \psi$ (ψ is *at least as epistemologically entrenched as* ϕ) says that ϕ, ψ are both logical truths, or ϕ is not believed at all, or a need to give up one of ϕ or ψ leads to discarding ϕ.

(EE1) If $\phi \leq \psi$ and $\psi \leq \chi$, then $\phi \leq \chi$
(EE2) If $\phi \vdash \psi$, then $\phi \leq \psi$
(EE3) $\phi \leq \phi \wedge \psi$ or $\psi \leq \phi \wedge \psi$
(EE4) If K is consistent, then $\phi \notin K$ iff $\phi \leq \psi$ for all ψ
(EE5) If $\phi \leq \psi$ for all ϕ, then $\vdash \psi$

Given a contraction relation, **C** below defines a relation of epistemic entrenchment. Conversely, given an entrenchment relation $\leq$, **E** defines a contraction relation:

C $\phi \leq \psi$ iff $\phi \notin K \dot{-} (\phi \wedge \psi)$ or $\vdash \phi \wedge \psi$

E $K \dot{-} \phi = K \cap \{\psi \mid \phi < \phi \vee \psi\}$ if not $\vdash \phi$, and K otherwise.

A third model for the AGM-postulates is Bayesian statistics. Conditional probability functions that were axiomatised by Popper give revision functions satisfying $(*1)$–$(*8)$ above, and again a representation theorem can be proved. Clearly, there are many analogies between this theory of belief revision and the various themes of this chapter.

General Upgrading. Dynamifying preference models for conditional logics was first proposed in Spohn 1988 as a possible worlds analogue to probabilistic updates of belief functions. The upgrade system presented here, geared toward default reasoning triggered by conditionals and adverbial modifiers, is from Veltman 1991. Similar ideas with a motivation from AI occur in Boutilier 1993, Boutilier and Goldszmidt 1993, and more elaborately, Pearl 1996. Jaspars and Kameyama 1996 is an up-to-date linguistic version, linking up with discourse processing strategies. An abstract account of changing preferences is in van Benthem et al. 1993, whose upgrade conditions add or remove preference arrows (in the sense of Arrow Logic: Chapter 8 below), with outcomes described in static preferential logics.

Questions and Queries. Another meeting ground for linguistics and computer science is the dynamics of questions. The pioneering work of Nuel Belnap and Jaakko Hintikka (surveyed in Groenendijk and Stokhof 1996) made it clear how *answers are updates* of the questioner's information state. This can be described, for instance, in dynamic modal logic and its quantificational versions. Asking "whether φ" invites an update to a state where I know φ or I know $\neg\varphi$ — while asking "who φ -s" invites an update to a state where I know of each individual whether it φ -s. Thus, answers have static *epistemic postconditions*, which can be used to classify natural types of question. Of course, the full setting must include 'higher-order' updates of knowledge that questioner and answerer have about each other. Somewhat different algorithmic concerns are found in the study of *query languages* over databases — and it remains to be seen how one can merge these two agendas in one dynamic logic.

Dynamic Parameters. Systematic 'Tarskian Variations' were proposed in van Benthem 1991a, and elaborated in van Benthem and Cepparello 1994, Cepparello 1995, who also consider 'Kripkean Variations' for intensional logics. Further parameters arise in the rich architecture for natural language processing of Kameyama 1993, 1994, which combines (amongst others) long-term default preferences from a knowledge base with short-term preferences arising from discourse processing. Cf. also the discussion of AI-style context change in Buvac and Fikes 1995. 'Language change' is also becoming a recognized desideratum in the AI theory of belief revision. As for computer science, Groenboom and Renardel de Lavalette 1994 define a dynamic modal logic for creation and modification of objects in the semantics of specification languages.

Semantic Tableaus as Dynamic Structures. This chapter does not exhaust the dynamic content of standard logic. To show this, we permit ourselves a little encore. An interesting alternative locus of dynamics is Beth's validity test of *semantic tableaus*. (E.g., Aliseda 1996 proposes new algorithmic uses of tableaus in defining abduction.) Tableau rules are instructions for stepwise construction of (counter-) models, witness

$\exists x\, \phi, \Sigma \bullet \Delta :$ introduce a new object d and continue with the new construction task $\phi(d)$, $\Sigma\Delta$.

This is a dynamic instruction adding a new object to the current domain of discourse. Alternatively, tableau rules are instructions for evaluation, with the 'introduction' of d a result of sampling from some given domain. Bottom up, tableau rules become left- and right-introduction rules in a Gentzen sequent calculus. But, one can also view tableaus as a record of moves in a *game* (Lorenzen). Consider a zero-sum game with player I claiming the existence of a counter-example, and player II denying this. Tableau rules then represent natural game conventions. A negation instruction

$\neg\phi, \Sigma \bullet \Delta :$ change to $\Sigma \bullet \Delta, \phi$

is a role change from defending $\neg\phi$ to attacking ϕ, and so is its converse. Rules for conjunction and disjunction may introduce multiple commitments or choices:

$\phi \wedge \psi, \Sigma \bullet \Delta :$ change to ϕ, ψ, $\Sigma \bullet \Delta$

$\Sigma \bullet \Delta, \phi \wedge \psi :$ continue with a choice for player I as to which component is to be taken: $\Sigma \bullet \Delta$, ϕ or $\Sigma \bullet \Delta$, ψ.

Winning positions for player II are atomic end sequents where some formula occurs on both sides. In textbooks, these three views on tableaus are engineered to produce the same valid sequents (Lorenzen and Lorenz 1979 lean hard on games to fit the mould). But in a dynamic perspective there may be different outcomes, reflecting decisions about the procedure executed, or the game played. E.g., consider the rules for existential quantifiers $\exists x\, \phi$. The above convention on the left lets player I choose a 'witness' d such that the game continues with $\phi(d)$. Dually, on the right, the choice of a 'challenger' d is the privilege of player II.

This plausible game will not lead to standard predicate-logical validity. As all proof rules corresponding to game rules increase complexity, proof search terminates in a finite number of steps. So this style of dynamic inference is *decidable*. The difference with the standard case is that, once treated, a formula disappears from the list of tasks. Many predicate-logical validities require more than one try on the same false existential (true universal) formula. Put differently, classical logic requires the structural rule of Contraction, which is invalid in this game. The latter rather suggests a finer systematic distinction. The primary case is an obligation which can be discharged in one single act. But there may also be 'standing obligations' which involve multiple discharges. Examples in ordinary argumentation are standing commitments to universal quantifiers or conditional statements ("whenever you challenge me with the antecedent, I will respond with the consequent"). So, one needs some explicit way of *distinguishing* one from the other. Again, this additional expressive power motivates the introduction of a new dynamic logical operator ! — this time encoding commitment. Given its intended interpretation, a formula $!\phi$ will be treated via the earlier rule for ϕ but with the difference that $!\phi$ itself is also inherited in the process. Corresponding proof rules are:

$$\frac{!\phi, \Sigma \bullet \Delta}{!\phi, \phi, \Sigma \bullet \Delta} \qquad\qquad \frac{\Sigma \bullet \Delta, !\phi}{\Sigma \bullet \Delta, !\phi, \phi}$$

E.g., the valid predicate-logical formula $\exists x\,((Pa \vee Pb) \rightarrow Px)$ is unprovable here, since it requires 'two tries', but one can prove its 'repetitive version' $!\exists x\,((Pa \vee Pb) \rightarrow Px)$. Full predicate logic can be embedded in the system with the new dynamic operator.

3

Technical Tools

In developing a general theory of logical dynamics we draw upon standard mathematical logic. This chapter provides basic background on *First-Order Logic*, *Modal Logic*, *Relational Algebra* and *Dynamic Logic*. Our exposition is not a manual for learning these topics, for which there exist excellent textbooks. Readers familiar with these matters may skip this chapter at first reading. Most of what follows is well-known, except for the view we take of modal languages and corresponding decidable fragments of first-order logic. This illustrates a larger theme in later chapters. Even though our point of departure lies in well-known systems, the *questions* that we are going to ask about them from the dynamic stance will be largely new — and this will show them in a fresh light.

3.1 First-Order Predicate Logic

We provide a quick tour of first-order logic, to outline the 'iron rations' for this book. For excellent introductions, see Enderton 1972, Hodges 1983.

3.1.1 Language and Semantics

The languages that we need have an alphabet containing *individual variables* x, y ..., *individual constants* c, d, ... *predicate symbols* P, Q, ... (arities indicated), an *identity predicate* $=$, *Boolean connectives* $\neg$, $\wedge$, $\vee$, $\rightarrow$, $\leftrightarrow$, *quantifiers* $\exists$, $\forall$, *brackets*), (.

Example 3.1 (Languages over Transition Models) First-order languages whose vocabulary fits labeled transition systems express various properties of computations by means of quantification over states:

- Action a is deterministic: $\forall xy\,(R_a xy \wedge R_a xz \rightarrow y = z)$
- Action a is confluent: $\forall xyz\,((R_a xy \wedge R_a xz) \rightarrow \exists u\,(R_a yu \wedge R_a zu))$
- Action a enables b: $\forall xy\,(R_a xy \rightarrow \exists z\, R_b yz)$.

Further possibilities exist. One can take a second domain of *actions* in addition to *states*, with action quantifiers and variables, and introduce a ternary predicate $x \xrightarrow{a} y$ (action a takes state x to state y). Labeled transition systems are also models for this first-order language, and one can express new computational properties:

- Endpoints exist: $\exists x \neg \exists a\, x \xrightarrow{a} y$
- Every action has a converse: $\forall a \exists b \forall xy\, (x \xrightarrow{a} y \leftrightarrow y \xrightarrow{b} x)$
- Actions are extensional: $\forall ab\, (\forall xy\, (x \xrightarrow{a} y \leftrightarrow x \xrightarrow{b} y) \rightarrow a = b)$.

Our syntax is as follows. *Terms* are individual variables and constants. (We disregard function symbols in this book.) *Atomic formulas* are of the form $Pt_1 \ldots t_k$ with P of arity k and $t_1, \ldots, t_k$ terms. *Formulas* are defined inductively as usual. We have the standard grammar, with *scope*, *binding*, *free* and *bound occurrence*, *substitution*, *substitutability*, *alphabetic variants*, *positive* and *negative occurrence*. A useful measure is the *quantifier depth* of a formula: i.e., the maximum length of a nest of quantifiers occurring in each other's scope within it. Recursively,

$$\begin{aligned} qd(\phi) &= 0 \text{ for atomic formulas } \phi \\ qd(\neg\phi) &= qd(\phi) \\ qd(\phi \# \psi) &= \max(qd(\phi), qd(\psi)) \text{ for all connectives } \# \\ qd(Qx\,\phi) &= qd(\phi) + 1 \text{ for both quantifiers } Q. \end{aligned}$$

Semantic interpretation uses structures $\mathbb{D} = (D, \mathbb{O}, \mathbb{P})$, with D a domain of objects, $\mathbb{O}$ a set of distinguished objects, and $\mathbb{P}$ a set of predicates. (Later Chapters will use ad-hoc notation for structures.) An *interpretation function* I maps individual constants c to objects $I(c) \in \mathbb{O}$, and k-ary predicate symbols Q to k-place predicates $I(Q) \in \mathbb{P}$. An assignment a maps individual variables x to objects $a(x) \in D$. Here, I is a more permanent linkage and a a more local 'dynamic' one. Next come term values :

$$\begin{aligned} value(x, \mathbb{D}, I, a) &= a(x) \\ value(c, \mathbb{D}, I, a) &= I(c). \end{aligned}$$

Now, Tarski's *truth definition* defines the central notion "*ϕ is true in $\mathbb{D}$ under I and a*":

$$\mathbb{D}, I, a \models \phi$$

through the following inductive clauses (the atomic case is an example):

$$\begin{aligned} \mathbb{D}, I, a \models Rt_1t_2 \quad &\text{iff} \quad I(R)(value(t_1, \mathbb{D}, I, a), value(t_2, \mathbb{D}, I, a)) \\ \mathbb{D}, I, a \models t_1 = t_2 \quad &\text{iff} \quad value(t_1, \mathbb{D}, I, a) \text{ equals } value(t_2, \mathbb{D}, I, a) \\ \mathbb{D}, I, a \models \neg\phi \quad &\text{iff} \quad \text{not } \mathbb{D}, I, a \models \phi \\ \mathbb{D}, I, a \models \phi \wedge \psi \quad &\text{iff} \quad \mathbb{D}, I, a \models \phi \text{ and } \mathbb{D}, I, a \models \psi \\ &\qquad \text{and likewise for the other propositional connectives} \\ \mathbb{D}, I, a \models \exists x\,\phi \quad &\text{iff} \quad \text{there exists some } d \in D \text{ with } \mathbb{D}, I, a^x_d \models \phi \\ &\qquad \text{and likewise for the universal quantifier.} \end{aligned}$$

Here a^x_d is the assignment b which is like a except for the possible difference that it assigns object d to the variable x. (The resulting relation is also written as $a =_x b$).

Often, the function I is taken for granted. We write $\mathbf{M}, a \models \phi$, where a *model* $\mathbf{M}$ is a pair of a structure and an interpretation function. Here are some useful basic properties of this semantics. The *Finiteness Lemma* says that, if two assignments agree on all free variables of a formula, then they give the same truth value to that formula (in some fixed model). Hence, for truth of 'sentences' without free variables, assignments may be disregarded. The *Substitution Lemma* is the following important equivalence, which holds provided that the term t is freely substitutable for x in the formula ϕ:

$$\mathbf{M}, a \models [t/x]\phi \text{ iff } \mathbf{M}, a^x_{value(t,\mathbf{M},a)} \models \phi.$$

Finally, we mention another equivalence between a syntactic and a semantic operation. Let $(\phi)^A$ be the formula arising from ϕ by replacing every quantifier $\exists x$, $\forall x$ in it by its *relativized* version $\exists x\,(Ax \wedge \ldots$, $\forall x\,(Ax \rightarrow \ldots$. Let $\mathbf{M} \upharpoonright A$ be the model obtained from $\mathbf{M}$ by taking the domain $I(A)$ and restricting every predicate interpretation to it. The *Relativization Lemma* states that, for assignments a taking their values in $I(A)$:

$$\mathbf{M}, a \models (\phi)^A \text{ iff } \mathbf{M} \upharpoonright A, a \models \phi.$$

These notions set a pattern for many other formal languages. One of these is *infinitary first-order logic* allowing conjunctions and disjunctions over arbitrary sets of formulas. (For the countable case, cf. Keisler 1971, and for the general one, Barwise 1975.) Infinitary languages will appear at several places in what follows.

3.1.2 Invariance and Games

Independently from any formal language, structures have their mathematical relations. Most basically, two models $\mathbf{M} = (D, \mathbb{O}, \mathbb{P}, I)$, $\mathbf{M}' = (D', \mathbb{O}', \mathbb{P}', I')$ are *isomorphic* if there exists an *isomorphism* between them, i.e., a bijection F between D and D' which respects distinguished objects in $\mathbb{O}$, $\mathbb{O}'$ (and operations, if these are relevant) and which also respects corresponding predicates in $\mathbb{P}$, $\mathbb{P}'$:

$$\begin{array}{rcl} F(I(c)) & = & I'(c) \text{ for all individual constants } c \\ I(Q)(\mathbf{d}) & \text{iff} & I'(Q)(F(\mathbf{d})) \text{ for all tuples of objects } \mathbf{d} \text{ in } D. \end{array}$$

Other important model relations and constructions are *submodel*, *direct product* and *ultraproduct*. A basic measure of expressive power of a formalism are its characteristic invariances across models, via such structural relations. The *Isomorphism Lemma* says that tuples related by an isomorphism cannot be distinguished by first-order formulas:

$$\mathbf{M}, a \models \phi \text{ iff } \mathbf{M}', F \circ a \models \phi.$$

In particular, isomorphic models are *elementarily equivalent*: they verify the same first-order sentences. The converse fails badly, except on *finite models*. Closer to elementary equivalence, two models $\mathbf{M}$, $\mathbf{M}'$ are *partially isomorphic* (also, 'potentially isomorphic') if there exists a non-empty *zigzag family* PI of finite *partial isomorphisms* (isomorphisms between finite submodels of $\mathbf{M}$, $\mathbf{M}'$) satisfying the following back-and-forth clauses:

- for any partial isomorphism $F \in \mathrm{PI}$ and any d in the domain of $\mathbf{M}$, there exists an object e in the domain of $\mathbf{M}'$ such that $F \cup \{(d, e)\} \in \mathrm{PI}$
- and analogously in the opposite direction.

The Isomorphism Lemma adapts at once to partial isomorphisms in zigzag families PI. Again, no converse holds for first-order logic — but we do have the following important *infinitary* equivalence, for tuples of objects in any two models (Karp's Theorem):

> $\mathbf{M}, \mathbf{d}$ and $\mathbf{N}, \mathbf{e}$ satisfy the same formulas of infinitary first-order logic iff there exists a zigzag family PI between $\mathbf{M}$, $\mathbf{N}$ containing some partial isomorphism F sending $\mathbf{d}$ (in that order) to $\mathbf{e}$.

To get finer semantic comparisons, one uses *Ehrenfeucht-Fraïssé Games*. The 'n-round comparison game' between two models $\mathbf{M}$, $\mathbf{N}$ is played by two players I and II in n successive rounds, each consisting of (i) selection of a model and an object in its domain by player I, (ii) selection of an object in the other model by player II. After n rounds, the partial map between the two domains created by these pairs is inspected. If it is a partial isomorphism respecting all predicates, then player II has won, otherwise the win is for player I . One writes $\mathrm{II}(\mathbf{M}, \mathbf{N}, n)$ if player II has a winning strategy in this model comparison game. The following theorem is basic:

> $\mathrm{II}(\mathbf{M}, \mathbf{N}, n)$ iff $\mathbf{M}$, $\mathbf{N}$ verify the same sentences of quantifier depth n.

It rests on a more general equivalence with assignments, easily proved by induction on first-order formulas. (Cf. Doets 1996, an excellent game-based course of mathematical logic.) This analysis can be refined even further. We shall often be concerned with *finite-variable fragments* of first-order logic, which restrict attention to formulas written with some fixed finite set $\{x_1, \ldots, x_k\}$ of variables (free or bound). The corresponding method is an Ehrenfeucht-Fraïssé game with *pebbling* (Immermann and Kozen 1987). There are k pairs of matched pebbles which players must put on objects to mark them. Thus, they can only create partial isomorphisms of size at most k. The above definitions remain the same, with $\mathrm{II}(\mathbf{M}, \mathbf{N}, n, k)$ stating a guaranteed win for player II in the k-pebble game over n rounds. Then, we get the following equivalence:

> $\mathrm{II}(\mathbf{M}, \mathbf{N}, n, k)$ iff $\mathbf{M}$, $\mathbf{N}$ verify the same first-order sentences up to quantifier depth n *inside the* $\{x_1, \ldots, x_k\}$-fragment of first-order logic.

These semantic games can be specialized to various other fragments of first-order logic (cf. the following section on Modal Logic). They also inspire the theory of process equivalences in Chapters 4, 5. Finally, another structural characterization of first-order predicate logic is *Keisler's Theorem*, of which we just state a simple version. A class of models is *elementary* — i.e., definable in the form $\{\mathbf{M} \mid \mathbf{M} \models \phi\}$ for some first-order sentence ϕ, iff it is closed under the formation of ultraproducts and zigzag families.

3.1.3 Model Constructions

Two central features of first-order predicate logic are the following results.

Compactness Theorem. *If each finite subset of a set of formulas Σ has a model plus assignment verifying it, then so does the whole Σ simultaneously.*

Löwenheim-Skolem Theorem. *If a set of formulas Σ is verified in some model plus assignment, then it is already verified in some countable model.*

These theorems have many applications in constructing new models. Also, non-first-orderness is often proved through them. (A standard example is the first-order undefinability of *finiteness* of the domain.) And indeed, they are even characteristic of the whole system. *Lindström's Theorem* says that first-order predicate logic is the strongest logic satisfying both Compactness and Löwenheim-Skolem. Next, we state a compactness-based technique that we will use repeatedly (cf. Chapters 4, 5, 10). It involves an auxiliary model relation. $\mathbf{N}$ is an *elementary extension* of $\mathbf{M}$ if (i) $\mathbf{M}$ is a submodel of $\mathbf{N}$ such that (ii) $\mathbf{M}$, $\mathbf{N}$ agree on the truth value of all first-order formulas over objects in $\mathbf{M}$. Now, we say that a model $\mathbf{M}$ is *ω-saturated* if, for each set Σ of formulas in a finite set of free variables x_1, $\ldots$, x_k and involving only finitely many objects from $\mathbf{M}$'s domain as fixed parameters, the following holds:

> if each finite subset of Σ has a k-tuple of objects satisfying it in $\mathbf{M}$, then there exists a k-tuple of objects satisfying the whole set Σ in $\mathbf{M}$.

Theorem 3.2 *Each model has an ω-saturated elementary extension.*

For a proof, cf. Chang and Keisler 1973, Keisler 1977. In particular, *finite* models must be ω-saturated, as they have no proper elementary extensions. (The statement of their cardinality is first-order definable, whence it would transfer to the extension.)

Survival of relations between models, even when not valid for all of predicate logic, may still be a good test for its useful *fragments*. Here are some preservation theorems. A first-order sentence ϕ is *preserved under submodels* if, for all models $\mathbf{M}$, $\mathbf{M} \models \phi$ implies $\mathbf{N} \models \phi$ for all submodels $\mathbf{N}$ of $\mathbf{M}$. The *Łoś-Tarski Theorem* says that a first-order sentence is preserved under submodels iff it is logically equivalent to a *universal* formula, made from a quantifier-free one by prefixing universal quantifiers. (Effective enumerability of these sentences may be predicted a priori. ϕ is preserved under submodels iff the semantic implication $\phi \models (\phi)^A$ is valid.) Next, a first-order sentence ϕ is *monotone* in the predicate symbol Q if, for each model $\mathbf{M}$, $\mathbf{M} \models \phi$ implies $\mathbf{N} \models \phi$ for all models $\mathbf{N}$ differing from $\mathbf{M}$ only in having a larger extension for the predicate $I(Q)$. *Lyndon's Theorem* says that a first-order sentence is monotone in Q iff it is logically equivalent to one with only *positive* occurrences of Q. Monotonicity is an important principle of reasoning in natural language (van Benthem 1986), and also in the mathematical theory of inductive definitions and fixed points. It will occur often in this book. A stronger requirement occurring occasionally is this. Sentence ϕ is *continuous* in the predicate symbol Q if, for each model $\mathbf{M}$ with $I(Q) = \bigcup\{Q_i \mid i \in I\}$, $\mathbf{M} \models \phi$ iff there exists some $i \in I$ such that $\mathbf{M}_i \models \phi$, where $\mathbf{M}_i$ is a model exactly like $\mathbf{M}$ except for having $I_i(Q) = Q_i$.

3.1.4 Validity and Axiomatization

Valid consequence in predicate logic is defined as 'transmission of truth':

> $\Pi \models \gamma$ if for all models $\mathbf{M}$ and assignments a, if $\mathbf{M}, a \models \pi$ for all π in Π, then $\mathbf{M}, a \models \gamma$.

On this notion of validity, statements in the following pairs are mutual consequences, and hence, logical synonyms: $\phi \vee \psi \,/\, \neg(\neg\phi \wedge \neg\psi)$, $\phi \rightarrow \psi \,/\, \neg\phi \vee \psi$, $\forall x\, \phi \,/\, \neg\exists x\, \neg\phi$. Therefore, attention may be restricted to just the logical constants $\neg$, $\wedge$ and $\exists$, as we shall often do in this Book. (This is only permitted on our classical semantics: these equivalences are invalid, e.g., in intuitionistic logic.) The quantification over the totality of all models makes validity a quite abstract notion. But, there exist more concrete methods of testing it, such as Beth's 'semantic tableaus'. More concrete from the start is the *deductive* approach, proceeding from combinatorial systems of inferential steps:

> $\Pi \vdash \gamma$ if there exists a *derivation* for γ from assumptions in Π using only permissible rules of some logical calculus.

In this book, we will mostly ignore deductive aspects (a good text on proof theory is Troelstra and Schwichtenberg 1996). But in Chapter 9, we shall have occasion to scrutinize one particular axiom system for predicate logic (Enderton 1972). These consists of all universal closures of arbitrary Boolean propositional laws (which we shall always take for granted here) plus the three quantifier axioms

$\forall x\, (\phi \rightarrow \psi) \rightarrow (\forall x\, \phi \rightarrow \forall x\, \psi)$

$\phi \rightarrow \forall x\, \phi$, provided that x do not occur free in ϕ

$\forall x\, \phi \rightarrow [t/x]\phi$, provided that t be free for x in ϕ.

Further from the mainstream, various authors have revived the original source of logic as a debating game. This gives a third notion of validity (Lorenzen and Lorenz 1979):

$\Pi \Vdash \gamma$ if a defender of γ has a guaranteed *winning strategy* against an opponent granting Π in a logical game of argumentation.

Hintikka 1973 even has a game-theoretical approach to semantic interpretation itself, between a 'verifier' and a 'falsifier', who draw objects from the domain of discourse which can be tested for atomic facts. Usually, semantic notions of validity serve as a touch-stone of adequacy for proof-theoretic or game-theoretic proposals. But the latter provide more vivid ideas about structuring of arguments and procedures for reasoning. The main technical result in this area is Gödel's *Completeness Theorem*, stating that (for suitable axiom systems, such as the above):

$$\Pi \models \gamma \text{ iff } \Pi \vdash \gamma \text{ for all } \Pi, \gamma \text{ in first-order logic.}$$

Probably the best-known proof method for this result uses *Henkin models* — but again, these will hardly be used in what follows.

3.1.5 Decidability and Undecidability

Validity in predicate logic is undecidable, by *Church's Theorem*. In this book, we shall often work with *fragments* of the full language, where matters are more complicated (cf. Dreben and Goldfarb 1979). Here are some facts of relevance to what follows. *Monadic* first-order logic (with only one-place predicates) is decidable, and so is valid consequence between *universal* formulas (as long as we have no function symbols). The *two-variable* fragment of first-order logic is decidable, but its *three-variable* fragment is already undecidable (this is the natural habitat of relational algebra). *Bounded fragments* with only *relativized* quantifiers of various sorts, are sometimes, though not always, decidable (Chapter 4). Decidability results naturally call for *complexity* results concerning degrees of 'feasibility', but we shall not pursue the latter line of investigation in this book.

3.2 Modal Logic

3.2.1 Basic Language and Semantics

The base language of modal propositional logic has the following formulas (stated in Backus-Naur Form):

$$\begin{array}{lcl} P & ::= & \text{propositional atoms } p, q, r, \ldots \\ F & ::= & P \mid \neg F \mid (F \wedge F) \mid \Diamond F \end{array}$$

Further standard propositional items, such as "false" ($\bot$), disjunction ($\vee$), implication ($\rightarrow$), are available by definition, as usual. Moreover, the modal necessity operator may be introduced as follows:

$$\Box\phi = \neg\Diamond\neg\phi.$$

This language is interpreted over possible worlds frames $\mathbf{F} = (W, R)$ where R is a binary relation over the 'worlds' W ('accessibility') enriched with a 'valuation' V encoding truth values for propositions at states, yielding possible worlds models

$$\mathbf{M} = \langle S, R, V\rangle$$

The truth definition now runs via the following recursion. Intuitively, $\mathbf{M}, s \models \phi$ says that ϕ is true at world s:

$$\begin{array}{rll} \mathbf{M}, s \models p & \text{iff} & s \in V(p) \\ \mathbf{M}, s \models \neg\psi & \text{iff} & \text{not } \mathbf{M}, s \models \psi \\ \mathbf{M}, s \models \phi_1 \wedge \phi_2 & \text{iff} & \mathbf{M}, s \models \phi_1 \text{ and } \mathbf{M}, s \models \phi_2 \\ \mathbf{M}, s \models \Diamond\phi & \text{iff} & \text{there exists } s' \text{ with } Rss' \text{ and } \mathbf{M}, s' \models \phi. \end{array}$$

In particular, modal possibility $\Diamond$ is a (restricted) existential quantifier. Dually, modal necessity $\Box$ becomes a restricted universal quantifier ('in all R-successors'). We shall be sloppy concerning the preferred primitive operators, with convenience as our lode-star. This simple system has served as a 'logical laboratory' for many years, showing the effects of working with relatively weak expressive power over interesting models allowing changing denotations across different indices (possible worlds, points in time, computational states). In the remainder of this book, we shall work mainly with an inessential (though interesting and suggestive) variation, viz. poly-modal logic. Its models live on frames which are nothing but the earlier *labeled transition systems*:

$$\mathbf{M} = \langle S, \{R_a\}_{a\in A}, V\rangle$$

whose various atomic actions R_a are described by corresponding modalities $\langle a\rangle$, $[a]$.

Example 3.3 (Computational Properties of Actions) Poly-modal logic expresses various types of statement about execution of programs.

1. *Partial Correctness* Precondition ϕ implies the truth of postcondition ψ following every successful execution of program a: $\phi \to [a]\psi$.
2. *Termination* Program a terminates from the current state: $\langle a\rangle\top$.
3. *Enabling* Program a 'enables' program b to produce effect ϕ: $[a]\langle b\rangle\phi$.

3.2.2 Invariance and Model Constructions

When analyzing the expressive power of a formalism, one seeks a semantic notion of equivalence between models providing a 'closest fit'. Isomorphism does guarantee invariance as for first-order logic, but it is way off the mark. The best candidate for modal logic is a well-known notion (cf. van Benthem

1976, Park 1981) relating worlds or states in two models in such a way that successor steps can always be imitated.

Definition 3.4 (Bisimulation) Let **M**, **N** be two possible worlds models. A *bisimulation* is a binary relation $\equiv$ between worlds in these two models satisfying the following two conditions:

'Atomic Harmony' If $x \equiv y$, then $\mathbf{M}, x \models p$ iff $\mathbf{N}, y \models p$, for all proposition letters p

'Modal Zigzag' If $x \equiv y$ and xR_az, then there exists u such that yR_au and $z \equiv u$.
If $x \equiv y$ and yR_au, then there exists z such that xR_az and $z \equiv u$.

Bisimulation resembles the earlier notion of a zigzag family, for the modal language. A straightforward induction shows that modal formulas are *invariant for bisimulation.*

Proposition 3.5 (Invariance Lemma) *Let* $\equiv$ *be a bisimulation between two models* **M**, **N** *with* $x \equiv y$. *Then, for all modal formulas* ϕ, $\mathbf{M}, x \models \phi$ *iff* $\mathbf{N}, y \models \phi$.

Proof. The atomic step uses precisely the first clause in the definition of bisimulation. The Boolean steps are for free, as usual. The key step is that for modalities $\langle a \rangle \phi$, where the zigzag conditions provide precisely what is needed. ⊣

Incidentally, this observation subsumes the usual facts in the modal literature about preservation under 'generated submodels', 'disjoint unions' and 'p-morphic images'. The latter again imply corresponding facts for pure frame constructions. Conversely, one can prove the following implication (van Benthem 1976), which makes the semantic fit between modal equivalence and bisimulation almost perfect:

Proposition 3.6 (From Invariance to Simulation) *Assume that* $\mathbf{M}, x \models \phi$ *iff* $\mathbf{N}, y \models \phi$, *for all modal formulas* ϕ. *Then there exist elementary extensions* $\mathbf{M}^+$, $\mathbf{N}^+$ *of* **M**, **N**, *respectively, and a bisimulation* $\equiv$ *between these such that* $x \equiv y$. *The converse implication also holds.*

For a proof, see Chapter 4. In particular, this implies that, with *finite* models **M**, **N**, two states x, y verify the same modal formulas if and only if they are connected by some bisimulation. This connection does not hold in general (cf. Hennessy and Milner 1985 on this). We shall return to these matters in the broader perspective of Chapter 4.

3.2.3 Translation into First-Order Logic

Modal languages may be viewed as fragments of standard first-order languages via the so-called 'standard translation' (cf. van Benthem 1976,

1985a). There is an effective map taking modal formulas ϕ to first-order formulas $\underline{\phi}$ with one free variable (for the 'current world' of evaluation) recording their truth conditions on transition systems.

Definition 3.7 (Standard Translation) The following recursion sends modal formulas to corresponding first-order ones:

$$\begin{aligned}\underline{p} &= Px \\ \underline{\neg\phi} &= \neg\underline{\phi} \\ \underline{\phi \wedge \psi} &= \underline{\phi} \wedge \underline{\psi} \\ \underline{\langle a\rangle\phi} &= \exists y\,(R_a xy \wedge \underline{\phi}(y)),\end{aligned}$$

where y is a fresh variable over worlds.

Example 3.8 (Translation of Modal Enabling)

$$[a]\langle b\rangle p \text{ translates to } \forall y\,(R_a xy \to \exists z\,(R_b yz \wedge Pz)).$$

Now, possible worlds models are also standard models for the first-order language employed in this translation. Evidently, a modal formula ϕ is true at state s in a model **M** if and only if its first-order translation $\underline{\phi}(x)$ is true in **M** under the variable assignment sending x to s. By exercising some care in our management of variables, an additional observation can be made (Gabbay 1981a):

Proposition 3.9 (Two Variables Suffice) *Every modal formula has a first-order equivalent employing only two variables over states (free or bound).*

E.g., in the previous example, the following would have worked just as well:

$$\forall y\,(R_a xy \to \exists x\,(R_b yx \wedge Px)).$$

Later on, we shall look at general finite-variable fragments of first-order languages. The translation can easily be extended to richer modal formalisms (cf. Chapter 5, or de Rijke 1993a). These may require essentially higher numbers of variables. E.g., standard temporal logic needs three for an adequate translation of its temporal 'betweenness'.

Example 3.10 (Translating "*Until*") The first-order translation for

Until pq

('q holds until p') is

$$\exists y\,(Rxy \wedge Py \wedge \forall z\,((Rxz \wedge Rzy) \to Qz)).$$

An embedding into predicate logic gives a number of facts about modal logic for free, namely the universal properties of first-order logic which are inherited by its fragments, such as the Löwenheim-Skolem Theorem, or by all its effective fragments, such as recursive axiomatizability for universal validity. The embedding need not give syntactic specifics. For instance, if a modal formula is preserved under submodels, it has a universal first-order equivalent by the Łoś-Tarski theorem: but we do not know a priori whether this equivalent must be *modal* itself. Thus, meta-properties making existential claims need more detailed analysis (cf. Andréka et al. 1995b for more thorough discussion). Here is a useful syntactic description. The class of translated modal formulas corresponds effectively to the following fragment of first-order logic.

Definition 3.11 (Modal Fragment of First-Order Logic) The *modal fragment* of the above first-order logic consists of all those formulas that can be formed using unary atoms, Boolean operations, and restricted existential quantifiers of the form

$$\exists z\,(R_a yz \wedge \ldots$$

(with y, z distinct variables).

Modal Logic is mainly concerned with unary first-order formulas $\phi(x)$ describing properties of states. But the Relational Algebra and Dynamic Logic of later chapters also involve binary first-order formulas $\pi(x, y)$ for program expressions denoting binary transition relations between states. The expressive power of the basic modal language with respect to classical logic is measured precisely by the following preservation result (van Benthem 1976, 1985a):

Theorem 3.12 (Invariance Theorem) *A first-order formula $\phi(x)$ is equivalent to the translation of a modal formula iff it is invariant for bisimulations between models.*

This result will be proved in Chapter 4 — where it will serve as a point of departure for a much broader analysis of simulation invariance and modal definability. Moreover, we shall find much richer fragments of first-order logic with pleasant 'modal' properties.

More finely-grained semantic analyses are possible. One can specialize the Ehrenfeucht-Fraïssé games of the preceding Section to the modal language. This time, players can only choose successors of worlds in the current match. A typical outcome is this:

> $\mathbf{M}, s$ satisfies the same modal formulas up to operator depth k as $\mathbf{N}, t$ iff there is a winning strategy for player II in the modal Ehrenfeucht-Fraïssé game over k rounds with respect to $\mathbf{M}$, $\mathbf{N}$ with a start from the match (s, t).

3.2.4 Frame Correspondence

One of the main attractions of possible worlds semantics for modal logic has been the correspondence between independently proposed modal axioms and natural structural mathematical properties of frames (or labeled transition systems). To bring these out, one defines frame truth of a modal formula as its truth on a frame *under all valuations*.

Example 3.13 (Frame Correspondences)

1. *K4 axiom.* The modal formula $\langle a\rangle\langle a\rangle p \rightarrow \langle a\rangle p$ is true on a frame $\mathbf{F} = (W, R)$ iff the accessibility relation R is transitive: i.e.,
$$\forall x \forall y\,(R_a xy \rightarrow \forall z\,(R_a yz \rightarrow R_a xz)).$$
2. *K.2 axiom.* The modal formula $\langle a\rangle[a]p \rightarrow [a]\langle a\rangle p$ is true on frame $\mathbf{F} = (W, R)$ iff accessibility is confluent: i.e.,
$$\forall x \forall yz\,((R_a xy \wedge R_a xz) \rightarrow \exists u\,(R_a yu \wedge R_a zu)).$$
Such principles have obvious poly-modal generalizations. For instance, the formula $\langle a\rangle[b]p \rightarrow [b]\langle a\rangle p$ expresses a 'Church-Rosser Property' of the two actions a, b:
$$\forall x \forall yz\,((R_a xy \wedge R_b xz) \rightarrow \exists u\,(R_b yu \wedge R_a zu)).$$

These equivalences are proved by straightforward semantic arguments. These are examples of first-order frame conditions corresponding to modal formulas. In general, however, frame truth is a second-order notion, described by the following monadic second-order closure of the earlier standard translation _ working on models. Let the modal formula ϕ have proposition letters $p_1, \ldots, p_n$:
$$\mathbf{F} \models \phi \text{ iff } \mathbf{F} \models \forall P_1 \ldots P_n \forall x\, \underline{\phi}.$$

Modal formulas often correspond to genuinely higher-order frame constraints. Thus, the Löb Axiom $\Box(\Box p \rightarrow p) \rightarrow \Box p$ expresses transitivity plus reverse *well-foundedness* of frames, and the so-called McKinsey Axiom $\Box\Diamond p \rightarrow \Diamond\Box p$ is higher-order, too. Some relevant mathematical results are provided by modal Correspondence Theory.

Theorem 3.14 (van Benthem 1976) *A modal formula corresponds to a first-order frame property iff it is preserved under ultrapowers.*

Theorem 3.15 (Goldblatt and Thomason 1974) *A first-order sentence has a corresponding modal formula iff it is preserved under the following operations on frames: (i) generated subframes, (ii) disjoint unions, (iii) p-morphic images, and its negation is preserved under (iv) ultrafilter extensions.*

We omit details, as frame constructions will be a minor concern in what follows. As for more constructive information, van Benthem 1976 presents preservation theorems stating (modulo logical equivalence) which first-order forms are preserved under the above constructions (i), (ii), (iii). But modal definability of first-order formulas is at least undecidable (and probably of much higher complexity). Conversely, first-orderness of modal formulas is undecidable, too (Chagrova 1991) — and probably much worse.

Van Benthem 1984 shows that first-order definability for arbitrary monadic second-order Π^1_1 sentences (having the prefix shape of the above translation, followed by arbitrary first-order matrices) is non-arithmetically definable.

These negative results do not preclude the existence of powerful algorithms computing useful first-order equivalences. E.g., the *Sahlqvist-van Benthem Algorithm* analyzes modal formulas through their second-order translations $\forall P_1 \ldots P_n \forall x\, \phi$, by making first-order substitutions for the second-order variables. (These represent suitable 'definable valuations'.) With modal formulas ϕ in so-called 'Sahlqvist form', this algorithm will produce frame equivalents. Many well-known modal axioms fall under this regime. Here is a simple formulation (cf. Venema 1991, de Rijke 1993a for more sophisticated modern versions, including completeness theorems for modal logics with these axioms).

Theorem 3.16 (Sahlqvist Theorem) *There exists an effective algorithm computing first-order frame correspondents for all modal formulas $\phi \rightarrow \psi$ satisfying the following syntactic conditions:*

1. *ϕ is constructed from proposition letters using only $\wedge$, $\vee$, $\Diamond$ and $\Box$ in such a way that no $\vee$ or $\Diamond$ occurs in the scope of a necessity $\Box$*
2. *ψ is wholly positive, constructed from atoms using $\wedge$, $\vee$, $\Diamond$, $\Box$.*

Instead of a proof, we offer a sample calculation.

Example 3.17 (Computing First-Order Frame Equivalents) We show how frame equivalents arise by effective transformation of modal axioms. Consider the axiom $\langle a\rangle[a]p \rightarrow [a]\langle a\rangle p$. Its first-order translation over models is

$$\forall x \exists y\, (R_a xy \wedge \forall z\, (R_a yz \rightarrow Pz)) \rightarrow \forall u\, (R_a xu \rightarrow \exists v\, (R_a uv \wedge Qv)).$$

This is equivalent to

$$\forall xy\, (R_a xy \wedge \forall z\, (R_a yz \rightarrow Pz)) \rightarrow \forall u\, (R_a xu \rightarrow \exists v\, (R_a uv \wedge Qv)).$$

Now we extract a 'critical valuation' describing a minimal zone in a model which would make the antecedent true, given worlds x and y:

$$Ps := R_a ys.$$

This is then substituted into our formula, to obtain its 'strongest instance'. The result is

$$\forall xy\, (R_a xy \wedge \forall z\, (R_a yz \rightarrow R_a yz)) \rightarrow \forall u\, (R_a xu \rightarrow \exists v\, (R_a uv \wedge R_a yv)).$$

Part of its antecedent is always true, and hence it may be dropped. The final result is

$$\forall xy\, (R_a xy \rightarrow \forall u\, (R_a xu \rightarrow \exists v\, (R_a uv \wedge R_a yv)),$$

which is indeed the frame property of confluence. The proof of the Sahlqvist Theorem is a model-theoretic correctness argument for this algorithm, where the stated syntactic restrictions on modal axioms like this all play an essential role.

The substitution method does not deliver all first-order equivalents. Van Benthem 1984 shows that the conjunction of the K4 axiom and the McKinsey Axiom has a first-order frame equivalent requiring an essential use of the Axiom of Choice. The computational literature has rediscovered this area, adding alternative algorithms (cf. Ohlbach 1991, Doherty et al. 1994, d'Agostino et al. 1995). In the following chapters, we often use frame correspondences. Here, the above methods are not special to the basic modal language, or to binary frames. Chapter 8 analyzes basic Relational Algebra via correspondences over frames with a ternary accessibility relation, for composition of 'transitions' or 'information pieces'. Analogously, Chapter 9 uses frame correspondences over transition systems to analyze first-order logic. Thus, predicate-logical axioms *themselves* become modal axioms. Completely like above, the standard axiom $\exists x \exists x\, \phi \to \exists x \phi$ then expresses transitivity, and $\exists x \forall y\, \phi \to \forall y \exists x\, \phi$ confluence over generalized DPL-style computational models for first-order logic. Finally, Chapter 13 uses correspondences to analyze principles of categorial grammar over frames modeling ternary combination of syntactic structures.

3.2.5 Proofs, Tableaux and Sequents

The following Section is a summary of some well-known facts about modal deduction. This is only a minor topic of this Book, which justifies our terseness.

Hilbert-Style Axiomatics

There are various ways of describing the universal validities of modal logic, being those formulas which are true in all models at all states (without imposing any condition on accessibility relations). The best-known axiomatic description in the standard Hilbert-style is the following deductive system, due to Kripke.

Definition 3.18 (Minimal Modal Logic) The axioms consist of all propositional tautologies, plus the following modal principles

Modal Distribution	$\Box(\phi \to \psi) \to (\Box\phi \to \Box\psi)$
Modal Duality	$\Box\phi \leftrightarrow \neg\Diamond\neg\Diamond$

To obtain the minimal poly-modal logic, one just adds indices to all the modalities. The two derivation rules of the minimal logic are Modus Ponens plus

Necessitation if ϕ is provable, then so is $\Box\phi$.

Theorem 3.19 (Basic Completeness Theorem) *A modal formula is universally valid iff it is provable in the minimal logic.*

On top of this result, many completeness results exist for stronger modal logics with additional axioms corresponding to more restrictive frame constraints. Examples are the well-known modal logics K4, S4 or S5 (whose axioms make accessibility into an equivalence relation). This is the main thrust of research in traditional modal logic. In this book, we shall be mostly concerned with another direction, namely, increasing expressive power of modal languages over the same broadest class of models.

Semantic Tableaux

A more constructive method for checking universal validity uses semantic tableaus. Its rules include the usual decomposition principles for Boolean operators on sequents, of which the following are samples. We take the usual semantic consequence between sets of formulas. That is, truth of *all* premises should imply that for *at least one* conclusion.

$$\Sigma, \neg A \Rightarrow \Delta \quad \text{iff} \quad \Sigma \Rightarrow A, \Delta$$
$$\Sigma \Rightarrow A \wedge B, \Delta \quad \text{iff} \quad \Sigma \Rightarrow A, \Delta \text{ and } \Sigma \Rightarrow B, \Delta.$$

In modal tableaus, the key rule is that for existential modalities, which are best treated in a bunch, when no further propositional reductions are possible:

true: $\Diamond\phi_1, \ldots, \Diamond\phi_n \quad \bullet_w \quad \Diamond\psi_1, \ldots, \Diamond\psi_m$:false
create new worlds $v_1, \ldots, v_n$ with Rwv_i $(1 \leq i \leq n)$ and start these with sequents $\phi_i \bullet_{v_i} \psi_1, \ldots, \psi_m$.

Theorem 3.20 *Modal semantic tableaus are an adequate test for valid consequence in the minimal modal logic (over the universe of all possible worlds models).*

Corollary 3.21 *Modal universal validity is decidable.*

The corollary follows since all tableau rules decrease formula complexity of sequents (even though they may temporarily increase the number of parallel tasks). That tableaus are adequate for validity hinges on the semantic validity of the above $\Diamond$-Rule. Let P, Q be disjoint sequences of proposition letters. Then we have the following equivalence:

Proposition 3.22 (Decomposition of Modalized Validity)

$$P, \Diamond\phi_1, \ldots, \Diamond\phi_n \models Q, \Diamond\psi_1, \ldots, \Diamond\psi_m \textit{ iff}$$
$$\textit{for some } i \ (1 \leq i \leq n),\ \phi_i \models \psi_1, \ldots, \psi_m.$$

Proof. The equivalence is immediate from right to left. The opposite part of the proof depends on bisimulation invariance, through a semantic construction of 'joint rooting': any family of models $\mathbf{M}_i$, $v_i \models \phi_i \wedge \neg\psi_1 \wedge \ldots \wedge \neg\psi_m$ $(1 \leq i \leq n)$ can be 'glued disjointly' under one new common root:

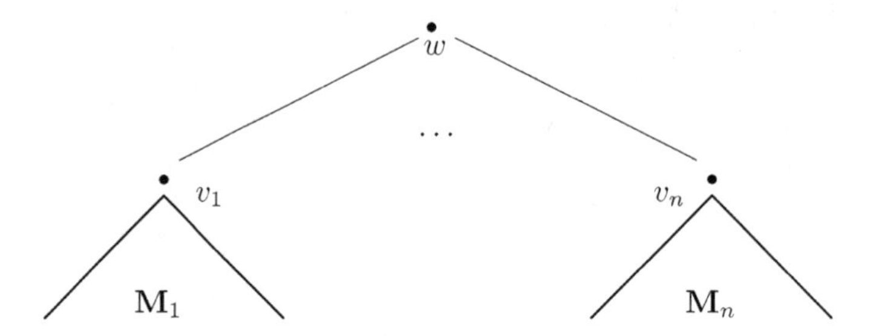

All models $\mathbf{M}_i$ are generated submodels (the identity is a bisimulation), whence no truth values change for modal formulas in their roots. So, the new root w verifies $\Diamond\phi_1 \wedge \ldots \wedge \Diamond\phi_n \wedge \neg\Diamond\psi_1 \wedge \ldots \wedge \neg\Diamond\psi_m$, thereby refuting the top sequent. $\dashv$

Sequent Calculus

Read bottom-up, tableau rules become introduction rules in a Gentzen-style calculus of sequents (cf. Fitting 1993). It sole axioms are sequents

$$\Sigma \Rightarrow \Delta \quad \text{with } \Sigma \cap \Delta \text{ non-empty.}$$

The following logical introduction rules are involved:

$$\frac{\Sigma, A \Rightarrow \Delta}{\Sigma \Rightarrow \neg A, \Delta} \qquad \frac{\Sigma \Rightarrow A, \Delta}{\Sigma, \neg A \Rightarrow \Delta}$$

$$\frac{\Sigma, A, B \Rightarrow \Delta}{\Sigma, A \wedge B \Rightarrow \Delta} \qquad \frac{\Sigma \Rightarrow A, \Delta \qquad \Sigma \Rightarrow B, \Delta}{\Sigma \Rightarrow A \wedge B, \Delta}$$

$$\frac{A \Rightarrow B_1, \ldots, B_m}{\Diamond A \Rightarrow \Diamond B_1, \ldots, \Diamond B_m}$$ where the part "$B_1, \ldots, B_m$" may be empty.

The rules for $\vee$, $\Box$ are analogous. Moreover, this calculus has two structural rules of inference:

Permutation	inside the premises and the conclusions
Monotonicity	from $\Sigma \Rightarrow \Delta$ infer $\Sigma', \Sigma \Rightarrow \Delta, \Delta'$.

These are needed to get the exact correspondence with closed semantic tableaus right. The classical structural rule of *Contraction* (deducing $\Sigma, A \Rightarrow \Delta$ from $\Sigma, A, A \Rightarrow \Delta$) is redundant for this completeness proof. In tableaus or sequent proofs for predicate logic, this rule ensures that false existential (and true universal) formulas can produce as many substitution instances as are required for the argument. (There is no a priori bound on this, given the undecidability of predicate logic.) With modal formulas, no such unbounded iteration is needed: we did all that is needed in one fell swoop. Hence, our calculus involves no shortening rules, and the proof search space is finite. (Thus, a form of 'linear logic' is complete for modal fragments of predicate logic.)

Once again, this section has been terse. We refer the reader to Goldblatt 1987, Blackburn et al. 1996 for good textbook expositions.

3.2.6 Meta-Theory

Basic modal logic shares many key meta-properties with first-order predicate logic. Here are some examples (Andréka, van Benthem and Németi 1995a present proofs, based on bisimulation arguments similar to those found in Chapter 4 below).

Theorem 3.23 (Interpolation Theorem) *Let $\phi \models \psi$. Then there exists a modal formula α whose proposition letters are included in both ϕ and ψ such that $\phi \models \alpha \models \psi$.*

Theorem 3.24 (Submodel Preservation Theorem) *A modal formula is preserved under model extensions iff it can be defined using only propositional atoms and their negations, $\wedge$, $\vee$, $\Diamond$.*

The similarities between modal logic and standard first-order logic call for explanation. One key analogy is that between bisimulation and 'partial isomorphism' $\cong_p$ between first-order models ('cut off' at length 2; cf. van Benthem 1991b). De Rijke 1995 is a systematic study of poly-modal model theory, revolving around the 'heuristic equation'

$$\text{Modal Logic} : \text{Bisimulation} = \text{Predicate Logic} : \text{Partial Isomorphism}.$$

Some key lemmas license 'transfer' between modal and classical reasoning. One is an earlier observation. Two models have the same modal theory iff they have bisimilar elementary extensions. (One can choose the latter to be countable ultrapowers.) Here is another recent result of this kind, again from Andréka et al. 1995a.

Proposition 3.25 (Bisimulation and Elementary Equivalence) *Two models $\mathbf{M}, x$ and $\mathbf{N}, y$ have the same modal theory iff they possess bisimulations with two elementarily equivalent models $\mathbf{M}^+, x$ and $\mathbf{N}^+, y$.*

The full meta-mathematical extent of these analogies and transfer principles has not yet been established.

3.2.7 Extended Modal Logics

One test of character for modal logic is how it survives generalization. At least for poly-modal languages, one can virtually copy the above, putting in indices. (Subtleties do arise — e.g., with the Interpolation Theorem.) Generalization also occurs for languages having k-ary *polyadic modalities*, referring to $(k+1)$-ary accessibility relations:

$$\Diamond\phi_1 \dots \phi_k \text{ translates into } \exists y_1 \dots y_k \, (R^{k+1}x, y_1 \dots y_k \wedge \bigwedge_{1 \leq i \leq k} \phi_i(y_i)).$$

Under this translation, modal formalisms end up in *bounded fragments* of first-order logic. The latter inherit all attractive properties of their modal originals (van Benthem 1991a, de Rijke 1993b; cf. the decidable 'guarded fragment' of Section 4.4.2). Another generalization strengthens a set of modalities over a fixed similarity type for frames, as in the passage from modal to temporal logic. Here, one adds backward operators, or stronger betweenness operators (essentially involving three state variables), such as

$$\mathbf{M}, x \models \mathit{Past}\,\phi \quad \text{iff} \quad \text{there exists } y \text{ with } Ryx \text{ and } \mathbf{M}, y \models \phi$$

$$\mathbf{M}, x \models \mathit{Until}\,\phi\psi \quad \text{iff} \quad \text{there exists } y \text{ with } Rxy,\ \mathbf{M},\ y \models \psi, \text{ and for all } z \text{ with } Rxz \text{ and } Rzy,\ \mathbf{M}, z \models \phi.$$

These have obvious first-order translations. Hierarchies of extended modal logics return in Chapter 4. Their model theory and proof theory are comparatively little explored.

Finally, we shall sometimes encounter *infinitary* modal languages, which allow infinite (set) conjunctions and disjunctions. Unlike the finitary modal language, equivalence of two worlds with respect to the infinitary modal language matches the existence of a bisimulation *precisely*. The invariance theorem for bisimulation generalizes to infinitary predicate logic, by removing any reliance on compactness (cf. Chapter 10). Infinitary modal correspondence theory is explored in Barwise and Moss 1996, who relate truth in models to frame truth in their collapses over the maximal bisimulation (cf. Chapter 4).

3.3 Relational Algebra

3.3.1 Language and Semantics

What are most general operations on actions, viewed as binary transition relations? Ubiquitous examples are *sequential composition* and *choice*. But others occur too, such as undoing an action (reversing its transitions). A convenient formalism for this repertoire is *Relational Algebra*, the study of logical operations on binary relations initiated by Schröder and Tarski (see Németi 1991, Maddux 1995). Its basic similarity type has the following operations, indicated with their set-theoretic interpretation:

Boolean operations	$-, \cap, \cup$	complement, intersection and union
Ordering Operations	$\circ, {}^{\smile}$	composition and converse
Identity Element	Δ	the identity relation

Intuitively, $\cup$ models 'choice', $\circ$ 'sequencing' and ${}^{\smile}$ 'reversal' for binary relations. Standard models for this algebraic formalism are families of binary relations over some fixed domain of objects. Interpretation of algebraic terms proceeds inductively, starting from any assignment of binary set relations to individual variables, with clauses

$$\begin{aligned}
-R &= \{(x,y) \mid \neg Rxy\} \\
R \cap S &= \{(x,y) \mid Rxy \wedge Sxy\} \\
R \cup S &= \{(x,y) \mid Rxy \vee Sxy\} \\
R \circ S &= \{(x,y) \mid \exists z\, Rxz \wedge Szy\} \\
R^{\smile} &= \{(x,y) \mid Ryx\} \\
\Delta &= \{(x,y) \mid x = y\}
\end{aligned}$$

These models form a class **SRA** of *set relation algebras*. This class may be extended by moving to abstract algebras of this similarity type — just as Boolean algebras have arisen through abstraction from concrete families of sets. Then, *relation algebras* become any abstract algebras satisfying 'enough postulates' valid on the original set interpretation. Excellent accounts of various algebraic possibilities here may be found in Németi 1991, Venema 1991, Marx 1995, Maddux 1990, 1995, Venema and Marx 1996a. A basic finite system of equational postulates is BRA (cf. Section 3.2.4 and Chapter 8). For a start, it contains all axioms of *Boolean Algebra*:

$$\begin{aligned}
x \cup (y \cup z) &= (x \cup y) \cup z & x \cap (y \cap z) &= (x \cap y) \cap z \\
x \cup y &= y \cup x & x \cap y &= y \cap x \\
x \cup x &= x & x \cap x &= x \\
x \cup (y \cap z) &= (x \cup y) \cap (x \cup z) & x \cap (y \cup z) &= (x \cap y) \cup (x \cap z) \\
x \cup (x \cap y) &= x & x \cap (x \cup y) &= x \\
-(x \cup y) &= -x \cap -y & -(x \cap y) &= -x \cup -y \\
x \cup 0 &= x & x \cap 0 &= 0 \\
x \cup 1 &= 1 & x \cap 1 &= x \\
x \cup -x &= 1 & x \cap -x &= 0 \\
--x &= x
\end{aligned}$$

Then it adds equations for composition, converse and diagonal:

$$\begin{aligned}
R \circ \Delta &= R = \Delta \circ R \\
R^{\smile\smile} &= R \\
(-R)^{\smile} &= -R^{\smile} \\
(R \cup S)^{\smile} &= R^{\smile} \cup S^{\smile} \\
(R \circ S)^{\smile} &= S^{\smile} \circ R^{\smile} \\
(R \circ S) \circ T &= R \circ (S \circ T) \\
R \circ (S \cup T) &= (R \circ S) \cup (R \circ T) \\
(R \cup S) \circ T &= (R \circ T) \cup (S \circ T) \\
(R \circ -(R \circ S)) \cup -S &= -S.
\end{aligned}$$

Of particular interest are the *representable relation algebras* in which the complete equational theory of all set relation algebras holds. (This theory is stronger than BRA.) By Birkhoff's Theorem (Grätzer 1968), these form the variety HSP(**SRA**), i.e., the closure of **SRA** under *homomorphic images*, *subalgebras* and *direct products*. Tarski has shown that this equational variety equals SP(**SRA**). Hence, it consists of all subalgebras of algebras which can be embedded in a relation algebra of the special form pow(U) (with the above operations) — where U is an equivalence relation over some underlying set of objects D, but not necessarily the full Cartesian product $D \times D$.

Digression. (Identities, Quasi-Identities, Universal Formulas) To put these results in perspective, we review some facts from Universal Algebra. Birkhoff's Theorem says that a class of algebras is definable by some set of *equations* (read with their universal closure) iff it is closed under homomorphic images, subalgebras and direct products. A class of algebras is definable by *quasi-equations* (universal Horn clauses made up of identities) iff it is closed under isomorphic images, subalgebras and direct products. Finally, a class of algebras is definable by a set of *universal formulas* (in the usual first-order sense, with identity atoms, and arbitrary disjunctions) iff it is closed under isomorphic images, subalgebras and ultraproducts. Sometimes, the pure identity theory of a class of algebras encodes its complete quasi-equational theory. For instance, on the above class **SRA** (though not on all Boolean algebras), non-identities can be expressed by identities. We first reduce to identities with 0, via $R = T$ iff $R \dot{-} T = 0$ (with $\dot{-}$ symmetric difference). Then we observe

$$R \neq 0 \text{ iff } 1 \circ R \circ 1 = 1.$$

Now, a quasi-equation $R = 0 \rightarrow S = 0$ is encoded by the identity $(1 \circ R \circ 1) \cup -S = 1$. (Conjunctions of equations $U = 0$, $V = 0$ can be reduced to single ones $(U \cup V) = 0$.) The trick extends to all 'subdirectly irreducible' algebras in the variety generated by set relation algebras. Tarski's quoted result now follows from another Birkhoff Theorem, viz. that any algebra is isomorphic to a subdirect product of subdirectly irreducible algebras satisfying its equational theory. More background on 'discriminator terms' allowing the above encoding is found in Németi 1991, Marx 1995, Mikulás 1995.

The pow(U) format leads to the modern algebraic perspective of *Relativization*. The above inductive value definition for algebraic terms will even work if the universe of available pairs is restricted to merely some subset of a full Cartesian product $D \times D$, encoded as a binary 'top relation' U which need not even be an equivalence relation. Then, one can reinterpret the above operations so as to stay within the restricted pair universe. One can think of this as having only a limited number of 'transitions' available:

the underlying states do not allow random mutual access:

$$\begin{aligned}
-R &= \{(x,y) \in U \mid \neg Rxy\} \\
R \cap S &= \{(x,y) \mid Rxy \wedge Sxy\} \\
R \cup S &= \{(x,y) \mid Rxy \vee Sxy\} \\
R \circ S &= \{(x,y) \in U \mid \exists z\, (x,z) \in U \wedge (z,y) \in U \wedge Rxz \wedge Szy\} \\
R^{\smile} &= \{(x,y) \in U \mid Ryx\} \\
D &= \{(x,y) \in U \mid x = y\}
\end{aligned}$$

This broader semantics will not validate the full equational theory of **SRA**. Indeed, it is instructive to consider individual relation-algebraic laws, and determine what they presuppose about availability of ordered pairs. E.g., the BRA axiom of Distributivity is valid as it stands, but Associativity of composition typically requires *transitivity* of U. Relativization to arbitrary top relations, or to top relations satisfying some subset of transitivity, symmetry and reflexivity has yielded several interesting classes of algebras. Sometimes, these techniques are even more abstract, relativizing via *ternary* relations:

$$R \circ S = \{(x,y) \mid \exists z\, U^3 xyz \wedge Rxz \wedge Szy\}.$$

These ideas lie behind the dynamic semantics that we shall propose in Chapters 8, 9, which use general techniques of relativization in a classical model-theoretic setting. In the field of *Cylindric Algebra*, which algebraizes all of predicate logic (cf. Henkin et al. 1985, Németi 1991, Marx 1995, Venema 1995a), the logic of the relativized model class corresponding to the above pair models is called **CRS** (an acronym for 'cylindric relativized set algebras'). The latter system will be very much in evidence in Chapter 9 below.

3.3.2 Permutation Invariance

The above algebraic operations O form a completely general procedural apparatus, which is 'logical' in that it is independent from any specific structure of states. Technically, this feature shows in so-called *invariance under permutations* of states. For instance, in the binary case, we have the following commutation property.

Proposition 3.26 (Permutation Invariance) *For every permutation π of the underlying state set S, $\pi[O(R,S)] = O(\pi[R], \pi[S])$.*

Here, the relevant permutation images $\pi[R]$, $\pi[S]$ are formed in the obvious way, permuting objects inside ordered pairs. An equivalent formulation runs as follows:

$$(x,y) \in O(R,S) \text{ iff } (\pi(x), \pi(y)) \in O(\pi[R], \pi[S]).$$

This formulation is easily derived from the Isomorphism Lemma for first-order logic, applied to the definitions of the algebraic operations. Thus, relational operators refer only to abstract arrow patterns, not to the individuals forming these arrows. This fact explains why operations from Relational Algebra return across many concrete dynamic systems with more specialized state spaces. This reflects a much more general notion of permutation invariance, applying to arbitrary set operators in finite-type domains (van Benthem 1986, 1989a, 1991b). Can we be more specific about logicality than this? All the above algebraic operations are permutation-invariant, but so are infinitely many others (see below). Following Tarski, algebraists view permutation invariance as a necessary condition for logicality (cf. Jónsson 1987, or Sher 1991). As for sufficient conditions, Chapter 5 will define much more restrictive notions of invariance for algebraic operators, involving modal bisimulation.

3.3.3 First-Order Translation

Relational Algebra started in the 19th century as an algebraic theory of binary relations, just as Boolean Algebra treated elementary operations on sets. Alternatively, it may be viewed as an algebraization of an ambitious non-monadic fragment of first-order predicate logic. This connection shows in an obvious first-order translation sending relation-algebraic terms R to first-order formulas $\underline{R}(x,y)$. Its clauses mimic the above inductive interpretation into **SRA**:

$$\begin{aligned}
\underline{-R}(x,y) &= \neg\underline{R}(x,y)\\
\underline{R\cap S}(x,y) &= \underline{R}(x,y)\wedge\underline{S}(x,y)\\
\underline{R\cup S}(x,y) &= \underline{R}(x,y)\vee\underline{S}(x,y)\\
\underline{R\circ S}(x,y) &= \exists z\,\underline{R}(x,z)\wedge\underline{S}(z,y)\\
\underline{R^{\smile}}(x,y) &= \underline{R}(y,x)\\
\underline{\Delta}(x,y) &= x=y
\end{aligned}$$

In principle, this translation has much the same uses as those given for Modal Logic. Known properties of first-order logic may be transferred. As it happens, however, Relational Algebra also inherits some of the bad. Notably, like predicate logic itself, the equational theory of full set relation algebras is *undecidable*. Also, not all important meta-properties go through: there are counter-examples to Interpolation and Submodel Preservation (cf. Németi 1991). These concerns will be largely irrelevant to our study. Of course, many other operations over relations can be defined in this first-order way. This translation explains an earlier observation, by an easy induction on formulas.

Proposition 3.27 (Definability Implies Invariance) *Each first-order definable operation $O(R, S, \ldots)$ is permutation invariant.*

But there is further fine-structure here. By exercising proper care, 'recycling variables' in the step for composition, one can make do with a rather simple first-order fragment.

Proposition 3.28 (Three Variables Suffice) *Each relation term has a corresponding first-order formula using only the three variables* $\{x, y, z\}$, *free or bound.*

There is also an interesting converse result, which may be shown by a straightforward analysis of inner quantifiers in first-order definitions:

Theorem 3.29 (Functional Completeness) *Each first-order definable relational operation using only three variables can be defined using the basic operations of Relation Algebra.*

Example 3.30 (Dynamic Negation) The strong test negation $\sim(R)$ of Chapter 2 was defined by the first-order formula $x = y \wedge \neg\exists z\, Rxz$. In fact, it can be defined using two variables: $x = y \wedge \neg\exists y\, Rxy$. This amounts to the relation-algebraic term $\Delta \cap -(R \circ \top)$ ('$\top$' for Boolean 'true'). (A functional completeness theorem for two variables is in van Benthem 1993b.)

Example 3.31 (Categorial Slashes) Two central operations in Categorial Grammar (Chapter 12) are directed implications

$$\begin{aligned} A \to B &= \{(x,y) \mid \text{for all } z, \text{ if } Azx, \text{ then } Bzy\} \\ B \leftarrow A &= \{(x,y) \mid \text{for all } z, \text{ if } Ayz, \text{ then } Bxz\}. \end{aligned}$$

Their first-order transcription uses only three variables. And indeed, they have relation-algebraic definitions which resemble the Boolean equivalence $(\phi \to \psi) \leftrightarrow \neg(\phi \wedge \neg\psi)$):

$$\begin{gathered} -(A \circ -(B)^{\smile}) \\ -(-(B)^{\smile} \circ A). \end{gathered}$$

We shall return to issues of functional completeness at finite-variable levels in Chapter 5.

Remark 3.32 (Relativization) When using the earlier-mentioned relativized semantics, the first-order translation must be modified. All translations will have their tuples relativized to the top relation Uxy. Then, the clause for composition will no longer have an unlimited existential quantifier z, as the relevant values must be U-related to x or y. This means that we end up, again, in *bounded fragments* of predicate logic (cf. Chapter 4 below), whose behaviour may be much better than that of the full system. This explains our improvements.

Remark 3.33 (Infinitary Operations) Relational Algebra in the present sense is a subsystem of first-order predicate logic. Thus, its algebraic calculations may be replaced by predicate-logical reasoning using explicit variables for states. This reduction from dynamic reasoning to standard logic is useful for technical purposes. But in addition to first-order operations, again, infinitary notions make sense. There are natural *infinitary* operations on binary relations. These encode unlimited action structures, such as endless repetition ('Kleene star'). The most prominent technical example of an infinitary operation is *reflexive transitive closure*:

$$\begin{aligned} R^* \;&=\; \{(x, y) \mid \text{some finite sequence of successive} \\ &\qquad R \text{ transitions links } x \text{ to } y\ \}. \end{aligned}$$

The Kleene star satisfies various principles of 'regular algebra', such as

$$\begin{aligned} R \cup R^* &= R^* \\ R^{**} &= R^* \\ (R \cup S)^* &= (R^* \circ S^*)^*. \end{aligned}$$

3.3.4 Equational Axiomatizations

The inferential properties of these relational algebraic operators are encoded in some well-known axiom systems. Recall the core system BRA of Basic Relational Algebra, which adds the following equations to those of Boolean Algebra:

$$\begin{aligned} R \circ \Delta &= R = \Delta \circ R \\ R^{\smile\smile} &= R \\ (-R)^{\smile} &= -R^{\smile} \\ (R \cup S)^{\smile} &= R^{\smile} \cup S^{\smile} \\ (R \circ S)^{\smile} &= S^{\smile} \circ R^{\smile} \\ (R \circ S) \circ T &= R \circ (S \circ T) \\ R \circ (S \cup T) &= (R \circ S) \cup (R \circ T) \\ (R \cup S) \circ T &= (R \circ T) \cup (S \circ T) \\ (R^{\smile} \circ -(R \circ S)) \cup -S &= -S. \end{aligned}$$

These principles are all valid under their intended interpretation. This is easy to check by noting the simple properties of ordered pairs expressed by them. The hardest case may be the final axiom. In Chapter 8, we provide a modal perspective for computing what these axioms mean via frame correspondences. It is known, however, that the above list does not capture all valid principles of Relational Algebra over set relations. The complete equational theory of **SRA** is *non-finitely-axiomatizable* (Monk 1969, Andréka 1991). The latter theory is effectively axiomatizable, of course, as it encodes a fragment of first-order logic. On the other hand, both BRA and SRA are *undecidable*. For an exhaustive treatment of the field, cf. the continuing survey Németi 1991, 1994. In particular, the situation may improve under the earlier technique of relativization, leading to decidable axiomatizations of subtheories.

Another direction of improvement concerns specialized operators. For instance, Hollenberg 1995a shows that the complete relational algebra of the repertoire $\{\circ, \sim\}$ for dynamic predicate logic (Chapter 2) is finitely axiomatizable, and it stays so when we add $\cup$. For the sake of concrete illustration, we list his axioms here:

$$\begin{aligned}
{\sim}R \circ R &= 0\\
R \circ 0 &= 0\\
0 \circ R &= 0\\
{\sim}0 \circ R &= R\\
(R \circ S) \circ T &= R \circ (S \circ T)\\
{\sim}R \circ {\sim}S &= {\sim}S \circ {\sim}R\\
R &= ({\sim}{\sim}R) \circ R\\
{\sim}{\sim}({\sim}R \circ {\sim}S) &= {\sim}R \circ {\sim}S\\
{\sim}(R \circ S) \circ R &= ({\sim}(R \circ S) \circ R) \circ {\sim}S\\
{\sim}(R \circ {\sim}({\sim}S \circ {\sim}T)) &= {\sim}{\sim}({\sim}(R \circ S) \circ {\sim}(R \circ T)).
\end{aligned}$$

In this book, we shall not perform any algebraic calculations using these equations. Instead, we will analyze Relational Algebra in the Arrow Logic of Chapter 8 from a modal perspective. (Venema and Marx 1996a demonstrate the great power of switching back and forth between algebraic logic and modal logic.) This is just one mathematical connection between the two fields. For a richer perspective, we refer to the general duality theory between 'complex algebras' and modal-style frame semantics (an up-to-date survey is Goldblatt 1988), as well as the detailed studies Venema 1991, Marx 1995, which provide many surprising mutual applications.

3.4 Dynamic Logic

3.4.1 Basic Language and Semantics

Our final tool was designed as a combination of Modal Logic and Relational Algebra. The core language of propositional dynamic logic has 'formulas' and 'programs' defined by the following mutual recursion (stated here in Backus-Naur Form):

$$\begin{aligned}
PA &::= \text{propositional atoms } p, q, r, \ldots\\
AA &::= \text{atomic actions } a, b, c, \ldots\\
F &::= PA \mid \neg F \mid (F \wedge F) \mid \langle P \rangle F\\
P &::= AA \mid (P\,;P) \mid (P \cup P) \mid P^* \mid (F)?.
\end{aligned}$$

This language is interpreted over polymodal Kripke models (labeled transition systems) with a valuation V encoding truth values for propositions at states, yielding models $\mathbf{M} = \langle S, \{R_a\}_{a \in A}, V\rangle$. The truth definition explains two notions in one double recursion. $\mathbf{M}, s \models \phi$ says that ϕ is true at state s, while $\mathbf{M}, s_1, s_2 \models \pi$ says that the transition from s_1 to s_2 corresponds to a successful execution for the program π:

- Propositions

$$\begin{array}{rll} \mathbf{M}, s \models p & \text{iff} & s \in V(p) \\ \mathbf{M}, s \models \neg\psi & \text{iff} & \text{not } \mathbf{M}, s \models \psi \\ \mathbf{M}, s \models \phi_1 \wedge \phi_2 & \text{iff} & \mathbf{M}, s \models \phi_1 \text{ and } \mathbf{M}, s \models \phi_2 \\ \mathbf{M}, s \models \langle\pi\rangle\phi & \text{iff} & \text{there exists } s' \text{ with } \mathbf{M}, s, s' \models \pi \\ & & \text{and } \mathbf{M}, s' \models \phi \end{array}$$

- Programs

$$\begin{array}{rll} \mathbf{M}, s_1, s_2 \models a & \text{iff} & (s_1, s_2) \in R_a \\ \mathbf{M}, s_1, s_2 \models \pi_1 ; \pi_2 & \text{iff} & \text{there exists } s_3 \text{ with } \mathbf{M}, s_1, s_3 \models \pi_1 \\ & & \text{and } \mathbf{M}, s_3, s_2 \models \pi_2 \\ \mathbf{M}, s_1, s_2 \models \pi_1 \cup \pi_2 & \text{iff} & \mathbf{M}, s_1, s_2 \models \pi_1 \text{ or } \mathbf{M}, s_1, s_2 \models \pi_2 \\ \mathbf{M}, s_1, s_2 \models \pi^* & \text{iff} & \text{some finite sequence of } \pi\text{-transitions in} \\ & & \mathbf{M} \text{ connects } s_1 \text{ with } s_2 \\ \mathbf{M}, s_1, s_2 \models (\phi)? & \text{iff} & s_1 = s_2 \text{ and } \mathbf{M}, s_1 \models \phi \end{array}$$

Thus, formulas have the usual Boolean operators, while the existential modality $\langle\pi\rangle\phi$ is a weakest precondition true at only those states where program π can be performed to achieve the truth of ϕ. The program operators are the usual regular operations of relational composition, Boolean choice, Kleene iteration, and tests for formulas. This system encompasses standard control operators on programs such as

Conditional Choice

$\texttt{if}\ \epsilon\ \texttt{then}\ \pi_1\ \texttt{else}\ \pi_2 \quad ((\epsilon)?\, ;\pi_1) \cup ((\neg\epsilon)?\, ;\pi_2)$

Guarded Iteration

$\texttt{while}\ \epsilon\ \texttt{do}\ \pi \quad ((\epsilon)?\, ;\pi)^*\, ;(\neg\epsilon)?.$

In this formalism, we can express all laws of modal logic and of relational set algebra.

3.4.2 Invariance and Model Constructions

The expressive power of propositional dynamic logic may be measured via the earlier invariance for bisimulations of a polymodal language. What about program operations? One can prove bisimulation invariance for all formulas in propositional dynamic logic — but there is a new aspect. Intertwined with the old inductive argument, one has to show that the zigzag clauses are inherited by the regular program constructions: each binary transition relation $[\![\pi]\!]$ shows this behaviour, upward from the atomic ones.

Proposition 3.34 *Let $\equiv$ be a bisimulation between two models* $\mathbf{M}$, $\mathbf{M}'$, *with $s \equiv s'$. Then*

1. *s, s' verify the same formulas of propositional dynamic logic*
2. *if $s[\![\pi]\!]^{\mathbf{M}}t$, then there exists t' with $s'[\![\pi]\!]^{\mathbf{M}'}t'$ and $s' \equiv t'$.*

Proof. This is a simultaneous induction on formulas and programs. Here is a typical step of the latter kind. If $s[\![\pi_1 ; \pi_2]\!]^{\mathbf{M}}t$, then, by the truth definition, there exists some state x such that $s[\![\pi_1]\!]^{\mathbf{M}}x$ and $x[\![\pi_2]\!]^{\mathbf{M}}t$. By the inductive hypothesis for π_1 with respect to $s \equiv s'$, there exists some x' with $x \equiv x'$ and $s'[\![\pi_1]\!]^{\mathbf{M}'}x'$. Next, by the inductive hypothesis for π_2 with respect to $x \equiv x'$, there exists some t' with $t \equiv t'$ and $x'[\![\pi_2]\!]^{\mathbf{M}'}t'$. But then, by the truth definition, $s'[\![\pi_1 ; \pi_2]\!]^{\mathbf{M}'}t'$ — and we are done. Similar arguments work for the other regular program constructions. The inductive step for tests will recurse toward the formula side. ⊣

This observation motivates the following notion of invariance for program operations (where we indulge in a slight abuse of notation, for greater readability).

Definition 3.35 (Safety for Bisimulation) An operation $O(R_1, \ldots, R_n)$ on programs is *safe for bisimulation* if, whenever $\equiv$ is a relation of bisimulation between two models for their transition relations $R_1, \ldots, R_n$, then it is also a bisimulation for $O(R_1, \ldots, R_n)$.

The core of the above program induction is that the three regular operations ;, $\cup$, * of PDL are safe for bisimulations. We study broader effects of this notion of invariance in Chapter 5. To show its impact, here is a non-example, of a non-regular program operation outside of standard propositional dynamic logic, which lacks safety.

Example 3.36 (Boolean Conjunction) Compare the following two labeled transition systems:

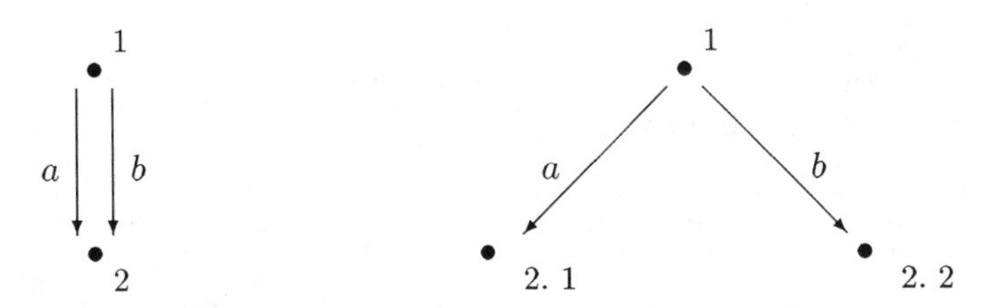

The numbers indicate an obvious bisimulation with respect to the atoms a, b — but zigzag fails for the intersection $R_a \cap R_b$.

3.4.3 Infinitary First-Order Translation

The first-order translation for Modal Logic may be extended to Dynamic Logic. Naturally, there are two parts to this procedure. * takes propositions to unary formulas $\phi(x)$, while $^{\#}$ takes programs to binary formulas $\phi(x, y)$, in a mutual recursion:

$$\begin{aligned}
(p)^* &= Px \\
(\neg\phi)^* &= \neg(\phi)^* \\
(\phi \wedge \psi)^* &= (\phi)^* \wedge (\psi)^* \\
(\langle\pi\rangle\phi)^* &= \exists y\, ((\pi)^{\#} \wedge [y/x](\phi)^*)
\end{aligned}$$

$$\begin{aligned}
(a)^{\#} &= R_a xy \\
(\pi_1 \cup \pi_2)^{\#} &= (\pi_1)^{\#} \vee (\pi_2)^{\#} \\
(\pi_1 \circ \pi_2)^{\#} &= \exists z\, ([z/y](\pi_1)^{\#} \wedge [z/x](\pi_2)^{\#}) \\
(\phi?)^{\#} &= x = y \wedge (\phi)^* \\
(\pi^*)^{\#} &= \bigvee_{n \in \mathbb{N}} (\underbrace{\pi \circ \ldots \circ \pi}_{n})^{\#}.
\end{aligned}$$

The latter clause, of course, makes the formalism infinitary.

Proposition 3.37 *The above is a correct translation from dynamic modal logic into (countably) infinitary first-order predicate logic over state transition models.*

This translation has various uses. For instance, without Iteration, we see that universal validity in dynamic logic is recursively axiomatizable. With Iteration, we are in a constructive fragment of the well-understood infinitary logic $L_{\omega_1\omega}$ (cf. Keisler 1971). For further discussion, cf. Harel 1984 — as well as Chapter 10 below.

3.4.4 Axiomatics and Proof Theory

The central mathematical result that put this formalism on the map was the following. Propositional dynamic logic has a natural minimal logic (the 'Segerberg Axioms'):

- All principles of the minimal modal logic for all modalities $[\pi]$
- Computation rules for weakest preconditions:
  $\langle\pi_1 ; \pi_2\rangle\phi \leftrightarrow \langle\pi_1\rangle\pi_2\phi$
  $\langle\pi_1 \cup \pi_2\rangle\phi \leftrightarrow \langle\pi_1\rangle\phi \vee \langle\pi_2\rangle\phi$
  $\langle\phi?\rangle\psi \leftrightarrow \phi \wedge \psi$
  $\langle\pi^*\rangle\phi \leftrightarrow \phi \vee \langle\pi\rangle\langle\pi^*\rangle\phi$
- Induction Axiom
  $(\phi \wedge [\pi^*](\phi \rightarrow [\pi]\phi)) \rightarrow [\pi^*]\phi$

This proof calculus generalizes regular algebra, propositional modal logic as well as Hoare Calculus for correctness assertions $\{\phi\}\, \pi\, \{\psi\}$. In particular, under the earlier transcription $\phi \rightarrow [\pi]\psi$, plus the above PDL definitions for the two basic program constructions `if ... then ... else` and `while`

... do ..., it derives all Hoare principles for sequencing, conditional choice and guarded iteration (Goldblatt 1987).

Theorem 3.38 *The Segerberg axioms plus the usual modal inference rules axiomatize universal validity in propositional dynamic logic.*

The completeness proof can be set up as a Henkin argument over a suitable finite universe of formulas, which yields *decidability* as a corollary. (The latter property was first established in Fisher and Ladner 1979 via modal filtration on arbitrary models.) As a consequence, we have the following observation:

Theorem 3.39 *The relational set algebra of* $\{\wedge, \sim\}$ *is decidable.*

Proof. Here is one argument. It is easy to show that one can extend PDL with the relational operator $\sim$ and still keep all of the above results. The characteristic axiom will read as follows (with $\neg$ standing for Boolean negation on the right-hand side):

$$\langle \sim R \rangle \phi \leftrightarrow \phi \wedge \neg \langle R \rangle \top .$$

The following effective reduction is virtually immediate. An algebraic identity $R = S$ in this language is valid iff, for some new proposition letter q, the poly-modal formula $\langle R \rangle q \leftrightarrow \langle S \rangle q$ is valid in propositional dynamic logic. Another proof reduces validity of equations like this directly into poly-modal logic, which is known to be decidable. Reduction steps are the above PDL axiom for composition plus the new one for $\sim$. ⊣

3.4.5 Extensions

The earlier modal apparatus (including frame correspondences) applies to propositional dynamic logic, both in its basic form and under expressive extensions. PDL is an open-ended enterprise. E.g., adding *converse* of programs, running them backwards, gives a 'tense-logical' variant of the system. Also, the missing Booleans $\cap$ and $-$ may be brought in after all. The former addition is known to preserve decidability, but the latter does not, as we can then encode full relational set algebra like in the preceding proof. Another useful extension has a *fixed point operator* (π) for programs π which holds at precisely those states where the transition relation $[\![\pi]\!]$ has a loop. PDL *with loops* remains effectively axiomatizable and decidable. We shall use some of these extensions in what follows (e.g., in the analysis of dynamic inference in Chapter 7).

3.5 Notes

We list some texts for the subjects of this chapter.

First-Order Logic. Good general introductions are Enderton 1972, Bell and Machover 1977, Hodges 1983. See also the *Handbook of Mathematical Logic* (Barwise 1977). An up-to-date introduction to logic and model theory based on Ehrenfeucht-Fraïssé games is Doets 1996. An advantage of the latter technique is that it also works in cases where compactness arguments fail, such as finite model theory. A good reference for the latter is Ebbinghaus and Flum 1995. A modern text book on Proof Theory is Troelstra and Schwichtenberg 1996.

Modal Logic. For a general picture, see the *Handbook of Philosophical Logic* (Gabbay and Guenthner 1983–1989). Van Benthem 1988c is a programmatic overview of intensional logic. Goldblatt 1987 is a compact textbook geared toward computational applications. Blackburn et al. 1996 is a sophisticated modern treatment, congenial to the spirit of this Book. For Correspondence Theory in particular: cf. van Benthem 1984, 1985a (and the addenda to its 1996 reprint).

Infinitary Versions. Barwise 1975 is a key source on general infinitary logic, Keisler 1971 on its countable fragment. Infinitary dynamic logic dates back to Salwicki 1970. Goldblatt 1982 is a textbook on such formalisms. For infinitary modal logic, see van Benthem and Bergstra 1994, de Rijke 1993b, Barwise and Moss 1996. An elegant form of infinitary logic for computer science is the so-called *μ-calculus*, which extends PDL with arbitrary fixed-point operators: cf. Harel et al. 1995, Janin 1996.

Relational Algebra. No good textbook exists yet. Useful surveys are Henkin, Monk, and Tarski 1985 (Section 5.3), Henkin, Monk, Tarski, Andréka, and Németi 1981, Németi 1991, 1994, Maddux 1995. Andréka et al. 1996 is a textbook in progress. For connections between algebraic logic and modal logic, see Venema and Marx 1996a.

Cylindric Algebra. The classic texts on this subject (rich, but slightly unreadable) are Henkin et al. 1971, 1985. See also Monk 1976, Németi 1991, 1994. Venema 1995a has a nice exposition for modal logicians.

Dynamic Logic. Classical texts are Harel 1984, Goldblatt 1987. Cf. also van Benthem and Meyer Viol 1993. A comprehensive recent textbook is Harel et al. 1995.

Infinitary Logic, Modal Logic and Set Theory. In particular, there are strong connections between the preceding formalisms and various *set theories*. Barwise and Moss 1996 (with contributions by Alexandru Baltag) develop non-well-founded set theory, exploiting the fact that infinitary modal formulas characterize sets up to bisimulation. This provides a purely mathematical connection for much of what follows in this Book, which still awaits further exploration. In particular, Barwise and van Benthem 1996 provide general tools for infinitary logic which make up (to some extent) for the loss of compactness. These include modified *interpolation* and *preservation* results of which the following is a sample. For any two infinitary formulas φ, ψ, there exists an interpolant α in $L_\varphi \cap L_\psi$ with $\varphi \models \alpha \models \psi$ iff, whenever $\mathbf{M} \models \varphi$ and $\mathbf{M}$ is partially L-isomorphic to $\mathbf{N}$ (cf. Section 3.1.2), then $\mathbf{N} \models \psi$. The latter condition is stronger than ordinary consequence, and of independent interest. The underlying proof method (combining a Lindström-type argument with the Barwise-Kunen Boundedness Theorem) may be used to derive infinitary versions of the modal invariance and safety theorems of Chapters 4, 5, 10 — thereby settling some conjectures in earlier versions of this book.

Part II

Logical Foundations

In this second part, we are going to study the fundamental questions identified so far, concerning repertoires of operators, dynamic inference, architecture and complexity. Our main tools come from modal logic, relational algebra, and their combination in propositional dynamic logic — all twisted somewhat to suit our present concerns. Recurrent technical themes are invariance for bisimulation, frame correspondence for modal axioms, and the switch between modal logics and fragments of first-order logic. In Chapter 4, we introduce our basic process structures, being possible worlds models ('labeled transition systems') with various semantic equivalences between them, and study a spectrum of matching logical languages, including new decidable 'bounded' first-order fragments. Chapter 5 studies logical spaces of natural process operations, mainly through invariances for process equivalences. In Chapter 6, we analyze systems of dynamic logic as a general two-level architecture for dynamic information flow. Chapter 7 uses this modal framework to give a general treatment of possible dynamic inference relations. Finally, we re-think classical relational algebra and predicate logic from a dynamic perspective. In Chapter 8, we give decidable versions of relational algebra viewed as theories of primitive transition arrows. In Chapter 9, we analyze predicate logic in the same style, letting standard validities become computational constraints on update patterns for variables. This strategy uncovers a whole landscape of decidable predicate logics with richer vocabularies underneath the standard system.

4

Process Simulation and Definability

Special systems of dynamic semantics exemplify general processes. We will study these using (poly-)modal logic, with a good deal of inspiration from computer science. Processes may be represented by possible worlds models, or equivalently, labeled transition systems. The underlying notion of process is then determined by imposing a semantic equivalence between different models — of which there is a whole hierarchy. Corresponding to such relations of process 'simulation', there are different logical languages of matching expressive power over our models. Our point of departure is an old characterization of the basic modal language as consisting of just those first-order formulas which are invariant for *bisimulation*. The proof of this result yields a general model-theoretic technique matching up process equivalences and fragments of predicate logic. In particular, we will consider finite-variable fragments and bounded fragments of predicate logic, which provide an interesting fine-structure to classical logic.

4.1 Transition Systems and Process Equivalences

4.1.1 Basic Structures

We start with a formal account of our dynamic structures and their semantic relations.

Definition 4.1 A *labeled transition system* (LTS) is a triple $\mathbf{M} = (S, \{R_a \mid a \in A\}, V)$ where S is a set of 'states', the R_a are binary relations describing atomic 'actions', and V is a valuation for proposition letters over states in the usual modal sense.

LTSs are, of course, the earlier Kripke models for polymodal logic. Equivalently, the frame part of these structures may be described as a triple $(S, A, \rightarrow)$ where A contains the atomic 'actions', and $\rightarrow$ is a ternary relation over $S \times A \times S$ providing 'labeled transitions' $\xrightarrow{a}$ between states for each atomic action a. In the computational literature, one finds several further elements to LTSs. For instance, there may be special arrows

$\xrightarrow{a} \surd$ indicating a successfully completed a-transition, or special markers for 'termination' or 'dead-lock'. This structure may be added as follows:

$\surd$ is a unary predicate holding at 'success' or 'termination' states
δ is a unary predicate holding at 'failure' or 'dead-lock' states.

If desired, these may be provided with suitable meaning postulates, such as

- $\forall x\,(\surd x \vee \delta x \to \neg\exists y\, R_a xy)$ (for all $a \in A$)
- $\forall x\,(\surd x \to \neg\delta x)$.

4.1.2 Process Equivalences

Two different LTSs may represent the same process. This observation suggests notions of equivalence across different semantic structures. Concerns of this kind are scarce in linguistic semantics or philosophical logic — even though appropriate criteria of process identity seem crucial to the dynamic turn. In the computational literature, notions of 'process equivalence' depend on one's practical aims, and therefore, there is a hierarchy from coarser to finer grain levels of comparison — all the way up to full LTS *isomorphism*. There are two notable choice points here. We shall mainly compare different LTSs, but the literature often compares states inside one 'super LTS'. These are equivalent views, as one can merge two different LTSs into their disjoint union. A genuine choice is the following. One can view process equivalence relations either as 'respecting' certain observational properties in two LTSs (this is the "if" direction in the definitions to follow) or as completely 'determined' by the latter properties (this is the "iff" version). Both alternatives occur in the literature.

Definition 4.2 Let $\mathbf{M} = (S, \{R_a \mid a \in A\}, V)$, $\mathbf{M}' = (S', \{R'_a \mid a \in A\}, V')$ be LTSs. A binary relation $\equiv$ over $S \times S'$ is a *finite trace equivalence* (FinTE) if, whenever $s \equiv s'$, then s, s' start the same finite traces, i.e., finite sequences of labeled transitions with their atomic markings. If the relation $\equiv$ consists of precisely the latter pairs of traces, then it is a *complete finite trace equivalence*.

This is about the coarsest notion of 'behavioral equivalence' for processes. We now consider some progressively more demanding comparisons.

Definition 4.3 A relation $\equiv$ as above is a *full trace equivalence* (FullTE) if any two states with $s \equiv s'$ start the same countable traces. The latter include all terminating finite traces (successful or not) and all countably infinite ones

$$s = s_1 \xrightarrow{a_1} s_2 \xrightarrow{a_2} s_3 \xrightarrow{a_3} \ldots.$$

Complete full trace equivalence is the case where the latter property even defines $\equiv$.

FIGURE 1

The next notion is basic in both the computational and the modal literature. Essentially, it adds respect for the 'choice structure' of processes (cf. Chapter 3):

Definition 4.4 A binary relation $\equiv$ between states in two LTSs is a *bisimulation* (BISIM) if the following back-and-forth conditions hold:

(i) if $s \equiv s'$ and $s \xrightarrow{a} t$, then there exists t' with $s' \xrightarrow{a} t'$ and $s' \equiv t'$
(ii) and vice versa.

Moreover, $\equiv$ should only connect states agreeing in their atomic markings. The *maximal bisimulation* between two LTSs is the union of all bisimulations between them (which is necessarily a bisimulation itself).

Finally, we consider one even stronger notion of process equivalence.

Definition 4.5 A relation $\equiv$ is a *generated graph isomorphism* (GGI) if, whenever $s \equiv s'$, then the 'generated subLTS's $\mathbf{M}_s$, $\mathbf{M}'_{s'}$ are isomorphic. (The latter are the submodels containing s, s' and all states that can be reached from these via some finite sequence of atomic transitions available in the model.) When used as a defining characteristic again, the latter relation gives *complete generated graph isomorphism*.

Here are some simple connections between these various notions, stated with some obvious abbreviations. The relevant proofs and counter-examples are straightforward.

Proposition 4.6

(i) GGI implies BISIM implies FullTE implies FinTE
(ii) None of these implications can be reversed.

Example 4.7 (Power of Distinction) We give some illustrations, with simulations only connecting roots (marked with $*$). Valuations are omitted. The structures in Figure 1 are finite trace equivalent, but not bisimilar. Those in Figure 2 are bisimilar, but not generated-graph equivalent.

FIGURE 2

One can vary the above notions of process equivalence in many ways. In what follows, we shall concentrate on the middle ground of bisimulation. But the eventual aim of our analysis is to provide a general model-theoretic perspective upon all of them.

4.2 Languages over Transition Systems

Labeled transition systems are ordinary mathematical graph structures, and as such, they may be described using standard logical formalisms. There is a great variety of suitable formal languages for this purpose, differing in expressive power.

4.2.1 Poly-Modal and Dynamic Logic

Given the needs of dynamic semantics (cf. Chapter 2) as well as mathematical reasons that will become clear below, we start with poly-modal and dynamic logics over such models, with the usual box and diamond notations. The basic theory of these useful formalisms was outlined in Chapter 3. They allow us to define many assertions concerning program execution. We recall an earlier illustration.

Example 4.8 (Basic Computational Assertions)

(i) Correctness: $\phi \to [\pi]\psi$; precondition ϕ implies postcondition ψ after successful execution of program π

(ii) Termination: $\langle\pi\rangle\top$; program π terminates

(iii) Weakest Precondition: $\langle\pi\rangle\phi$; running program π can produce effect ϕ

(iv) Enabling: $[\pi_1]\langle\pi_2\rangle\phi$; program π_1 enables π_2 to produce ϕ.

Poly-modal logic lies at the bottom of a landscape of formalisms describing transition systems. Higher up lies their full first-order language, and even further extensions. To see this, recall the 'standard translation' from Chapter 3 sending modal and dynamic languages to fragments of standard first-order and higher-order languages.

Example 4.9 (Translation of Modal and Dynamic Formulas)

$$[a]\langle b\rangle p \quad \text{translates to} \quad \forall y\,(R_a xy \rightarrow \exists z(R_b yz \wedge Pz))$$

$$\langle a^*\rangle[b \cup c]q \quad \text{translates to} \quad \exists y(\bigvee_{n\in\mathbb{N}} R_a^n xy \wedge \forall z\,((R_b yz \vee R_c yz) \rightarrow Qz))).$$

4.2.2 Finitary and Infinitary First-Order Logic

The full first-order language expresses various natural properties of computations.

Example 4.10 (First-Order Computational Notions)

'Action a is deterministic'	$\forall xy\,((R_a xy \wedge R_a xz \rightarrow y = z)$
'Action a is confluent'	$\forall xyz\,((R_a xy \wedge R_a xz) \rightarrow \exists u\,(R_a yu \wedge R_a zu))$
'Action a enables b'	$\forall xy\,(R_a xy \rightarrow \exists z\, R_b yz)$

Some computational notions require infinitary first-order formalisms — a move which has been advocated by many authors in the literature on dynamic logic.

Example 4.11 (Infinitary Computational Notions) (One has to assume here that there are only countably many atomic actions.)

'Acyclic LTS'

$$\bigwedge_{n\geq 1,\, a_1,\ldots,a_{n-1}\in A} \neg(\exists x_1 \ldots x_n\,(\bigwedge_{1\leq i<n} R_{a_i} x_i x_{i+1} \wedge x_1 = x_n))$$

'Action a terminates'

$$\forall x\,(\surd(x) \vee \bigvee_{n\geq 1} \exists x_1 \ldots x_n(\bigwedge_{1\leq i<n} R_a x_i x_{i+1} \wedge \surd(x_n))).$$

Remark 4.12 (First-Orderization) In Logic, things are not always what they seem. Higher-order formulations may often be 'first-orderized' by additional vocabulary. For instance, there is a *two-sorted* first-order version of Termination taking 'paths' or 'branches' as a new sort of object (in addition to states), with some plausible predicates about them, and stating that every state lies on a successful a-path. Such a technical move may even have intrinsic interest as an explicit geometrical theory of computation paths (cf. Chapter 13).

4.2.3 Fine-Structure of Expressive Power

In between modal logic and first-order logic, there are natural intermediate formalisms for describing transition systems. Such *extended modal logics* have been studied extensively (starting with the work of the logic group at Sofia University in the 80s). A simple nice example is modal logic extended with a 'universal modality':

$$\mathbf{M}, x \models \Diamond\phi \text{ iff } \mathbf{M}, y \models \phi \text{ for some state } y.$$

There is a more general notion of fine-structure here. Both the basic modal language and its universal extension lie within the *two-variable* fragment of predicate logic, with formulas using only two state variables x, y (free or bound). This fragment also expresses further properties of LTSs, referring to predecessors rather than successors in labeled transitions. One can think of its two variables as the only available 'registers' for computation. Thus, the two-variable fragment states those computational properties that involve no more than comparison of two states at any time. The next level up in this hierarchy has three variables. Here lie the usual languages of Temporal Logic, which allow one to also express properties of intermediate stages of labeled transitions.

Example 4.13 (Temporal Betweenness) Three variables are involved essentially in the well-known temporal operator "*Until*":

$Until_a pq \quad \exists y\,(Py \wedge \forall z\,((R_a xz \wedge R_a zy) \rightarrow Qz))$.

The general picture is of 'finite-variable fragments', to which we shall return later on.

4.2.4 Alternative Formalisms over Transition Systems

A related, though not quite equivalent, approach is the tradition of *Process Algebra*, which considers LTSs consisting of processes with additional structure. Here is one typical axiom system (Chapter 10 below provides further background). There is one binary operation of addition or 'choice' (+) plus a family of unary operations a; ('prefixing action a') , satisfying algebraic identities like the following:

Example 4.14 (Elementary Process Algebra (EPA))

(1) $x + x = x$ — idempotence
(2) $x + y = y + x$ — commutativity
(3) $x + (y + z) = (x + y) + z$ — associativity
(4) $0 + x = x$ — zero element

Process Algebra can be seen as a calculus for combining whole LTSs into new ones, via operations of 'choice', 'product' or 'merge' mirroring the algebraic operations. Thus, Modal Logic is an 'internal' description language for LTSs, which is in harmony with the latter 'external' calculus. (Chapter 10 presents more detailed comparisons.)

Other logical possibilities exist. *First-order Action Logic* describes operational structure via a second domain of *actions* in addition to states, with action quantifiers and variables. (Specific atomic actions have constant names.) Labeled transition systems are still models for this language, with the domains S and A now treated on a par. The earlier arrow $\rightarrow$ becomes a ternary predicate letter relating actions and states, and one can express new computational properties, both first-order and infinitary:

- 'Endpoints exist': $\exists x \neg\exists a\, x \xrightarrow{a} y$,
- 'Every action has a converse': $\forall a \exists b \forall xy\, (x \xrightarrow{a} y \leftrightarrow y \xrightarrow{b} x)$,
- 'Actions are extensional': $\forall ab\, (\forall xy\, (x \xrightarrow{a} y \leftrightarrow x \xrightarrow{b} y) \rightarrow a = b)$.

Example 4.15 (Preservation Under Changing Repertoires) Assertions in pure state languages refer to some fixed set of atomic actions. Hence they are invariant for all changes in the action domain of LTSs that leave the relevant actions undisturbed. This yields a semantic characterization of the 'pure state fragment' of the full two-sorted state-action logic. A related issue is which formulas are preserved when actions are *added* to an LTS (over the same states). The answer is in van Benthem 1985a (Lemma 17.19). Essentially, these are all formulas constructed using atoms and their negations, the connectives $\wedge$ and $\vee$, quantifiers $\forall$, $\exists$ over states, and only $\exists$ over actions. Many further semantic questions extend to first-order action logic.

4.3 Bisimulation Invariance and Modal Definability

A striking difference between the logical and the computational literature on transition systems is one of emphasis. On the former approach, one tends to start from a given formalism, and then study its expressive power by inventing some notion of 'semantic equivalence' between models for which the language is invariant. A typical example are *Ehrenfeucht Games* in first-order logic (Section 3.1). Indeed, such model comparison games and their natural modifications provide an attractive alternative take on general process equivalences as studied in Section 4.1. The computational literature, however, starts from intuitions concerning processes and their equivalences, and then designs an appropriate language respecting the latter. Sub specie aeternitatis, both approaches are two sides of the same coin, and they often produce the same results. Thus, the computational notion of bisimulation (Park 1981) had already been proposed as the preferred invariance for modal logic in Van Benthem 1976 — and in fact, it may also be seen as a clipped version of Ehrenfeucht Games, appropriate to the modal fragment of our first-order logic of LTSs. We shall encounter this duality repeatedly in this Book, starting with the following typical case.

4.3.1 The Paradigm: Basic Modal Logic

We recall some basic results from Chapter 3. To begin with, modal formulas are *invariant for bisimulation* in the following sense.

Proposition 4.16 (From Simulation to Invariance) *Let $\equiv$ be a bisimulation between two models* $\mathbf{M}$, $\mathbf{N}$ *with $x \equiv y$. Then, for all modal formulas ϕ,* $\mathbf{M}, x \models \phi$ *iff* $\mathbf{N}, y \models \phi$.

Conversely, we had the following connection:

Proposition 4.17 (From Invariance to Simulation) *Let* $\mathbf{M}, x \models \phi$ *iff* $\mathbf{N}$, $y \models \phi$, *for all modal formulas* ϕ. *Then there exist elementary extensions* $\mathbf{M}^+$, $\mathbf{N}^+$ *of* $\mathbf{M}$, $\mathbf{N}$, *respectively, and a bisimulation* $\equiv$ *between these such that* $x \equiv y$.

In particular, for *finite* models $\mathbf{M}$, $\mathbf{N}$ (though not in general!), two states x, y verify the same modal formulas iff they are connected by some bisimulation. (This is shown by a straightforward argument, revolving around description of finite branchings by conjunctions of existential modalities.) Now, we come to our main result, telling us that poly-modal logic and bisimulations are a perfect match in a first-order perspective.

Theorem 4.18 (Invariance Theorem) *A first-order formula* $\phi(x)$ *is equivalent to one in the modal fragment iff it is invariant for bisimulation.*

Proof. (van Benthem 1976) That all modal formulas are invariant was shown before. Conversely, suppose that $\phi = \phi(x)$ is an invariant first-order formula. Let $\mathbf{mod}(\phi)$ be the set of all modal consequences of ϕ. We prove the following implication:

Claim. $\mathbf{mod}(\phi) \models \phi$.

From this, by Compactness, ϕ must be equivalent to some finite conjunction of its modal consequences. The proof of the claim is as follows. Let $\mathbf{M}, x$ be any model for $\mathbf{mod}(\phi)$. Now consider the complete modal theory of x in $\mathbf{M}$ together with $\{\phi\}$. This set of formulas is finitely satisfiable. (Otherwise, ϕ implies some disjunction of negations of formulas in $Th(\mathbf{M}, x)$ — and as the latter is a modal formula in $\mathbf{mod}(\phi)$, it should have been true at x in $\mathbf{M}$.) By Compactness, the whole set therefore has a model $\mathbf{N}, y$. Now, take two *ω-saturated elementary extensions* $\mathbf{M}^+, x$ and $\mathbf{N}^+, y$ of $\mathbf{M}, x$ and $\mathbf{N}, y$, respectively. (These exist by earlier observations in Section 3.1.)

Claim. *The relation of modal equivalence between worlds is a bisimulation between the two models* $\mathbf{M}^+$ *and* $\mathbf{N}^+$, *which connects* x *with* y.

Proof. Of course, the key observation lies in the back-and-forth clauses. If some world u in $\mathbf{M}^+$ is modally equivalent with v in $\mathbf{N}^+$, and Rus holds, then the following set of formulas is finitely satisfiable in $\mathbf{N}^+$:

$$\{Rvt\} \text{ plus the full modal theory of } s \text{ in } \mathbf{M}^+.$$

But then, by ω-saturation, some world t exists satisfying all of this in $\mathbf{N}^+$: which is the required match. The converse argument is symmetric. ⊣

Having thus proved the second claim, we return to the first, and clinch the argument by 'diagram chasing'. For a start, $\mathbf{N}, y \models \phi$, and hence $\mathbf{N}^+, y \models \phi$ (by elementary extension), whence $\mathbf{M}^+, x \models \phi$ (by bisimulation invariance), and so $\mathbf{M}, x \models \phi$ (passing to an elementary submodel). ⊣

This style of argument can be extended in many directions, by modulating its key connection between establishing zigzag clauses in saturated models and the presence of corresponding restricted quantifier patterns in the formal language under investigation. We shall encounter several applications in this Chapter and the next.

Remark 4.19 (Complexity of Fragments) Natural fragments of first-order logic may be undecidable. The modal formulas are a case in point, because of the following effective reduction. Let $\alpha(x)$ be any first-order formula, and P (unary), R (binary) new predicate letters outside of α.

Claim. *The following two assertions are equivalent:*

(i) α *is universally valid*

(ii) $\exists y\, Py \wedge (\neg\alpha)^\rho$ *is bisimulation invariant, where* $(\neg a)^\rho$ *is the formula* $\neg\alpha$ *syntactically relativized to the subdomain* $\lambda z.\, Rxz \vee x = z$.

Proof. If α is universally valid, then $\exists y\, Py \wedge (\neg\alpha)^\rho$ is equivalent to a contradiction, which is trivially invariant. If α is not universally valid, then $\neg\alpha(x)$ is true in some LTS **M** at x. Now let R be the universal relation in **M** — and add one unrelated point where P holds to obtain a model $\mathbf{M}^+$:

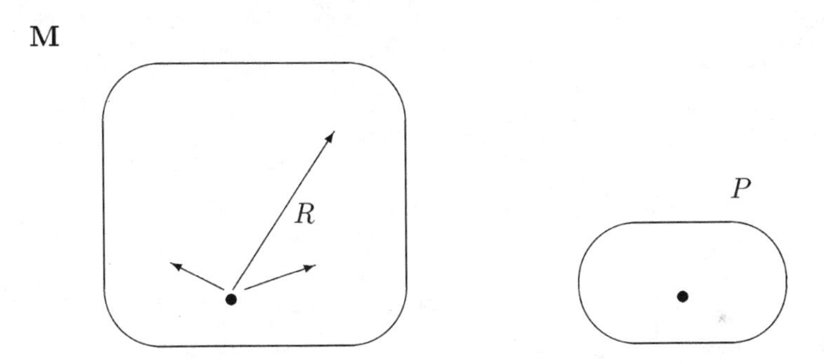

In the new model, x validates $\exists y\, Py \wedge (\neg\alpha)^\rho$. But this formula fails in the R-closed submodel generated by x, thereby violating invariance for bisimulation. ⊣

4.3.2 From Languages to Simulations

The above paradigmatic result suggests two routes. Given some formalism, one can seek a matching notion of process equivalence. But also, given some notion of process equivalence, one can seek an appropriate language with just this discriminatory power. We shall illustrate both directions. Here is a simple warm-up case. The earlier modal logic with a universal modality can be analyzed by a simple modification of the earlier bisimulation. One needs zigzag clauses for the (unrestricted) universal modality, stating that arbitrary choices of a state on either side can be matched with arbitrary choices on the other side. This is equivalent to requiring that the bisimulation be *total*, making its domain and range the full state sets of the two models being compared. Now, the earlier analysis of bisimulation is easily extended to the following result.

Theorem 4.20 *A first-order formula is definable by a modal formula with additional universal modalities iff it is invariant for total bisimulations.*

A similar analysis works for the *two-variable fragment* of predicate logic containing the preceding formalism. Its proper notion of invariance involves an interesting extension of the notion of bisimulation, from a binary relation between single states to one between *pairs of states*. Here is the right formulation for this extended equivalence.

Definition 4.21 (Pair Simulation) A *pair simulation* is a relation C between states in S and S' as well as between pairs of states in S and pairs of states in S' satisfying the following three conditions:

(i) matches in C are partial isomorphisms w.r.t. all predicates of our language
(iia) if sCs', then for all states t in S, there is a t' in S' with $(s,t)C(s',t')$
(iib) and vice versa
(iii) C is closed under *restrictions*: if $(s,t)C(s',t')$, then sCs' and tCt'.

This notion is stronger than bisimulation. It allows arbitrary choices in the zigzag clauses (not just successor states of the previous one). An easy induction shows that C-matched states or state pairs in two LTSs satisfy the same formulas from the two-variable fragment of first-order logic. A similar analysis works for the temporal three-variable fragment. Again, we encounter k-variable levels of first-order logic containing modal operators of ascending strength, computed over a growing number of 'registers':

$k = 1$ Boolean operators (negation, conjunction)
$k = 2$ modal operators (Necessary, Possible)
$k = 3$ temporal operators (Since, Until)
$k = 4$ lattice operators (supremum, infimum)
etcetera.

This Finite Variable Hierarchy will be analyzed in more detail in Section 4.4.1 below.

4.3.3 From Simulations to Languages

By arguments similar to that proving the modal Invariance Theorem, one can analyze the other notions of process equivalence introduced in Section 4.1. Here are some relevant results from van Benthem and Bergstra 1994. Incidentally, all these will go through if one imposes arbitrary first-order conditions on our model class, such as the earlier meaning postulates on success and dead-lock.

Theorem 4.22 *A first-order formula is invariant for Finite Trace Equivalence iff it is definable as a Boolean combination of successful path formulas.*

Proof. From right to left, this was already observed above. Conversely, suppose that $\phi(x)$ is invariant for FinTE . Consider the set $BPF(\phi)$ of all Boolean combinations of successful path formulas that follow logically from $\phi(x)$. The desired definability assertion is then an obvious consequence of the following

Claim. $BPF(\phi) \models \phi(x)$.

Proof. (of the Claim) Let $\mathbf{M}, s \models BPF(\phi)$. Then, by a simple argument, the set consisting of $\phi(x)$ together with all Boolean combinations of successful path formulas true at $\mathbf{M}, s$ is finitely satisfiable. Therefore, by Compactness, this set must be simultaneously satisfiable, say in some LTS $\mathbf{N}, s'$. This means that the relation of complete finite trace equivalence between these two models relates s to s'. But then, by the given invariance of $\phi(x)$, this implies that $\mathbf{M}, s \models \phi(x)$. ⊣

Indeed, the same syntactic class captures invariance with respect to Complete Finite Trace Equivalence. In general, of course, invariance for a completed equivalence notion is implied by, but need not imply, that under the original one. Next, we move from successful path formulas to arbitrary ones, being existential descriptions of finite chains of transitions with arbitrary success or dead-lock behaviour in their states.

Theorem 4.23 *A first-order formula is invariant for Full Trace Equivalence iff it is definable as a Boolean combination of arbitrary path formulas.*

Proof. The argument is similar to the preceding one. This time, however, one passes again from $\mathbf{M}, s$ and $\mathbf{N}, s'$ to two ω-saturated elementary extensions. Note that, in the latter LTSs, coincidence of arbitrary path formulas at two states is in fact the same as full trace equivalence. The reason is that any countably infinite chain on one side can be described using some countable set of formulas which will be finitely satisfiable on the other side, after which saturation gives us the whole chain there, too. ⊣

Next comes the earlier modal Invariance Theorem for bisimulation. Again its proof also characterizes another invariance. For, its key bisimulation between saturated models $\mathbf{M}^*$, $\mathbf{N}^*$ is in fact the *maximal* one between them. This shows that the modal fragment of first-order logic also captures invariance with respect to maximal bisimulations. A slight refinement of this argument shows the next result. Let the *restricted fragment* of first-order logic be constructed like the modal fragment, but allowing binary atoms Ruv, $u = v$ in addition to unary ones. (Section 4.4.2 has several such formalisms.)

Theorem 4.24 *A first-order formula is invariant for Generated Graph Isomorphism iff it is definable as a restricted formula.*

Proof. We show that the restricted consequences of an invariant formula ϕ imply it. In an obvious notation: $restr(\phi) \models \phi$. It suffices to make a few changes in the above argument, without loss of generality. First, the models **M**, **N** can be chosen to be countable by the Löwenheim-Skolem Theorem. Then, we pass on to countable *recursively saturated elementary* extensions $\mathbf{M}^*$, $\mathbf{N}^*$ (cf. Keisler 1977 for this refined technique), which only have saturation with respect to RE sets of first-order formulas. (The reason is this. With arbitrary ω-saturated elementary extensions as before, we cannot keep the cardinality down to countable infinity.) Equality of restricted theories yields the extra information that there is an isomorphism between the generated submodels of x in $\mathbf{M}^*$ and y in $\mathbf{N}^*$. This may be obtained as follows. Start from complete enumerations $x = a_1, a_2, \ldots$ and $y = b_1, b_2, \ldots$ of the generated submodels. Then use a Cantor-style zigzag argument to extend the initial match to one between progressive sequences, maintaining the following invariant:

$$\mathbf{M}, a_1, \ldots, a_n \text{ satisfies the same restricted formulas as } \mathbf{N}, b_1, \ldots, b_n.$$

With each extension with a new object in the enumeration, use the RE set of formulas

$$\phi(a_1, \ldots, a_n, x_{n+1}) \leftrightarrow \phi(b_1, \ldots, b_n, x_{n+1}) \text{ with } \phi \text{ restricted,}$$

to locate a matching object on the other side. Here, one may use restricted quantifiers only to describe the behaviour of some object x_{n+1} sharing finitely many restricted properties with, say, a_{n+1}. This is possible, as each object in a generated submodel is reachable from its origins via some finite sequence of labeled transitions. $\dashv$

Again, by the same argument, the above syntactic class also captures invariance with respect to Complete Generated Graph Isomorphism.

4.4 Modal Fine-Structure of First-Order Logic

In this final Section, we consider the 'modal fine-structure' of first-order predicate logic which is coming to light in our analysis. It makes sense to bring this out by itself, even along two different lines — as in Andréka et al. 1995a, 1995b, 1996.

4.4.1 Finite Variable Hierarchy

Finite-variable fragments of first-order logic were introduced in Henkin 1967. These consist of all formulas using only some fixed finite set of variables (free or bound): say $\{x\}$, $\{x, y\}$, $\{x, y, z\}$, etcetera — while allowing arbitrary quantifiers and connectives. This restricts predicate logic to 'semantic computation' over some fixed finite number of registers (cf. Chapters 2, 9). A nice connection with modal logic was pointed out in Gabbay 1981b. Finite operator sets generate modal languages whose transcriptions involve only a fixed finite number of variables (free and bound). The basic modal language needs *two* state variables (cf. Chapter 3), while temporal logic with 'Since' and 'Until' uses essentially *three* variables. Thus, there is a *Finite Variable Hierarchy*, inside whose levels one finds modal logics of ascending expressive strength:

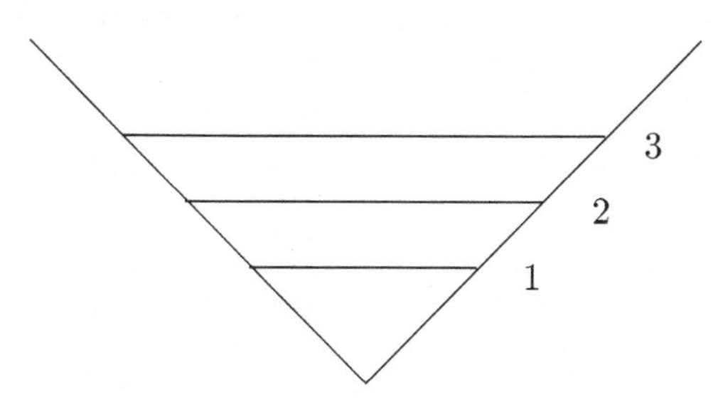

Like the basic modal language, these successive levels can be characterized via a natural semantic invariance (van Benthem 1991b). For this result, recall the usual convention that $\phi(x_1, \ldots, x_k)$ is a formula whose free variables are all among $\{x_1, \ldots, x_k\}$. Also, *k-variable formulas* will be first-order formulas whose variables (free or bound) are all among $\{x_1, \ldots, x_k\}$. Now, we consider a cut-off version of a well-known notion from Abstract Model Theory (cf. Section 3.1, Barwise 1975).

Definition 4.25 (k-Partial Isomorphism) A *k-partial isomorphism* is a non-empty zigzag family $\mathbf{I}$ of partial isomorphisms of size at most k between two models $\mathbf{M}$, $\mathbf{N}$, which is closed under taking restrictions to smaller domains, while the Back-and-Forth properties hold only up to length k.

A formula $\phi(x_1, \ldots, x_k)$ is *invariant for k-partial isomorphism* if, for every such $\mathbf{I}$, its truth value is invariant between any pair of k-tuples of objects $(\mathbf{a}, \mathbf{b})$ in $\mathbf{I}$.

Theorem 4.26 *A first-order formula $\phi(x_1, \ldots, x_k)$ is equivalent to a k-variable formula if and only if it is invariant for k-partial isomorphism.*

Proof. All k-variable formulas are preserved under k-partial isomorphism, by a simple induction. More precisely, one proves, for any assignment A and any partial isomorphism $I \in \mathbf{I}$ which is defined on the A-values for all variables $x_1, \ldots, x_k$, that

$$\mathbf{M}, A \models \phi \text{ iff } \mathbf{N}, I \circ A \models \phi.$$

The crucial step in this induction is the quantifier case. Quantified variables are irrelevant to the assignment, so that the relevant partial isomorphism can be restricted to size at most $k-1$, whence a matching choice for the witness can be made on the opposite side. This proves "only if". The converse direction "if" has a proof analogous to that of the Invariance Theorem for modal formulas. One shows that an invariant formula $\phi(x_1, \ldots, x_k)$ must be implied by the set of all its k-variable consequences. The key step in this argument is as before. We find two models which are elementarily equivalent for all k-variable formulas. These then possess ω-saturated elementary extensions — for which the relation of k-elementary equivalence itself defines a family of partial isomorphisms between tuples of objects up to length k, which satisfies all the above requirements for being a k-partial isomorphism. $\dashv$

Finite-variable fragments have many uses in mathematical logic and computer science. But they have some drawbacks. E.g., important meta-properties of full first-order logic fail inside them, including the Łoś-Tarski Submodel Preservation Theorem and Craig's Interpolation Theorem (cf. the Notes following this Chapter). Whatever the precise verdict, there exists another powerful perspective on modal fine-structure, which seems preferable from our point of view. We now turn to this alternative.

4.4.2 Bounded Fragments

The basic modal fragment is only a subset of the full two-variable fragment, since its syntax satisfies additional constraints. In particular, all quantifiers in translated modal formulas occur 'restricted' or 'bounded', in the forms

$$\exists y\,(Rxy \wedge \phi(y)), \quad \forall y\,(Rxy \rightarrow \phi(y)).$$

The latter form motivates the definition of bisimulation, explaining its zigzag clauses. This observation suggests another classification, in terms of *quantifier restrictions*, which may be varied along various dimensions. The general schema here is as follows:

$\exists \mathbf{y}(R\mathbf{xy} \wedge \phi(\mathbf{x}, \mathbf{y}, \mathbf{z}))$, where $\mathbf{x}$, $\mathbf{y}$, $\mathbf{z}$ are finite sequences of variables.

And the question is how much can be allowed without losing the attractive features of basic modal logic, including its decidability. This perspective suggests a hierarchy of 'bounded' fragments of predicate logic. Initial stages occurred in earlier sections. Polymodal logic shows that families of restricting predicates R_i are admissible. And polyadic modal logics showed the same for restrictions $\exists \mathbf{y}\,(Rx, \mathbf{y} \wedge \bigwedge_i \phi_i(y_i))$. We shall be concerned mainly with the following schemata in what follows:

Fragment 1	$\exists \mathbf{y}\,(R\mathbf{yx} \wedge \phi(\mathbf{y}))$
Fragment 2	$\exists \mathbf{y}\,(R\mathbf{yx} \wedge \phi(\mathbf{x}, \mathbf{y}))$
Fragment 3	$\exists \mathbf{y}\,(R\mathbf{yx} \wedge \phi(\mathbf{x}, \mathbf{y}, \mathbf{z}))$.

These fragments of a standard first-order predicate language start with arbitrary atoms, and allow further constructions with Boolean operators as well as the above bounded quantifiers, where the R can be any relation symbol (this atom is called the 'guard' of the formula), whose variables may appear in any order. Identity atoms are allowed. There are other plausible bounded fragments — but these will do. In particular, Fragment 2, the *Guarded Fragment* of predicate logic, combines nice modal behaviour with expressive power. Crucial for these fragments is the atomic nature of guards: Boolean combinations of atoms are not permitted. E.g., symmetry of a relation is in Fragment 2, but transitivity is not. These fragments may be understood as follows. Model-theoretically, we will extend modal *bisimulation* to describe them, which shows there is a genuine upward hierarchy of expressive strength. There is also a more combinatorial approach, focusing on their 'looseness' and decidability. Fragment 1 is decidable, being close to modal logic. Also, Fragment 3 is easily shown undecidable. Our main result is that the powerful intermediate Guarded Fragment 2 is decidable.

Bounded Quantifiers and Bisimulation

Bounded fragments may be analyzed semantically in terms of suitable bisimulations. As before, a *partial isomorphism* is a finite one-to-one partial map between models which preserves relations both ways. In any model $\mathbf{M}$, call a set X of objects *guarded* if there exists a relation symbol R, say k-ary, and objects $a_1, \ldots, a_k \in \mathbf{M}$ (possibly with repetitions) such that $R^{\mathbf{M}}(a_1, \ldots, a_k)$ and $X = \{a_1, \ldots, a_k\}$. Here is the correct bisimulation for the Guarded Fragment, the others involve obvious variations.

Definition 4.27 (Guarded Bisimulations) A *guarded bisimulation* is a non-empty set $\mathbf{F}$ of finite partial isomorphisms between two models $\mathbf{M}$ and $\mathbf{N}$ which satisfies the following back-and-forth conditions. Given any $f : X \to Y$ in $\mathbf{F}$,

(i) for any guarded $Z \subseteq M$ there is a $g \in \mathbf{F}$ with domain Z such that g and f agree on the intersection $X \cap Z$

(ii) for any guarded $W \subseteq N$ there is a $g \in \mathbf{F}$ with range W such that the inverses g^{-1} and f^{-1} agree on $Y \cap W$.

The point of this definition shows in semantic invariance for guarded bisimulation, which is proved by a straightforward induction on the construction of Fragment 2-formulas. The two zigzag conditions take care precisely of bounded existential quantifiers.

Proposition 4.28 *Let* $\mathbf{F}$ *be a guarded bisimulation between models* $\mathbf{M}$ *and* $\mathbf{N}$ *with* $f \in \mathbf{F}$. *For all guarded formulas* ϕ *and all variable assignments* α *into the domain of* f, *we have* $\mathbf{M}, \alpha \models \phi$ *iff* $\mathbf{N}, f \circ \alpha \models \phi$.

A straightforward analogue of the earlier modal proof yields a full preservation result.

Theorem 4.29 *Let ϕ be any first-order formula. Then ϕ is invariant for guarded bisimulations iff ϕ is equivalent to a Fragment 2-formula.*

One can adapt these notions to Fragments 1 and 2 in a straightforward manner. The modified back-and-forth clauses will now match the relevant quantifier restrictions.

Theorem 4.30 *The three fragments form a properly ascending hierarchy.*

Proof. We sketch the gist of the counter-examples. (1) The formula $\exists y\,(Rxy \wedge Sxy)$ is in Fragment 2, but not in Fragment 1. For, the following two models have different truth values for this formula — though a Fragment 1-bisimulation runs between them, consisting of the following matches between single objects: (x, x), (y, y_1), (y, y_2).

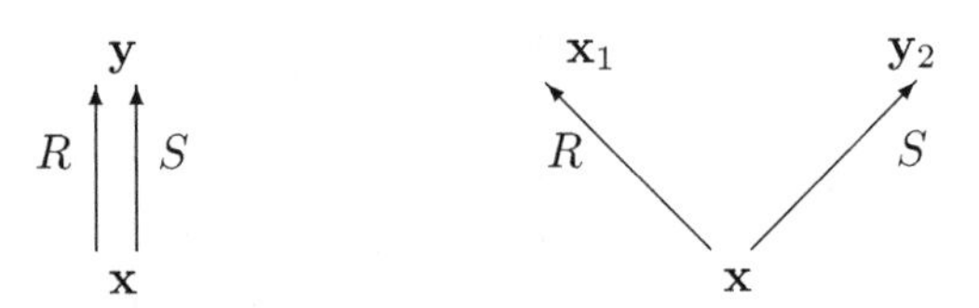

(2) The formula $\exists y\,(Ay \wedge \neg Rxy)$ is in Fragment 3, but not in Fragment 2 . For, it can distinguish between the following two models, even though they admit a Fragment-2 bisimulation, consisting of the partial isomorphisms $\{(y, y)\}, \{(z, y)\}, \{(x, x), (y, y)\}, \{(u, x), (z, y)\}$:

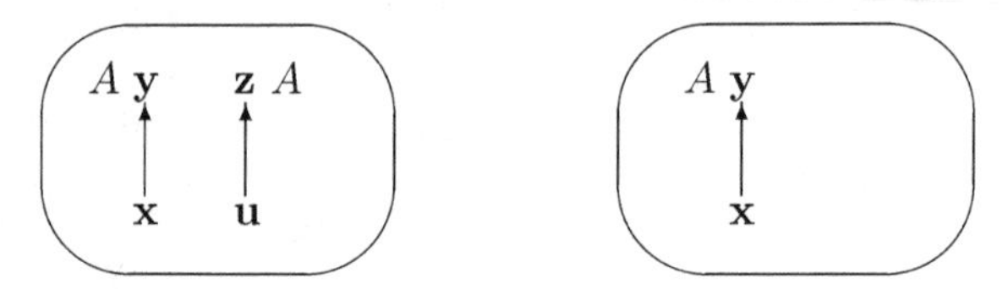

(3) Finally, Fragment 3 is still somewhat poorer than predicate logic as a whole. For instance, the formula $\forall x\, Ax$ is beyond it. This may be shown by the Fragment-3 bisimulation (x, x) between the following two models:

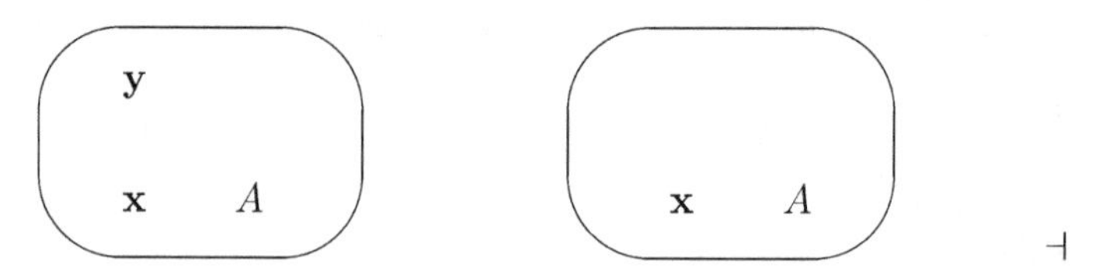

⊣

More generally, the bounded fragments serve as a point of departure for a new hierarchy in predicate logic, orthogonal to the finite-variable levels:

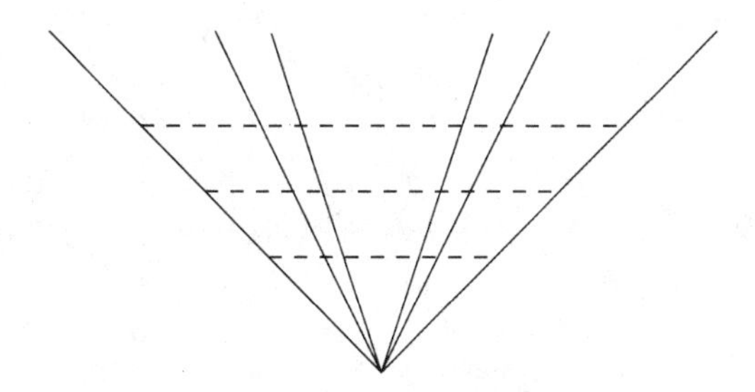

Decomposition of Universal Validity

To complete our analysis of bounded fragments, we now turn to their decidability. One way of proving this is via modal-style semantic tableaus, seeking some direct reduction for modalized sequents as in Chapter 3. Recall the relevant observation.

Fact 4.31 *The following are equivalent:*

1. $\exists y_1(Rxy_1 \wedge \phi_1(y_1)), \ldots, \exists y_k\,(Rxy_k \wedge \phi_k(y_k)) \models$
   $\exists y_1(Rxy_1 \wedge \psi_1(y_1)), \ldots, \exists y_m\,(Rxy_m \wedge \psi_m(y_m))$
2. *for some* i $(1 \leq i \leq k)$ $\phi_i \models \psi_1, \ldots, \psi_m$.

Here is a straightforward extension, which we only sketch here.

Theorem 4.32 *Validity of formulas in Fragment 1 is decidable.*

Proof. (Sketch) First, we perform all possible propositional reductions in a sequent, so that only atoms and existential quantifiers remain on both sides. Then we prove a reduction to matrix formulas like above. Again, we glue together counter-examples for sequents below where the quantifiers have been stripped off, so as to refute the original sequent with quantifiers. The longer sequent arguments **x**, **y** do not make an essential difference to this modal construction. And neither does the presence of arbitrary arguments **y** in the matrix formula, provided that the former have the required invariance property. Even when starting with 'mixed' initial formulas like $\exists y\,(Rx_1x_2, y \wedge \phi_1(y))$, $\exists y\,(Rx_3x_1, y \wedge \phi_2(y))$ and $\exists y\,(Rx_2, y \wedge \phi_3(y))$ followed by similar heterogeneous conclusions, one just matches up premise/conclusion pairs with identical sequences of **x**-parameters, as no semantic dependencies hold between behaviour of R-successors for sequences and their subsequences. ⊣

The preceding argument establishes a bit more than was stated.

Corollary 4.33 *Fragment 1 has the Finite Model Property.*

This direct approach does not work for the Guarded Fragment. Before trying another, let us give an easy result about Fragment 3. The latter turns out to be *undecidable*. Bounded quantification $\exists y\,(R\mathbf{xy} \wedge \phi(\mathbf{x}, \mathbf{y}, \mathbf{z}))$ is as powerful as full predicate logic.

Fact 4.34 *Predicate-logical satisfiability is effectively reducible to satisfiability in Fragment 3.*

Proof. The reduction takes any predicate-logical sentence ϕ to its relativization $\rho(\phi)$ to some unary predicate U not occurring in ϕ. $\rho(\phi)$ lies inside the parameterized restriction language. It is easy to see that ϕ is satisfiable if and only if $\rho(\phi)$ is. $\dashv$

Decidability of the Guarded Fragment

Our final theorem generalizes several results from the modal and algebraic literature.

Theorem 4.35 *Universal validity in the Guarded Fragment is decidable.*

Proof. Our general strategy is as follows. We show that any satisfiable formula ϕ in the Guarded Fragment has a finite 'quasi-model' of types (to be described below) of a size effectively computable from that of ϕ, which can be used conversely to generate a model for ϕ. (This is like modal filtration: which finds finite counter-models over sets of subformulas (Goldblatt 1987) — though we do not present our procedure as defining a model for ϕ.) Thus, the question whether a guarded formula is satisfiable is equivalent to whether it has a finite quasi-model — and from the effective description below, it can easily be seen that the existence of such a structure is decidable.

From Standard Models to Finite Quasi-Models. Suppose that a formula ϕ is satisfiable in some standard model $\mathbf{M}$. Let V be the set of variables occurring in ϕ (free or bound). Henceforth, we restrict attention to the finite set Sub_ϕ consisting of ϕ and all its subformulas (and closed under alphabetic variants using only variables in V, as explained below). Each variable assignment realizes a 'type' Δ consisting of finitely many formulas from this set. Types satisfy some closure conditions, which will emerge in due course in the proofs that follow. Our quasi-model has a universe consisting of the finitely many types realized in $\mathbf{M}$. Furthermore note that, for each guarded formula $\exists \mathbf{y}\,(Q\mathbf{xy} \wedge \psi(\mathbf{x}, \mathbf{y})) \in \Delta$ (no special order for the variables intended), there exists a type Δ' with (i) $Q\mathbf{xy}$, $\psi(\mathbf{x}, \mathbf{y}) \in \Delta'$ and (ii) Δ, Δ' agree on all 'unaffected' formulas with free variables contained in $\mathbf{x}$. Let us sum this up in the following notion, which may be viewed as an abstract version of the 'mosaics' of Andréka and Németi 1994. Assume in what follows that a guarded Fragment 2-formula ϕ is given, where V is the set of variables occurring in ϕ.

Definition 4.36 (i) Let F denote the set of all guarded formulas of length $\leq |\phi|$ that use only variables from V. Note that $\phi \in F$ and

F is closed under taking subformulas as well as 'alphabetic variants'. Also, F is finite.

(ii) An F-*type* is a subset Δ of F for which (a), (b), (c) below hold:
   (a) $\neg\psi \in \Delta$ iff not $\psi \in \Delta$ whenever $\neg\psi \in F$
   (b) $\psi \wedge \xi \in \Delta$ iff $\psi \in \Delta$ and $\xi \in \Delta$ whenever $\psi \wedge \xi \in F$
   (c) $[\mathbf{u}/\mathbf{y}]\psi$ implies $\exists y\, \psi \in \Delta$ whenever $\exists y\, \psi \in F$.

   Here $[\mathbf{u}/\mathbf{y}]\psi$ is the formula obtained from ψ by replacing each free variable in $\mathbf{y}$ with the corresponding variable in $\mathbf{u}$, simultaneously.

(iii) Let $\mathbf{y}$ be a sequence of variables, and let Δ, Δ' be types. We say that Δ and Δ' are $\mathbf{y}$-*close*, in symbols, $\Delta =_{\mathbf{y}} \Delta'$, if Δ and Δ' have the same formulas with free variables disjoint from $\mathbf{y}$.

(iv) A *quasimodel* is a set of F-types S such that, for each $\Delta \in S$ and each guarded formula $\exists \mathbf{y}\,(Q\mathbf{xy} \wedge \psi) \in \Delta$, there is a type $\Delta' \in S$ with $Q\mathbf{xy}$ and $\psi(\mathbf{x}, \mathbf{y})$ in Δ' and $\Delta =_{\mathbf{y}} \Delta'$. We say that ϕ *holds in a quasi-model* if $\phi \in \Delta$ for some Δ in this model.

Clearly, if ϕ is satisfied by some model, then ϕ also holds in a natural associated quasi-model. The converse holds as well:

From Quasi-Models to Standard Models. Given a quasi-model $\mathbf{M}$ of the above kind, we can again define a standard model $\mathbf{N}$. We say that π is a *path* if

$$\pi = \langle \Delta_1, \phi_1, \ldots, \Delta_n, \phi_n, \Delta_{n+1} \rangle.$$

where Δ_1, D_{n+1} are types in $\mathbf{M}$, each formula ϕ_i is of the form $\exists \mathbf{y}\,(Q\mathbf{xy} \wedge \psi) \in \Delta_i$ and Δ_{i+1} is an alternative type as described above (i.e., $Q\mathbf{xy}$, $\psi(\mathbf{x}, \mathbf{y})$ in Δ_{i+1} and $\Delta_{i+1} =_{\mathbf{y}} \Delta_i$). We say that the variables in $\mathbf{y}$ *changed their values* from Δ_i to Δ_{i+1}, whereas the others did not. (We can also encode these 'modified variables' in our definition of paths.) Finally, a variable z is called *new in a path* π if either $|\pi| = 1$ or z's value was changed at the last round in π. Now we are ready to define our model $\mathbf{N}$.

Objects in $\mathbf{N}$ are all pairs (π, z) where π is a path, and z is new in π. Next, we *interpret predicates* over these objects. We say that $I(Q)$ holds of the sequence of objects $\langle(\pi_j, x_j)\rangle_{j \in J}$ iff the paths π_j fit into one linear sequence under inclusion, with a maximal path π^* such that (i) the atom $Q\langle x_j \rangle_{j \in J} \in \Delta^*$ (the last type on π^*) and for no (π_j, x_j) does x_j change its value on the further path to the end of π^*. Finally, we also define a *canonical assignment* s_π for each path. We set $s_\pi(x) =_{\text{def}} (\pi', x)$ where π' is the unique subpath of π^* at whose end x was new, while it remained unchanged afterwards. Now, we formulate correctness of this construction via the following assertion — where $last(\pi)$ is the last type on the path π.

Lemma 4.37 *For all paths π in $\mathbf{N}$, and all formulas $\psi \in F$, $\mathbf{N}, s_\pi \models \psi$ iff $\psi \in last(\pi)$.*

Proof. Induction on relevant formulas ψ. *Boolean cases* are immediate, using the standard closure conditions for $\neg$ and $\wedge$ on types.

Atoms: here we do an example. (i) Suppose that $Qxyx \in last(\pi)$. The objects $s_\pi(x)$, $s_\pi(y)$ were introduced at some maximal subsequence π^* of π. Note that no x- or y-values changed on π after their introduction. Therefore, by the above transfer of 'unaffected' formulas across successor types in these sequences, $Qxyx \in last(\pi^*)$. Then by the above definition, $I(Q)$ holds for the objects $s_{\pi^*}(x)$, $s_{\pi^*}(y)$, $s_{\pi^*}(x) = s_\pi(x)$, $s_\pi(y)$, $s_\pi(x)$. This means that $\mathbf{N}, s_\pi \models Qxyx$. (ii) Next, suppose that $\mathbf{N}, s_\pi \models Qxyx$. This means that $I(Q)$ holds for the objects $s_\pi(x)$, $s_\pi(y)$, $s_\pi(x) = s_{\pi^*}(x)$, $s_{\pi^*}(y)$, $s_{\pi^*}(x)$. The picture is the same as in the previous case. By the definition again, we have that $Qxyx \in last(\pi^*)$. And then, once again by transfer of unaffected formulas, $Qxyx \in last(\pi)$. (With identity atoms in the language, an argument like this is to be complicated — allowing for the same object being marked by different variables on different paths.)

Finally, consider the case of bounded *Existential Quantifiers* $\exists \mathbf{y}\,(Q\mathbf{xy} \wedge \psi(\mathbf{x},\mathbf{y}))$. (i) First, suppose that $\exists \mathbf{y}\,(Q\mathbf{xy} \wedge \psi(\mathbf{x},\mathbf{y})) \in last(\pi)$. Then there is an extended path $\pi^+ =_{\text{def}} \pi$ concatenated with $\langle \exists \mathbf{y}\,(Q\mathbf{xy} \wedge \psi(\mathbf{x},\mathbf{y})), \Delta' \rangle$, where Δ' is a successor type for Δ chosen as above with $Q\mathbf{xy}$, $\psi(\mathbf{x},\mathbf{y}) \in \Delta'$ (and satisfying the transfer condition for unaffected formulas with free variables $\mathbf{x}$). All objects (π^+, ψ_i) with y_i in $\mathbf{y}$ are new here. By definition, the atomic guard predicate $I(Q)$ holds for the object tuples $s_{\pi^+}(\mathbf{y})$, $s_{\pi^+}(\mathbf{x})$ $(= s_\pi(\mathbf{x}))$. Next, by the inductive hypothesis, we must have $\mathbf{N}, s_{\pi^+} \models \psi(\mathbf{x},\mathbf{y})$. Therefore, $\mathbf{N}, s_{\pi^+} \models \exists \mathbf{y}\,(Q\mathbf{xy} \wedge \psi(\mathbf{x},\mathbf{y}))$. From this we see, by standard $\mathbf{x}$-invariance in the model $\mathbf{N}$, that indeed also $\mathbf{N}, s_\pi \models \exists \mathbf{y}\,(Q\mathbf{xy} \wedge \psi(\mathbf{x},\mathbf{y}))$.

(ii) Conversely, suppose that $\mathbf{N}, s_\pi \models \exists \mathbf{y}\,(Q\mathbf{xy} \wedge \psi(\mathbf{x},\mathbf{y}))$. By the truth definition, there exist objects $d_i = (\pi_i, u_i)$ such that $\mathbf{N}, s_\pi{}^{\mathbf{y}}_{\mathbf{d}} \models Q\mathbf{xy} \wedge \psi(\mathbf{x},\mathbf{y})$. (Here, $s_\pi{}^{\mathbf{y}}_{\mathbf{d}}$ is the assignment which is like s_π except for setting all y_i to d_i.) In particular, $I(Q)$ holds of the objects $s_\pi(x)$, d_i. This leads to a simple picture of forking paths. The objects $s_\pi(x)$ were all introduced by stage π^* inside π, and then the objects d_i were (either interpolated among them or) added to form a maximal sequence π^+ where the true atom $Q\mathbf{xy}$ holds at the end. The fork is such that $\mathbf{x}$-values do not change any more from π^* onward, whether toward π or π^+. (This is the only relevant situation where the atomic guard on our quantifier comes in essentially.) Now, a minor complication. Note that the variables u_i do not have to be the y_i. We can be sure that they are not x_i, though, as the original objects $s_\pi(x) = s_{\pi^*}(x)$ were involved in the true atom $Q\mathbf{xy}$. Also, π^+ is such that $s_{\pi^+}(u_i) = (\pi_i, u_i) = d_i$. Thus, the two assignments $s_\pi{}^{\mathbf{y}}_{\mathbf{d}}$ and s_{π^+} agree on $\mathbf{x}$, and for all $y_i \in \mathbf{y}$ we have $s_\pi{}^{\mathbf{y}}_{\mathbf{d}}(y_i) = d_i = s_{\pi^+}(u_i)$. Then, by $\mathbf{N}, s_\pi{}^{\mathbf{y}}_{\mathbf{d}} \models Q\mathbf{xy} \wedge \psi$ and the above observations, we have $\mathbf{N}, s_{\pi^+} \models [\mathbf{u}/\mathbf{y}]Q\mathbf{xy}$ and $\mathbf{N}, s_{\pi^+} \models [\mathbf{u}/\mathbf{y}]\psi$. By the inductive hypothesis then, $[\mathbf{u}/\mathbf{y}]\psi \in last(\pi^+)$. (Here we assume that our set of relevant formulas is closed under simultaneous substitutions, that do not increase syntactic complexity. For a proof, see the remark below.) Also, from the initial description of π^+, we see at once that $[\mathbf{u}/\mathbf{y}]Q\mathbf{xy} \in last(\pi^+)$ (by the interpretation of atomic predicates). By closure conditions (b), (c) in the definition of a type, one gets $\exists\mathbf{y}\,(Q\mathbf{xy} \wedge \psi(\mathbf{x},\mathbf{y})) \in last(\pi^+)$. Finally, since no changes in **x**-values occurred on the fork from π^*, the transfer condition on successor types along paths ensures that this same formula is in $last(\pi)$. ⊣

Remark 4.38 (Finite Variable Fragments are Closed under Simultaneous Substitutions) Our proof assumes the finite set of relevant formulas is closed under simultaneous substitutions — without enlarging the set V of relevant variables. To show this, consider any substitution $[\mathbf{x} := f(\mathbf{x})]\phi$ in a k-variable fragment with variables $\mathbf{x} = x_1, \ldots, x_k$. Atomic replacements are straightforward. Also, we can push substitutions inside over Booleans. The only interesting case is when we encounter an existential quantifier: $[\mathbf{x} := f(\mathbf{x})]\exists x_j\,\psi$. Then, the assignment clause $x_j := f(x_j)$ has no effect, and so it can be omitted. Hence, in the remaining substitution σ, at least one variable x_k is not used at all on the right-hand side in any assignment. But then, the following formula is easily shown to be equivalent to the original one: $\exists x_k[x_j := x_k, \sigma]\psi$. This gives a simple recursive algorithm computing substitutions inside our fragment. (With function symbols, the result fails: witness the case of $[x := fxy]\exists y\, Rxy$.)

From the Lemma, the Theorem is immediate. Our original formula ϕ is satisfied in a standard model iff it has a quasi-model, and it is decidable whether ϕ has the latter property. ⊣

This argument may be extended to deal with the Guarded Fragment with identity. What it leaves open is the Finite Model Property, which may be proved by an additional combinatorial argument. There are also other proofs, which make a detour through the generalized modal semantics ('**CRS**-models') of Chapter 9.

4.5 Notes

Further Simulations. The notion of 'bisimulation' is essentially that of a modal 'p-relation' (van Benthem 1976), which generalized modal 'p-morphisms' (due to Segerberg around 1970). The latter are predated by the 'strongly isotonic functions' of de Jongh and Troelstra 1966. Notions of bisimulation are springing up around the clock. The survey van Glabbeek 1990 has 256 items... Newcomers include the *directed bisimulations* of Kurtonina 1995 (these consist of directed arrows pointing in either direction), which will be used in Chapter 12 to analyze categorial grammar logics. Directed bisimulations are also useful in the analysis of temporal or intuitionistic logics (Kurtonina and de Rijke 1996).

Simulations and Games. Model comparisons can also be made with Ehrenfeucht-Fraïssé games. Bisimulations are such games over ω rounds with a restricted choice for successive objects. Like games, they also have cut-off versions with pebbling. Games have their own style of argument, which may sometimes replace ordinary model-theoretic arguments on special model classes and with non-first-order languages (cf. Barwise and van Benthem 1996).

More Modal Logic of Transition Systems. This is a vast field of research. Amongst many others, cf. Hennessy and Milner 1985 (a classic paper among computer scientists), Stirling 1989, Harel et al. 1995, van Benthem et al. 1994, van Benthem and Bergstra 1994, and Hollenberg 1996b, as well as various contributions to the volume Ponse, de Rijke and Venema, eds., 1995.

Alternative Models. First-order remodeling strategies for labeled transition systems leading to two-sorted geometrical theories of 'states' and 'paths' may be found in van Benthem et al. 1994, van Benthem 1996a. Classics of Process Algebra are Milner 1980, Bergstra and Klop 1984. For detailed comparisons with modal logic, see van Benthem et al. 1994, Hollenberg 1995b.

Extended Modal Logics. For some samples of this program (initiated at Sofia University), cf. Goranko 1990, 1995, Blackburn and Seligman 1995, de Rijke 1993b.

Modal Fragments of First-Order Languages. Modally inspired fragments of first-order languages abound. Cf. Blackburn and Seligman 1995 on 'hybrid languages', or Ohlbach et al. 1995 on 'terminological languages'.

The Modal Invariance Theorem. More elaborate discussion of this result and its generalizations to richer modal languages may be found in van Benthem and Bergstra 1994, de Rijke 1995. (These also discuss connections with the work by Hennessy and Milner 1985 on modal process equivalences.) The complicated original proof used unions of elementary chains. By now, several alternative proofs are available. One goes via Ehrenfeucht games, and proves the theorem also in finite model theory (Rosen 1995). Another runs via 'consistency families' (van Benthem 1996e), avoiding compactness and saturation, and also works for infinitary languages. Yet another route is via interpolation (Barwise and van Benthem 1996).

Infinitary Versions. Infinitary dynamic logics are proposed in Salwicki 1970, Goldblatt 1982, Harel 1984. Cf. also the references given at the end of Chapter 3, or Sturm 1996. In particular, one may mention the very recent results obtained by Janin and Walukiewicz 1996, who extend the modal Invariance Theorem to the following powerful language over labeled transition systems. A formula of *monadic second-order logic* is invariant for bisimulation if and only if it is definable in the μ-calculus (PDL with fixed-points). The techniques used involve tree automata combined with modal unraveling.

Modal Filtration and Decidability. Standard modal filtration as a method for obtaining finite counter-examples of effectively bounded size to non-valid modal formulas is explained in many textbooks, including Goldblatt 1987, Blackburn et al. 1996. Powerful generalizations dealing with extended modal logics are in Marx 1995, Venema and Marx 1996a. More detailed complexity analysis of various filtration arguments is found in Spaan 1993.

Finite Variable Fragments. For further results and critical discussion of pros and cons, see Andréka et al. 1995a. This paper shows failures of the Łoś-Tarski and Craig theorems inside finite-variable fragments. But it also proposes modified 'modal' versions of such failed model-theoretic results which do go through. Moreover, 'cutting off' bisimulations at some fixed length, one can get many positive properties of bounded fragments to carry over to their *internal* finite-variable hierarchy.

Bounded Fragments. Many further results show that bounded fragments have nice 'modal' behaviour. Andréka et al. 1996 generalize the modal unraveling construction (which is implicit in the decidability proof of our main text), and use it to prove a Łoś-Tarski theorem for preservation of guarded formulas under submodels. As for decidability strategies, direct reduction does work after all for subfragments of the Guarded Fragment. The decidability of the Guarded Fragment has various applications to algebraic logic. For instance, it implies the decidability of the modal generalized assignment semantics proposed in Chapter 9 for predicate logic. Bounded quantification is also essential to 'safe queries' on databases (Badia 1996).

Acknowledgment

The work reported here on the Guarded Fragment is a typical specimen of joint research with *Hajnal Andréka* and *István Németi* at the Mathematical Institute, Hungarian Academy of Sciences, Budapest. Our larger ongoing program, pursued together with a group of colleagues and students, is the systematic interfacing of modal logic and algebraic logic, with an emphasis on repercussions for logic in general.

4.6 Questions

Simulations and Languages. Provide a general account of the correspondence between simulations and logical formalisms that fit each other most closely. Consider the temporal language of "Since", "Until". Its main operators are not existential quantifiers generating immediate zigzag clauses. What is its appropriate bisimulation? (Kurtonina and de Rijke 1996 have announced a solution in the meantime.)

Decidable First-Order Fragments. How far can the decidability of the Guarded Fragment be extended? In particular, which composite atomic guards are acceptable? (Andréka et al. 1995a show that composite guards may produce undecidable logics.) Provide a generalization which explains, at least, the decidability of the minimal logic of "Since" and "Until". Also, does the Guarded Fragment remain decidable over special model classes where atomic guards satisfy additional conditions? For instance, what about the case where all guards are binary *transitive* relations?

Complexity Results. What is the precise complexity class of the various decidable logics in this Chapter? Basic modal logic is PSPACE-complete (Spaan 1993): how does complexity increase along the hierarchy of bounded fragments? Can one also locate the exact complexity of determining bisimilarity between finite models, for our various notions of simulation? (Grohe 1996 shows the task is polynomial-time complete for k-partial isomorphism, or modal bisimulation.)

Specializations. Extend all the above results to special model classes of computational interest. In particular, when specializing to finite models, one enters the area of *Finite Model Theory* (Gurevich 1985). For instance, Rosen 1995 extends the modal Invariance Theorem to this case, and Hollenberg (p.c.) observed the same for modal Interpolation. Chapter 12 conjectures that, indeed, most of the meta-theory of modal logic goes through for finite models. This would be quite unlike the situation for general first-order predicate logic — where the finite-model versions of Compactness, Łoś-Tarski and Interpolation fail. (This need not be the last word. As with the above finite-variable fragments, one can then ask for suitably *modified versions* of these classical results which do go through over the universe of finite models.)

Alternative Modelings. Redo the above theory without ordered pairs, using the 'arrows' of Chapter 8. (Bisimulations can then relate 'states' and 'arrows' directly.) Also, develop the theory of first-order action logic, and two-sorted state/path logics.

5

Relational Algebra of Process Operations

This chapter investigates natural repertoires of operations for combining processes. We seek functional completeness theorems which characterize such repertoires through appropriate logical behaviour. The two main approaches here are semantic invariances, and syntactic definability. We start with some classical notions, namely, 'permutation invariance', as well as various forms of first-order and higher-order definability. Classification techniques are presented for permutation-invariant operators satisfying natural semantic properties — which are often of a Boolean nature, reminiscent of the theory of generalized quantifiers. But eventually, we propose a more restrictive form of semantic invariance, namely, 'safety for simulations'. Our main mathematical result is then a complete classification of all process operations that are safe for bisimulation. These are essentially the regular dynamic ones encountered in Chapters 2 and 3.

5.1 Spaces of Logical Operations

Logical constants in standard logic are the key operators forming new propositions out of old ones. In dynamic logic, logical constants are the key operators of control, combining procedures. Now, much recent literature has a conservative bias, as the only issue raised is 'what the standard logical constants mean' dynamically. But the latter setting allows for finer distinctions than the standard one, so there may not be a clear sense to this issue. Standard 'conjunction' really collapses different notions: sequential composition, but also various forms of parallel composition. Likewise, standard 'negation' may be either some test, or an invitation to make a move refraining from some forbidden action ("anything, as long as you leave your father alone"). And natural logical operators in the dynamic setting may even lack classical counterparts altogether, such as conversion or iteration of procedures. To study these phenomena, we start with the Relational

Algebra of Chapters 2, 3. Its repertoire included the ubiquitous *sequential composition* and *choice*, as well as *reversal*. More precisely, we had

$-, \cap, \cup$	set-theoretic Booleans	complement, intersection and union
$\circ, \breve{}$	ordering operations	composition and converse
Δ	one constant entity	the identity relation.

5.1.1 Logicality as Permutation Invariance

There is a general perspective on logicality relating many different domains. Intuitively, 'logical' operators do not care about specific individual objects inside their arguments. This is also true for procedural operators. What makes a complement $-R$ a logical negation is that it works uniformly on all 'arrow patterns' R, in contrast to a social operator like 'Dutch' whose action depends on the content of its relational arguments ("Dutch dining" means making one's guests pay for themselves, "Dutch climbing" is running to the top ahead of one's companions just at the finish.) The common mathematical generalization behind this involves *invariance under permutations* π of the underlying set of relevant individuals (that is, here, states in transition systems).

Definition 5.1 (Permutation Invariance) Let π be a permutation of the base set S of states. This can be lifted to sets, relations, indeed, to all functions in a finite type hierarchy over S in a canonical way. An object F in the hierarchy is permutation-invariant if $\pi(F) = F$ for all permutations π.

Working out this definition for particular cases, one finds that permutation invariance means the following for classical operations on declarative propositions (sets of states):

$$\pi[O(X, Y, \dots)] = O(\pi[X], \pi[Y], \dots).$$

For procedural operations over dynamic propositions (i.e., binary relations over states), one gets a similar commutation property:

$$\pi[O(R, S, \dots)] = O(\pi[R], \pi[S], \dots).$$

A useful equivalent formulation recalls invariance under isomorphisms (Chapters 3, 4):

$$(x, y) \in O(R, S, \dots) \text{ iff } (\pi(x), \pi(y)) \in O(\pi[R], \pi[S], \dots).$$

It is easy to check the following by a direct set-theoretic argument.

Fact 5.2 *All the above operators of relational algebra are permutation-invariant.*

Further sources of permutation invariance are Tarski 1986 (relational algebra), van Benthem 1986, 1989a (generalized quantifiers, categorial operators). In general, any operation which is defined in some logical formalism that does not refer to specific objects in a domain (such as first-order logic with identity, higher-order logic, or even finite type theory) will be permutation invariant. Dual view-points on definability and invariance under suitable transformations date back to the 19th century. For instance, in the ground-breaking work by Helmholtz and Heymans on perception, invariants of movement correspond to primitives for geometrical theories. Similar ideas have been proposed in this century by Weyl for analyzing the primitives of physical space-time.

As before, a natural question is when invariants are *definable* in some logical language. We provide a glimpse of the much larger context of our discussion. Consider terms in a *typed language* L_ω over finite-type hierarchies, with function application, lambda abstraction and identity. This formalism is stronger than anything so far. In particular, it defines all standard first-order operators $\neg$, $\wedge$, $\exists$. In an ad-hoc notation, we then have the following invariance.

Proposition 5.3 (Invariance for Type-Theoretic Objects) *For every term* τ_b *with* n *free variable occurrences of types* $a_1, \ldots, a_n$, *and all permutations* π *of the individuals, suitably lifted to higher types,* $\pi_b([\![\tau(u_1, \ldots, u_n)]\!]) = [\![\tau(\pi_{a_1}(u_1), \ldots, \pi_{a_n}(u_n)]\!]$.

Conversely, there may be many more permutation invariants than logical definitions (with infinite base domains, the former are an uncountable set, unlike the latter). But, using model-theoretic arguments, we do have the following result.

Theorem 5.4 (From Invariance to Definability) *Over* finite *base domains, all permutation invariants in their finite type function hierarchy are explicitly definable in the type theory* L_ω.

Proof. Cf. van Benthem 1991b, Chapters 10 and 11, for this and related results. $\dashv$

Specializing to Relational Algebra, its operations are all permutation-invariant, being first-order definable (cf. Chapter 3). This explains their return across many concrete systems of dynamics making more specialized choices for their state space.

5.1.2 A Procedural Hierarchy: Finite Variable Levels

Mathematical permutation invariance still leaves a host of possible relational operators. A finer view is afforded by a 'linguistic' perspective, scrutinizing syntactic forms of definition for these. For instance, the earlier examples lived in a first-order language with variables over states and binary predicates for procedures. Now, as was observed in Chapter 4, a reasonable measure of complexity is the number of variables essentially employed in a definition — which gives the largest configuration of states involved in the action of the operator. Boolean intersection of relations employed two variables, while composition involved three. 'Finite variable levels' form a hierarchy of definitory complexity against which we can measure proposed procedural operations. (Some infinitary first-order version will be needed to include program iteration and its ilk.) We state some basic facts. (Cf. Chapters 15, 16, 17 of van Benthem 1991b for details.)

Theorem 5.5 (Hierarchy of Relational Operators)

(i) *The usual similarity type of Relational Algebra is functionally complete for all relational operators with a three-variable defining schema.*

(ii) *Each n-variable level has a finite functionally complete set of operators.*

(iii) *There is no finite functionally complete set of algebraic operators for the whole procedural hierarchy at once.*

Proof. (i) Consider all first-order forms, working inside out. By standard logic, innermost quantifiers in first-order formulas need only occur in disjunctions of forms

$$\exists x \bigwedge (\neg) Ruv, \quad \text{where } x \in \{u, v\} \subseteq \{x, y, z\}.$$

These can be written explicitly using composition, complement and converse. To show what is going on, a concrete example says more than an elaborate syntactic procedure. Consider the left-implication slash defined as follows (cf. Section 3.3):

$$R \to S = \{(x, y) \mid \text{ for all } z, \text{ if } Rzx, \text{ then } Szy\}.$$

We can transform its first-order definition via the algebraic operations as follows:

$$\forall z\, (Rzx \to Szy),\ \neg\exists z\, (Rzx \wedge \neg Szy),\ \neg\exists z\, ((R)^{\smile}xz \wedge (-S)zy),\ -(R \circ -(S)^{\smile}).$$

(ii) A proof may be found in Gabbay 1981b. The latter paper proposed an interesting general program of identifying 'finite variable fragments' with 'modal logics'.

(iii) This can be shown by *Ehrenfeucht-Fraïssé games* with *pebbles* for the available variables (cf. Section 3.1). Recall that two models are indistinguishable by first-order sentences up to depth n employing some fixed set of k variables iff there is a winning strategy for the second player in a game over n rounds, where each player receives k pebbles at the start, and can only select objects by putting one of these pebbles on them. Now, any finite operator formalism has a first-order transcription involving only some fixed finite number k of variables over states. Therefore, if any such formalism were functionally complete, the full first-order language over our models would be logically equivalent to one of its k-variable fragments. But such a reduction is impossible. Let the first-order sentence ϕ state the existence of a top node having at least $k+1$ distinct immediate successors.

Consider two models consisting of a top node with k and $k+1$ immediate successors, respectively, and the same valuation at all states. The second player has an evident winning strategy in the Ehrenfeucht comparison game between such models with k pebbles, over an arbitrary finite number of rounds. Hence, no k-variable sentence distinguishes between these two models: whereas ϕ can. ⊣

5.2 Boolean Structure

Finite-variable levels may contain implausible operators with contrived definitions. To exclude these, further constraints may be imposed, of some computational import. Such constraints often involve *Boolean structure*, using the fact that set-theoretic relations form (at least) a Boolean algebra.

5.2.1 Continuity and Homomorphism

The following well-known mathematical condition forces operations to determine their values 'locally' at singleton arguments.

Example 5.6 (Continuity) An operation $O(\ldots, X, \ldots)$ is *continuous* in a position X if it commutes with arbitrary unions of arguments:

$$O(\ldots, \bigcup_{i \in I} R_i, \ldots) = \bigcup_{i \in I} O(\ldots, R_i, \ldots).$$

'Locality' then follows by the singleton decomposition

$$R = \bigcup \{\{(x, y)\} \mid Rxy\}.$$

Continuous operations include Boolean intersection and union, relational composition and converse. A non-example is Boolean complement. Simple checking of possibilities (cf. van Benthem 1986) establishes the following finite functional completeness result, which is the relational version of a more general type-theoretic observation.

Proposition 5.7 (Continuous Logical Operations)

(i) Each continuous operation can be written in an existential form (displayed here for the two-argument case $O(R, S)$ only)

$$\lambda xy.\, \exists zu\, (Rzu \wedge \exists vw\, (Svw \wedge$$

'Boolean combination of identities in $\{x, y, z, u, v, w\}$.'

(ii) For each fixed arity, there are only finitely many continuous permutation-invariant relational operators.

Proof. (i) All forms listed define permutation-invariant operations, as they involve the identity predicate only. They are also continuous, because of the special existential context for their argument predicates. Conversely, the above singleton decomposition already explains the existentially quantified form: since $f(R, S)$ must equal the union of all values $f(\{(u, v)\}, \{(w, r)\})$

with Ruv, Swr. On top of this, permutation invariance induces the given Boolean matrix of identity statements, by imposing a certain 'uniformity'. For instance, if a value (x, y) is admissible with x distinct from y, u, v, w, r, then any such distinct object x' must be admissible in this position. To see this, consider any permutation of individuals interchanging x and x', but leaving all other objects fixed. More precisely, the Boolean matrix is constructed as follows. Argument objects u, v, w, r can exhibit only one of a finite number of (non-)identity patterns. For each of these, choose some pair $\{(u,v)\}, \{(w,r)\}$ exemplifying it, and consider the value set $f(\{(u,v)\}, \{(w,r)\})$. Then, for each pair (x, y) occurring inside the latter, write down the conjunction of all (non-)identity statements that are true for the objects x, y with respect to u, v, w, r. The result is a finite disjunction of conjunctions of identities and their negations. The correctness of this definition requires a double uniformity. First, for any tuple of argument objects, the value of f may be correctly described as the set of all pairs satisfying the stated Boolean condition, by the earlier observation. But also, descriptions obtained in this way are uniformly valid for all argument tuples having the same pattern of (non-)identities: since any two instances of the same pattern can be mapped onto each other by a permutation of individuals.

(ii) This is a simple consequence of (i). ⊣

Continuity in this strong form rules out too much, although it does describe a 'natural kind' of logical operator. (More liberal versions would only require continuity over ascending countable chains of arguments, as in Scott's theory of information domains.) An even smaller interesting class is that of *Boolean homomorphisms* commuting with arbitrary unions but also with complements of their arguments:

$$O(-R) = -O(R).$$

We cite another classification result.

Proposition 5.8 (Classifying Logical Homomorphisms) *Let A, B be arbitrary sets in some finite type hierarchy. There exists a canonical isomorphism between the following two function spaces:*

(1) the space of permutation-invariant Boolean homomorphisms from $\text{pow}(A)$ *to* $\text{pow}(B)$

(2) the space of permutation-invariant maps from B to A.

Proof. The isomorphism even works without the restriction to permutation invariance (van Benthem 1986). A most general type-theoretic formulation of this kind of connection is in van Benthem 1991b, Chapter 11. ⊣

Specialized to binary relational operators, the permutation-invariant homomorphisms match precisely the permutation-invariant maps from ordered pairs of states (s_1, s_2) to pairs of ordered pairs $((s_3, s_4), (s_5, s_6))$. It is easy to check all possibilities here, as permutation invariance enforces that $\{s_3, \ldots, s_6\} \subseteq \{s_1, s_2\}$. Another example occurs in Chapter 6, when analyzing logical projections from dynamic to static propositions.

5.2.2 Analyzing Special Repertoires

One can also start from specific dynamic operators in the literature, and study their semantic characteristics. An example is the procedural repertoire $\{\sim, \circ\}$ of dynamic predicate logic and its ilk (Chapter 2). This is more special than the regular operations of dynamic logic, as it lacks Boolean union of procedures. But the matter is delicate. This is best seen in a two-level dynamic logic, having both propositions and programs:

Fact 5.9 *The operations $\sim$, $\circ$ are both regular, and they suffice to embed the propositional component into the procedural one via the test mode ?:*

$$\begin{aligned} ?(\phi \wedge \psi) &= ?(\phi)\circ ?(\psi) \\ ?(\neg\phi) &= \sim ?(\phi) \\ ?(\langle\pi\rangle\phi) &= \sim\sim(\pi\circ ?(\phi)). \end{aligned}$$

Thus, at least at the level of propositions or their corresponding tests, this repertoire does provide all Boolean operations. Moreover, we have this observation (Chapter 2):

Fact 5.10 *Adding an explicit union $\cup$ to $\{\sim, \circ\}$ results only in addition of outermost unions of $\{\sim, \circ\}$ programs, because of the valid equivalences*

$$\begin{aligned} (R \cup S) \circ T &= (R \circ T) \cup (S \circ T) \\ R \circ (S \cup T) &= (R \circ S) \cup (R \circ T) \\ \sim(R \cup S) &= \sim(R) \circ \sim(S). \end{aligned}$$

Digression. (Zooming in on the DPL Repertoire) Here is a more radical semantic way of excluding unions (i.e., procedural 'choice'). Let the *direct product* of two transition models have for its domain the Cartesian product of the two state sets, with the following stipulation for its atomic repertoire:

$$(x, x')R(y, y') \text{ iff } xRy \text{ and } x'Ry'.$$

Arbitrary products may be defined in the same manner.

Fact 5.11 *All procedures formed from atomic ones using only the repertoire $\{\sim, \circ, \cap\}$ are invariant for direct products in the sense of the above equivalence; whereas the latter may fail for $\cup$ and $\sim$.*

The reason is that only first-order formulas constructed from atoms Rxy and $x = y$ using $\wedge$, $\exists$ are invariant for direct products. Here is a sample observation.

Fact 5.12 *The logical operators satisfying Continuity plus Product Invariance are those definable by the following schema:*

$$\lambda xy.\, \exists zu\, (Rzu \wedge \exists vw\, (Svw \wedge\, ,$$

$$\text{'}\textit{conjunction of identities in } \{x, y, z, u, v, w\}\text{'}.$$

Finally, there is also semantic analysis of special atomic repertoires. E.g., in dynamic predicate logic, basic actions π are all 'propositional tests' or 'random assignments', satisfying the identity $\pi \circ \pi = \pi$, and both are symmetric relations. (These properties are not preserved by the operations $\{\sim, \circ\}$.) Algebras over special kinds of binary relation have also been studied in the recent mathematical literature.

5.3 Safety for Bisimulation

We now present our final analysis of functionally complete natural dynamic repertoires. The main result of this Section is a new kind of Safety Theorem characterizing the program operations associated with the process equivalence of bisimulation.

5.3.1 From Invariance to Safety

The starting point is the earlier invariance for bisimulations of the poly-modal language, and its extension to Dynamic Logic (Chapter 3). We repeat the relevant result.

Proposition 5.13 *Let $\equiv$ be a bisimulation between two models* **M**, **M**$'$, *with $s \equiv s'$.*

(i) s, s' verify the same formulas of propositional dynamic logic
(ii) if $s[\![\pi]\!]^{\mathbf{M}}t$, then there exists t' with $s'[\![\pi]\!]^{\mathbf{M}'}t'$ and $s' \equiv t'$

This observation motivated the following notion of invariance for program operations (where we indulge in a slight abuse of notation, for greater readability).

Definition 5.14 (Safety for Bisimulation) An operation $O(R_1, \ldots, R_n)$ on programs is *safe for bisimulation* if, whenever $\equiv$ is a relation of bisimulation between two models for their transition relations $R_1, \ldots, R_n$, then it is also a bisimulation for $O(R_1, \ldots, R_n)$.

Safety is a drastic strengthening of the earlier invariance for permutations.

Fact 5.15 *Algebraic operators which are safe for bisimulation are permutation-invariant.*

The converse is false: only suitably 'modally definable' operations turn out to be safe. Independently, safety for bisimulation is a natural criterion for program operations. The regular operations of relation composition ; and choice $\cup$ (Boolean union) are safe — and so are tests (ϕ)? for modal ϕ

(cf. the inductive proof of the above proposition.) For later use, we also mention the safety of DPL negation (Chapter 2):

'counter-domain' $\quad \sim(R) = \{(x, y) \mid x = y \text{ and for no } z : xRz\}$.

Now, the following question arises. Is there some companion to the earlier preservation theorem for modal invariance, characterizing all programming operations that guarantee safety for bisimulation? We might even ask whether these are just the *regular* programming operations that are prominent for independent computational reasons. To make the question precise, we use the standard first-order language over labeled transition systems — adding arbitrary binary relation symbols for transition relations.

Definition 5.16 (First-Order Program Operations) A programming operation is *first-order* if it can be defined using a first-order formula $\theta(x, y)$ with two free variables. All earlier operations are first-order in this sense:

$$\begin{array}{rl} (R_1\,;R_2) & \exists z\,(R_1xz \wedge R_2zy) \\ (R_1 \cup R_2) & R_1xy \vee R_2xy \\ \sim(R) & x = y \wedge \neg\exists z\, Rxz \\ (P)? & x = y \wedge Px. \end{array}$$

Then, we have the following main result, providing a complete semantic characterization:

Theorem 5.17 (Safety Theorem) *Any first-order relational operation* $O(R_1, \ldots, R_n)$ *is safe for bisimulation iff it can be defined using atomic relations* $R_a xy$ *and atomic tests* $(q)?$ *for propositional atoms* q*, using the three operations* ;, $\sim$ *and* $\cup$.

To prove this result, we need a further excursion into the model theory of Modal Logic.

5.3.2 Continuous Modal Formulas

First, we need the following semantic notion, familiar from earlier Sections.

Definition 5.18 (Continuous Formulas) A modal formula $\phi(p)$ is *continuous* in the proposition letter p if (with some abuse of notation), the following equivalence holds in each model:

$$\text{for each family of subsets } \{P_i\}_{i\in I},\ \phi(\bigcup_{i\in I} P_i) \leftrightarrow \bigvee_{i\in I} \phi(P_i).$$

Examples of continuous formulas are $p \wedge q$, $\langle a\rangle p \wedge \langle b\rangle \neg q$, $p \vee \langle a\rangle p$ — and non-examples are $\neg p$, $[a]p$. We want to find a syntactic characterization for this notion. Some precedents exist already. Continuity (properly) implies the well-known property of semantic *monotonicity*, whose syntactic correlate is definability of ϕ using only positive occurrences for the proposition letter p. (For first-order predicate logic, this is the Lyndon Theorem, which also holds for basic modal logic.) Thus, we expect some even stricter syntactic constraint in the present case, which is provided by the following result:

Theorem 5.19 (Preservation Theorem for Continuity) *Modulo logical equivalence, the p-continuous modal formulas $\phi(p)$ are all those that can be written as disjunctions of formulas of the 'existential forms'*

$$\alpha_0 \wedge p, \alpha_0 \wedge \langle a_1 \rangle(\alpha_1 \wedge p), \alpha_0 \wedge \langle a_1 \rangle(\alpha_1 \wedge \langle a_2 \rangle(\alpha_2 \wedge p)), \text{ etcetera,}$$

where all formulas α_i are p-free.

First-order predicate logic has an easy syntactic characterization of Continuity (cf. van Benthem 1986), but this argument cannot be reproduced inside the modal fragment.

Proof. Clearly, all forms described are continuous with respect to their proposition letter p. The hard part, as usual, is the converse, which we approach as follows:

Claim. *A continuous modal formula ϕ implies the infinite disjunction of all modal existential forms which themselves imply ϕ as a consequence.*

Then, by Compactness for the modal language, ϕ will imply some finite disjunction of these forms, and hence it is equivalent to the latter, as it follows from each disjunct. Thus, the real work is in the following

Proof of the Claim. Assume that $\mathbf{M}, w \models \phi$. By Continuity in its downward direction, using the fact that the set $V(p)$ is the union of its singletons, we have $\mathbf{M}', w \models \phi$, where p holds at only one world. ($V(p)$ can never be empty: since $\phi(\emptyset)$ would then imply an empty disjunction, which is a clear contradiction.) We may assume that $\mathbf{M}', w$ is a transition model generated from w (since ϕ is modal, this makes no difference in truth value). Thus, let there be a finite path with labeled transitions

$$w = w_0 R_1 w_1 R_2 \dots R_n w_n \models p$$

from the root to the unique world where p holds. Let Φ_i be the set of all p-free modal formulas true at w_i. Via our first-order translation, one can regard all modal formulas here as first-order ones. Now, the following valid semantic consequence holds in first-order logic:

Subclaim. $\Phi_0(x_0), R_1 x_0 x_1, \Phi_1(x_1), \dots, R_n x_{n-1} x_n, \Phi_n(x_n), P x_n \models \phi.$

From this, by Compactness again, some finite subsets of the Φ_i will do for a premise. This yields the desired existential form implying ϕ by straightforward predicate logic. Generalizing over all models, we have shown that ϕ locally implies some such existential form everywhere: whence it implies their infinite disjunction globally.

Proof (of the Subclaim). Consider any model $\mathbf{N}$ for the following set of formulas: $\Phi_0(x_0), R_1 x_0 x_1, \Phi_1(x_1), \dots, R_n x_{n-1} x_n, \Phi_n(x_n)$, with the required n-sequence $v_0, v_1, \dots, v_n$ for its free variables starting at v. As in the above proof of the Invariance Theorem for modal formulas, plus the usual unraveling techniques of modal logic (cf. Chapters 3, 4), we may suppose without loss of generality that

(i) $\mathbf{M}'$, $\mathbf{N}$ are ω-saturated (we are only dealing with first-order formulas)
(ii) $\mathbf{M}'$, $\mathbf{N}$ are intransitive trees via some 'unraveling bisimulation': (we are in fact only dealing with modal formulas). ⊣

We can draw these models as indicated, each with a distinguished branch of length $n+1$, and the rest of the models lying in disjoint subtrees branching off from these:

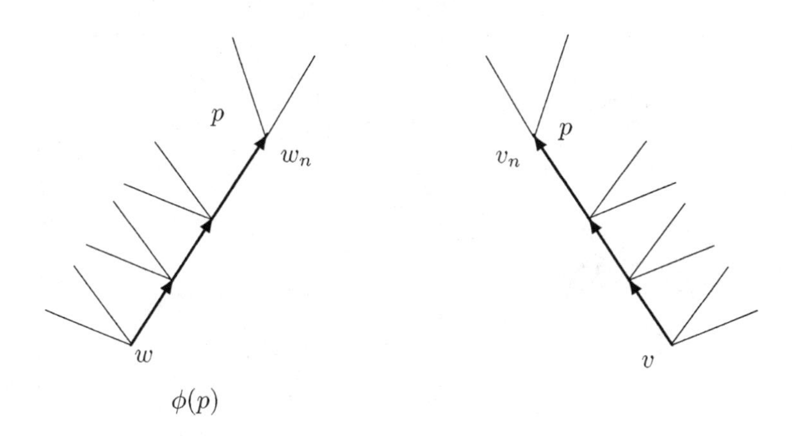

By a zigzag argument on saturated models, as in the proof of the Invariance Theorem, modal invariance between states defines an L-bisimulation between these two models:

$$x \equiv u \text{ iff } \mathbf{M}', x \text{ and } \mathbf{N}, u \text{ verify the same } L\text{-modal formulas.}$$

In particular, this stipulation matches corresponding points on the above two special branches, by the definition of the sets Φ_i. Now, our purpose is to further improve this L-bisimulation to a $(L+P)$-bisimulation respecting also the special propositional atom p, so that we can transfer the truth of the initial formula $\phi(p)$ from left to right. For this purpose, judicious geometrical rearrangement will be used on labeled transition graphs, to produce a situation where the matching on the two branches is unique (in particular, the final world w_n corresponds only to v_n), while subtrees on the left only contain matches in their corresponding subtrees on the right — and vice versa.

For a start, without loss of generality, we may assume that all bisimulation links occur between worlds at the same levels in the trees (others may be omitted without losing the bisimulation property). Here are the main steps in our procedure, which works upward along the special branches. As often in possible worlds semantics, it helps to visualize what is going on using suitable graph pictures:

(1a) Start with the match w_1, x_1. Suppose that world w_1 has links with other level-one worlds in **N**, as in the following picture:

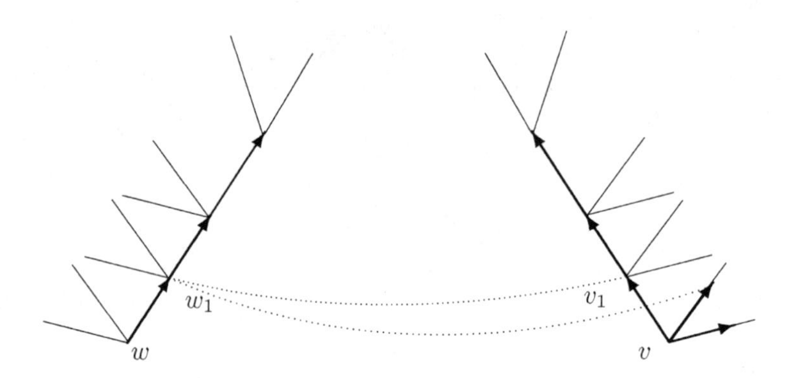

Then 'de-couple' by making a copy of the subtree T, w_1 on the left, attaching this to the root w, and matching the former mates of w_1 on the right to its copied companion w'_1 (and so on upward in the graph for the relevant relational successors). This transformation works out as follows:

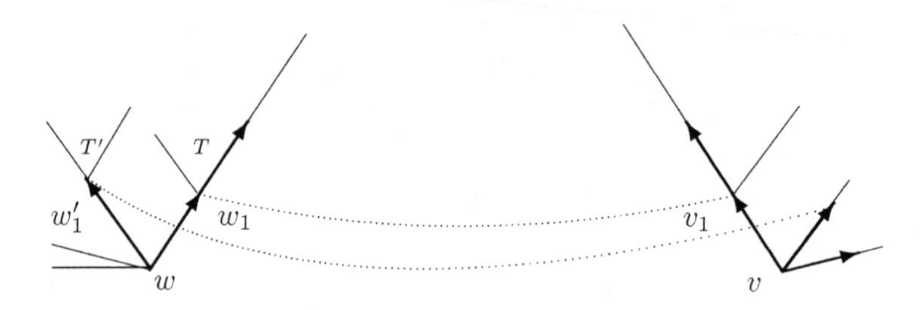

The result is still an L-bisimulation between the model on the right and the new enlarged one on the left, while the latter also L-bisimulates with the original **M**. Now, in the enlarged model, a duplication occurs of the single world w_n with p, which is not what we want. But applying the downward half of continuity again, we see that $\phi(p)$ will still hold at w if we have p true at just one of its two locations. Moreover, by the isomorphism of copies, we may suppose that this happens at the original location, without loss of generality.

(1b) Likewise, matches between v_1 and sisters of w_1 can be decoupled as shown in the following picture, which again involves copying, but now without any special moves for economizing on p-worlds:

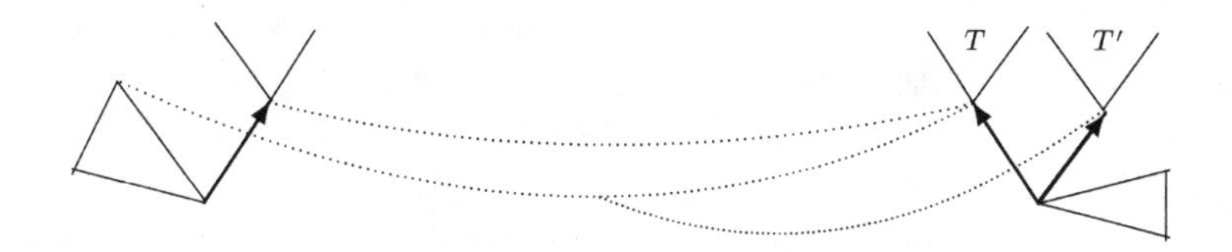

(2) The effect of the previous two moves is that the initial match w_1, v_1 has become unique — and we can now repeat the procedure, moving upward along the special branches until we reach the final match w_n, v_n, and make that unique too.

Now, let $\mathbf{M}^*$ be the enlarged model on the left-hand side (which has p true only at the end of the special branch), and $\mathbf{N}^*$ its companion on the right-hand side with p only true at the end of its special branch. The original enlarged $\mathbf{N}^+$ is this same model, but with possibly more occurrences of p in other worlds. By the construction, the latter model $(L+P)$-bisimulates with $\mathbf{M}$, so that v matches in both models. Then,

- $\phi(p)$ still holds at w in the enlarged model $\mathbf{M}^*$ (because of the construction)
- via the $(L + P)$-bisimulation between $\mathbf{M}^*$ and $\mathbf{N}^*$, it transfers to w in $\mathbf{N}^*$
- by upward monotonicity (the other side of continuity, which has not been used so far in this proof), $\phi(p)$ will also hold at w in $\mathbf{N}^+$
- by $(L+P)$-bisimulation, it will also hold at w in $\mathbf{N}$: which was to be shown. ⊣

5.3.3 Proof of the Safety Theorem

The argument for our main result may now be broken up into the following steps.

The Given Format Is Safe. By earlier observations, each operation of the described kind is safe for bisimulation. Here are two examples.

Composition. Let $xR;Sy$ in $\mathbf{M}_1$, and xCu, with u in $\mathbf{M}_2$. There exists z in $\mathbf{M}_1$ with xRz and zSy. By zigzag for R, one finds v in $\mathbf{M}_2$ such that uRv and zCv. Then, by bisimulation for S, one finds a w in $\mathbf{M}_2$ such that vSw and yCw. The latter is the required $R\,;S$-successor of u in $\mathbf{M}_2$.

Counterdomain. Let $x{\sim}(R)y$ in $\mathbf{M}_1$: that is, $x = y$ and x has no R-successors. Let xCu in $\mathbf{M}_2$. Now, suppose that u had some R-successor in $\mathbf{M}_2$. By bisimulation for R, there would also be a corresponding R-

successor for x in $\mathbf{M}_1$: quod non. Thus, (u, u) is the required matching $\sim(R)$-successor step in $\mathbf{M}_2$.

Safety Stays Inside The Format. Next, let $\theta(x, y)$ be a first-order operation in a language L, which is safe for bisimulation. Note that invariance for bisimulation is a language-dependent notion. What matters is if the bisimulation goes back-and-forth with respect to the atomic binary relations, and whether its matchings of states respect the relevant unary predicates. Choose a new unary predicate letter P.

Claim. *The formula* $\exists y\,(\theta(x, y) \wedge Py)$ *is invariant for* $(L+P)$*-bisimulation.*

Proof. Immediate from the safety of θ for L-bisimulation, plus the obvious fact that $(L + P)$-bisimulations are also L-bisimulations. $\dashv$

By the earlier Invariance Theorem for modal formulas, this means that the first-order formula $\exists y\,(\theta(x, y) \wedge Py)$ must be equivalent to some modal formula $\phi(p)$. Moreover, given the shape of our first-order formula, this $\phi(p)$ will be continuous in the proposition letter p. But then, it will be definable using a disjunction of formulas

$$\alpha_0 \wedge \langle a_1 \rangle (\alpha_1 \wedge \ldots \wedge \langle a_n \rangle (\alpha_n \wedge p) \ldots),$$

where the α_i do not contain the proposition letter p.

Now, it is easy to check the following assertion.

Claim. θ *will be definable using the corresponding union of relations of the form* $(\alpha_0)?\,;a_1\,;(\alpha_1)?\,;\cdots;a_n\,;(\alpha_n)?$.

Finally, it remains to remove possibly complex modal tests in the latter schema:

Claim. *All complex modal tests* $(\phi)?$ *can be reduced to atomic ones using only the regular operations* ; *and* $\sim$.

Proof. Use the following three valid identities to push all tests inward:

$$\begin{aligned} (\phi \wedge \psi)? &= (\phi)?\,;(\psi)? \\ (\neg\phi)? &= \sim(\phi)? \\ (\langle a \rangle \phi)? &= \sim\sim(a\,;(\phi)?). \quad \dashv \end{aligned}$$

5.3.4 Extensions and Variations

Infinitary Languages. The above analysis stayed inside a first-order formalism for programming operations. But many useful programming constructions involve *infinitary operations*, definable in the logic $L_{\infty\omega}$, such as regular Kleene Iteration. This is an instance of general Fixed-Point Recursion $\mu p.\,\phi(p)$ for formulas ϕ that are monotone in the proposition p. (In each model, the latter have explicit solutions by infinite disjunctive 'unwinding' — even though there need not be one uniform definition in $L_{\infty\omega}$.) First, we state an infinitary analogue of the modal Invariance Theorem.

Theorem 5.20 *The $L_{\infty\omega}$-formulas that are invariant for bisimulation are precisely the infinitary modal ones, allowing arbitrary conjunctions and disjunctions.*

For proofs and details, cf. Chapter 10. The point is to replace applications of first-order Compactness in the previous arguments by infinitary techniques of model construction, such as 'consistency families'. As for our Safety Theorem, we only state a

Conjecture 5.21 *A relational operation $O(R_1, \ldots, R_n)$ definable in $L_{\infty\omega}$ is safe for bisimulation if and only if it can be defined using atomic relations $R_a xy$ and atomic tests $(q)?$ for propositional atoms q in our models, using only the three operations ;, $\cup$ and $\sim$, where the unions may now be infinitary.*

By more delicate reasoning, these results may be specialized to the language $L_{\omega_1\omega}$ with countably infinite conjunctions and disjunctions only. Even so, all these results leave a substantial question unanswered. In the realm of infinitary programming operations, what is the semantic extra of the *regular program operations*, over and above their safety for bisimulation? Safety for bisimulation gives us a total semantic space of reasonable program operators, where infinite union is natural, while further restrictions to regular programs motivate a different subhierarchy inside this semantic space, by additional criteria of computational complexity. Here is another way to go (Gerard Renardel, p.c.). First restrict attention to some suitable *effective fragment* of $L_{\omega_1\omega}$, and then characterize the regular operations as the safe ones definable inside that fragment. Finally, we mention again the *μ-calculus*, i.e., propositional dynamic logic with arbitrary fixed-points (Harel et al. 1995). This is a powerful recursive programming formalism, which can be analyzed via modal techniques (Janin and Walukiewicz 1996). Can we also analyze the fine-structure of its repertoire via notions of safety?

Invariance and Safety with State Parameters. Often, labeled transition systems come with a distinguished 'root' s_0 standing for the starting state of the process:

$$(S, \{R_a\}_{a \in A}, s_0).$$

In the definition of 'bisimulation', one then adds the requirement that the roots be related. The previous results can be extended to this case, too, with the following modifications. First, the modal language is to be enriched to deal with this new feature, by introducing an operator $\text{GOTO}_{\text{root}}\phi$ 'resetting' evaluation to the root:

$$\mathbf{M}, s \models \text{GOTO}_{\text{root}}\phi \text{ iff } \mathbf{M}, s_0 \models \phi.$$

There are some obvious valid interchange principles governing the use of this operator (as with the similar "Now" operator in temporal logic), which validate certain normal forms for its occurrence. An easy adaptation of the proof for the Invariance Theorem to this enriched modal language gives the following result:

Theorem 5.22 (Invariance with State Parameters) *A first-order formula $\phi(x)$ over labeled transition systems is invariant for root-to-root bisimulation if and only if it is definable by a modal formula using ordinary modalities as well as* $\mathrm{GOTO}_{\mathrm{root}}$.

The new formulas may be used to force two roots to have the same modal types, so that they will stand in the bisimulation constructed. Next, concerning the Safety Theorem, the crucial new addition to our repertoire is the following binary relation ('root resetting'), which is always safe for root-to-root bisimulations:

$$\lambda xy.\, y = s_0.$$

Evidently, this is the natural binary relation corresponding to the modality $\mathrm{GOTO}_{\mathrm{root}}$. All earlier arguments go through with this addition, which will allow us to enforce a root-to-root bisimulation in the crucial part of the Continuity Lemma of Section 2. Thus, mutatis mutandis, we obtain:

Theorem 5.23 (Safety with State Parameters) *A first-order relational operation $O(R_1, \ldots, R_n)$ is safe for root-to-root bisimulation iff it can be defined using atomic relations $R_a xy$ as well as root resetting and atomic tests $(q)?$ for propositional atoms q, using the three operations* ;, $\cup$ *and* $\sim$.

These results can be generalized to admit further distinguished states as parameters in labeled transition systems, leading to additional fixed links in our bisimulations. In this case, for each of these states y, one adds a modal operator GOTO_y and a resetting relation RES_y to the syntactic repertoire, just as for the above root.

This concludes our discussion of safety for bisimulation. We shall return to these matters in Chapter 10 below, extending our analysis to operations in Process Algebra. Another possible direction would return to the various notions of process equivalence defined in Chapter 3. All of them have corresponding notions of safety, which can be determined syntactically by model-theoretic methods.

5.4 Notes

Logicality and Permutation Invariance. Permutation invariance is one of the most frequently rediscovered notions of logicality. Tarski 1986 (a 1966 lecture) may be the oldest source, in the context of Relational Algebra. Van Benthem 1986 proposes it for generalized quantifiers — and for logical operators in general, the notion is found in van Benthem 1989a, 1991b, Sher 1991. A historical discussion of logical invariance and definability in the context of philosophy of science occurs in van Benthem 1982.

Further Intuitions of Logicality. Other intuitive features of 'logical' operations include their *uniformity* (van Benthem 1986) or their *monotonicity* (cf. Keenan and Faltz 1985). Concrete implementations of these ideas have been proposed in the literature on *generalized quantifiers*: cf. Westerståhl 1989, Keenan and Westerståhl 1996.

Boolean Structure. Keenan and Faltz 1985 investigate general Boolean structure for natural language. Type-theoretic generalizations of Boolean structure are found in van Benthem 1991b. As for special constraints, Belnap 1977 proposes update logics whose operators satisfy only 'Scott continuity' (as in the information domains of Scott 1982).

Semantic Automata. Especially relevant to the dynamic paradigm of this Book is the procedural approach to logical operations in terms of so-called *semantic automata* (van Benthem 1986), which classifies them by the complexity of their associated procedural descriptions in the Automata Hierarchy. Thus, first-order quantifiers turn out to involve *finite state machines*, whereas higher-order quantifiers such as "most" essentially employ *push-down store automata* with unbounded memory.

Type-Theoretic Invariance and Definability. Van Benthem 1991b has details of earlier results, for lambda calculi and type theories with added Boolean operations. Plotkin 1980 strengthens permutation invariance to invariance under all binary relations (suitably defined). Barendregt 1992 provides logical background on all these matters.

Generalized Safety. Notions of Safety will reappear in Chapter 10, in an analysis of process algebra. A powerful generalization is given in Hollenberg 1996a. Call a first-order formula $\phi(x_1, \ldots, x_k, y_1, \ldots, y_m)$ (k, m)*-safe* if the following holds. Whenever C is a bisimulation between two models $\mathbf{M}$, $\mathbf{N}$ and we have object tuples $s_1, \ldots, s_k \in \mathbf{M}$, $t_1, \ldots, t_k \in \mathbf{N}$ with $s_i C t_i$ $(1 \leq i \leq k)$, then $\mathbf{M}, s_1, \ldots, s_k, u_1, \ldots, u_m \models \phi$ only if there exist objects $v_1, \ldots, v_m \in \mathbf{N}$ with $\mathbf{N}, t_1, \ldots, t_k, v_1, \ldots, v_m \models \phi$ and $u_j C v_j$ $(1 \leq i \leq k)$. This is the 'Zig' clause for (k, m)-safety: the 'Zag' clause is symmetric. The special case $k = 1$, $m = 0$ is in fact our earlier notion of invariance for bisimulation. This format allows operations beyond relational set algebra, which may create new state spaces in the process (e.g., direct products of arguments). Modal languages matching this notion of safety manipulate arbitrary n-ary relations. Not unnaturally, their repertoire is essentially the following: (i) atomic operations permuting and identifying argument positions, (ii) analogues for our three safe core operations $\{;, \sim, \cup\}$, and (iii) a concatenation operation performing 'disjoint addition' of argument places.

Specialized Safety. One can specialize the above to important non-first-order model classes (where earlier arguments need not go through). Hollenberg 1995b proves the Safety Theorem on *finite models*, by the game techniques of Rosen 1995.

5.5 Questions

Permutation Invariance in Finite Types. There are several open questions in this area. For instance, no explicit *counting formula* is known for obtaining all invariant objects in a finite-type hierarchy over finite base domains. For further questions, cf. the Addenda to the 1995 paperback reprint of van Benthem 1991b.

Boolean Preservation Theorems. Find an appropriate extension to a typed lambda calculus of *Lyndon's Theorem* characterizing the first-order operations that are monotone in a predicate P in terms of positive occurrence for P in their syntactic definition. (Cf. van Benthem 1996g for some partial results, and pitfalls.)

Connections with Semantic Automata. Find an *automata-theoretic* perspective on process operations similar to that developed for generalized quantifiers.

Safety in Games. Give a safety analysis of first-order quantifiers, in terms of Ehrenfeucht-Fraïssé games, or partial isomorphisms (Chapter 3).

Infinitary Safety. Prove the Safety Theorem for $L_{\infty\omega}$ and $L_{\omega_1\omega}$. (It appears possible to adapt the techniques of Barwise and van Benthem 1996 to this effect.) Characterize the *regular* operations precisely, perhaps as safety for the effective fragment of $L_{\omega_1\omega}$.

Recursive Program Operations. Prove a Safety Theorem for the *μ-calculus*. (Recently, such a result has been announced by Marco Hollenberg.) Analyze its operations that are *Scott-continuous* (commuting with unions of countable chains) and hence reach a fixed-point after ω rounds. (Section 11.2.2 below motivates these in logic programming, and gives a syntactic characterization in first-order logic.)

Safety for Other Process Equivalences. Safety may also be defined for the bounded fragments of Chapter 4, and their associated notions of guarded bisimulation. How is this to be characterized? What is the connection with Hollenberg's results? Answers may involve standard preservation theorems for fragments of first-order logic.

6

Two-Level Static-Dynamic Architecture

In this chapter, we want to show how logical interactions occur across multiple levels of a dynamic architecture. We are interested in this general schema, which was motivated in Chapter 1:

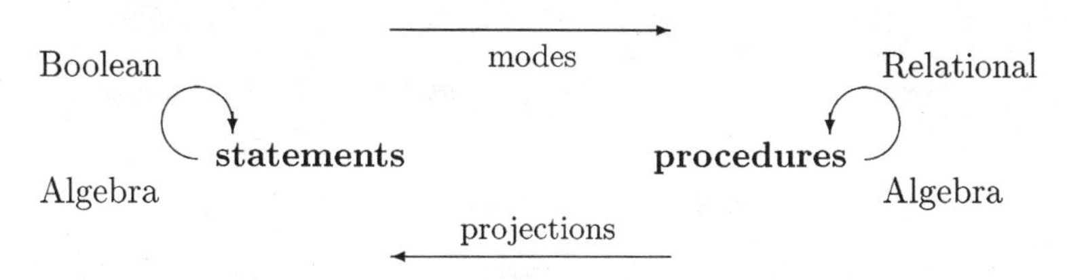

Our technical paradigm for this is Propositional Dynamic Logic (cf. Chapter 3), not as one specific calculus, but as a vehicle for conceptual experimentation. In such a layered setting, earlier questions of definability or inference occur across traditional boundaries. We start with a most abstract version, which allows us to classify possible modes and projections set-theoretically through invariance properties — but also to define a practical static-dynamic calculus of pre- and post-conditions. Next, we consider a slightly more structured dynamic modal logic of information states under inclusion (cf. Chapter 2), and investigate its expressive power, axiomatization and complexity.

6.1 Abstract Dynamic Logic

6.1.1 The Core System

We recall the system of propositional dynamic logic, presented in Chapter 3, with its dual language of declarative propositions and dynamic programs that can interact:

$$\begin{array}{rcl} PA & := & \text{propositional atoms } p, q, r, \ldots \\ AA & := & \text{atomic actions } a, b, c, \ldots \\ F & := & PA \mid \neg F \mid (F \wedge F) \mid \langle P \rangle F \\ P & := & AA \mid (P \,;\, P) \mid (P \cup P) \mid P^* \mid (F)? \end{array}$$

Its basic truth definition over labeled transition systems read as follows.

- formulas

$$\begin{array}{rll} \mathbf{M}, s \models p & \text{iff} & s \in V(p) \\ \mathbf{M}, s \models \neg\psi & \text{iff} & \text{not } \mathbf{M}, s \models \psi \\ \mathbf{M}, s \models \phi_1 \wedge \phi_2 & \text{iff} & \mathbf{M}, s \models \phi_1 \text{ and } \mathbf{M}, s \models \phi_2 \\ \mathbf{M}, s \models \langle\pi\rangle\phi & \text{iff} & \text{there exists } s' \text{ with } \mathbf{M}, s, s' \models \pi \text{ and } \mathbf{M}, s' \models \phi \end{array}$$

- programs

$$\begin{array}{rll} \mathbf{M}, s_1, s_2 \models a & \text{iff} & (s_1, s_2) \in R_a \\ \mathbf{M}, s_1, s_2 \models \pi_1 \,;\, \pi_2 & \text{iff} & \text{there exists } s_3 \text{ with } \mathbf{M}, s_1, s_3 \models \pi_1 \\ & & \text{and } \mathbf{M}, s_3, s_2 \models \pi_2 \\ \mathbf{M}, s_1, s_2 \models \pi_1 \cup \pi_2 & \text{iff} & \mathbf{M}, s_1, s_2 \models \pi_1 \text{ or } \mathbf{M}, s_1, s_2 \models \pi_2 \\ \mathbf{M}, s_1, s_2 \models \pi^* & \text{iff} & \text{some finite sequence of } \pi\text{-transitions} \\ & & \text{in } \mathbf{M} \text{ connects } s_1 \text{ with } s_2 \\ \mathbf{M}, s_1, s_2 \models (\phi)? & \text{iff} & s_1 = s_2 \text{ and } \mathbf{M}, s_1 \models \phi. \end{array}$$

And its complete axiomatization was provided by the 'Segerberg Axioms':

- All principles of the minimal modal logic for all modalities $[\pi]$
- Computation rules for 'weakest preconditions':

$$\begin{array}{ll} \langle\pi_1 \,;\, \pi_2\rangle\phi \leftrightarrow \langle\pi_1\rangle\langle\pi_2\rangle\phi & \langle\pi_1 \cup \pi_2\rangle\phi \leftrightarrow \langle\pi_1\rangle\phi \vee \langle\pi_2\rangle\phi \\ \langle\phi?\rangle\psi \leftrightarrow \phi \wedge \psi & \langle\pi^*\rangle\phi \leftrightarrow \phi \vee \langle\pi\rangle\langle\pi^*\rangle\phi. \end{array}$$

- An Induction Axiom

$$(\phi \wedge [\pi^*](\phi \rightarrow [\pi]\phi)) \rightarrow [\pi^*]\phi.$$

6.1.2 Characterizing Modes and Projections

First, we look at the new components of 'modes' and 'projections' in an earlier light. There is only a limited repertoire of structure-preserving logical switches between our two levels. E.g., by the earlier methods of Chapter 5, we can observe the following.

Proposition 6.1 (Homomorphic Logical Modes and Projections) *Only one projection is a Boolean homomorphism, viz. the diagonal* $\lambda R.\, \lambda x.\, Rxx$. *There are exactly two homomorphic modes, viz.* $\lambda P.\, \lambda xy.\, Px$, $\lambda P.\, \lambda xy.\, Py$.

Proof. We only consider the first assertion. Its ground is an earlier 'deflation analysis'. The diagonal function has a semantic recipe

$$\lambda R_{(s,(s,t))}.\, \lambda x_s.\, R(x)(x).$$

This is a Boolean homomorphism — with type $((s,(s,t)),(s,t))$, or, when curried $((s \cdot s,t),(s,t))$ — which is clearly permutation-invariant. Vice versa, these semantic features determine reflexivization completely. Such logical homomorphisms correspond to permutation-invariant maps in type $(s, s \cdot s)$. And of the latter, there is only one, the 'duplicator'

$$\lambda x_s.\, \langle x, x\rangle. \quad \dashv$$

Relaxing the condition of Boolean homomorphism, one obtains further natural modes and projections. For instance, typically, the permutation-invariant test mode ? of PDL is *continuous* (cf. Chapter 5), it preserves arbitrary unions, but it does not preserve negations. And the same holds for the domain projection $\langle\cdot\rangle$. Again by earlier results, we know that there exist only finitely many permutation-invariant continuous modes and projections, whose shapes are all enumerable using the methods of Chapter 5.

6.1.3 Digression on Projections in Update Semantics

The same set-theoretic analysis works in more complex settings. Our purpose is to show that it may then yield non-trivial results. (This section can be skipped, though.) We digress on eliminative update semantics (Chapter 2). Its dynamic propositions are functions from sets of objects (valuations) to sets of objects, and projections map these to static propositions, i.e., sets of objects. This is more complicated in terms of types than the projections of dynamic logic. Nevertheless, we can characterize the logical homomorphic projections like before. First, here is an obvious projection (which drove the earlier reduction of 'continuous introspective functions', cf. Chapter 2).

$$T^*(F) = \{s \mid F(\{s\}) = \{s\}\}.$$

By earlier results, the map T^* is permutation-invariant, as it has a purely set-theoretic definition. Now consider the Boolean algebra of all eliminative dynamic propositions, given a set of states S. This is not the full function space $\mathrm{pow}(S)^{\mathrm{pow}(S)}$, but a relativized space obtained by taking only the functions $F \leq \mathrm{Id}$. (Recall that the negation $-F$ in update semantics was really '$\mathrm{Id} - F$'). Then a simple calculation shows the following

Fact 6.2 *T^* is a Boolean homomorphism from update functions to classical propositions.*

(i) $T^*(-F) = -T^*(F)$
(ii) $T^*(\bigcup_i F_i) = \bigcup_i T^*(F_i)$

Thus, T^* is a logical projection respecting Boolean structure. Our main result is that, conversely, only two mappings have this behaviour.

Theorem 6.3 *There are only two permutation-invariant Boolean homomorphisms from eliminative dynamic propositions to declarative classical ones.*

Proof. Let P be such a logical Boolean homomorphism. It is determined by its values on Boolean atoms in the 'update algebra', being all functions $\alpha_{W,w}$ $(w \in W)$ such that

$$\alpha_{W,w}(X) = \begin{cases} \{w\} & \text{if } X = W \\ \emptyset & \text{otherwise.} \end{cases}$$

This is so because $P(F) = \bigcup_{\alpha \leq F} P(\alpha)$, for all F and atoms α (by the second homomorphism clause). Now we need an auxiliary observation.

Claim. *The values $P(\alpha)$ for all atoms α form a complete partition of S.*

Proof. Distinct atoms have $\alpha_1 \cap \alpha_2 = 0$, whence $P(\alpha_1 \cap a_2) = P(\alpha_1) \cap P(\alpha_2) = 0$: so they are disjoint. Moreover, $\bigcup \alpha_i = 1$, whence $P(1) = S = \bigcup_i P(\alpha_i)$. ⊣

Now define a map P* from S to such atoms, by setting

$$P^*(s) \text{ is the unique atom } \alpha \text{ with } s \in P(\alpha).$$

Note that $P(\alpha) = P^{*-1}(\{\alpha\})$. It is easy to show that P^* is permutation-invariant. Now our classification problem reduces to proving the following

Proposition. *There are only two permutation-invariant maps P^* sending states to atoms in the eliminative update algebra.*

Proof. We have $P^*(s) = \alpha_{W,w}$, for some state w and set of states W. If P^* is logical, then familiar reasoning about permutations tells us that w can only be s itself. But W might be any of the four sets $\{s\}$, $S - \{s\}$, $\emptyset$, S. Here, the requirement '$w \in W$' rules out two, and we retain just

$$\begin{aligned} P_1^*(s) &= \alpha_{\{s\},s} \\ P_2^*(s) &= \alpha_{S,s}. \end{aligned}$$

Calculating backwards, we get the maps P induced by these.

$$\begin{aligned} P_1(F) &= \{w \mid \exists\alpha \leq F{:}\, w \in P_1(\alpha)\} \\ &= \{w \mid \exists W, w{:}\, \alpha = \alpha_{W,w} \wedge w \in F(W) \wedge \alpha = \alpha_{\{w\},w}\} \\ &= \{w \mid F(\{w\}) = \{w\}\} \end{aligned}$$

$$\begin{aligned} P_2(F) &= \{w \mid w \in F(S)\} \\ &= F(S). \end{aligned}$$

It is easy to check that both these maps are permutation-invariant Boolean homomorphisms. ⊣

Thus, eliminative update semantics has two logical homomorphic projections, where relational update semantics had just one. Without the restriction to *eliminative* updates, the preceding argument produces exactly four such projections.

6.1.4 Inverse Logic on Specific Operators

The earlier methods can also be used to determine characteristic properties of *specific* modes and projections in semantic terms, avoiding deductive detail (assuming that the operators on propositions and procedures already have their standard interpretation). This 'inverse logic' was first studied for generalized quantifiers in van Benthem 1986.

Proposition 6.4 *'Test' is the only permutation-invariant continuous operator satisfying*

(i) $?(P) \leq \Delta$
(ii) $?(\neg P) = \Delta \cap -?(P)$

Proof. By Continuity, it suffices to determine the behaviour of 'test' on singleton arguments $\{x\}$. And since test values are subrelations of the diagonal, by (i), it suffices to specify the individual states in the image. By permutation invariance, there are only four basic options here: $\{x\}$ itself (1), $\{y \mid y \neq x\}$ (2), the unit set (3), and the empty set (4). Moreover, the choice will be made uniformly for all states x, again by permutation invariance. Now, outcome (4) would result in any test relation being empty: which contradicts (ii). Outcomes (3) and (2) may also be ruled out, by observing that they would allow for two distinct sets $\{x\}$, $\{y\}$ to have overlapping test values, whence some value $?(\neg P)$ would not be disjoint from $?(P)$: another contradiction with (ii). Thus, only the standard interpretation (1) remains. ⊣

A similar kind of argument characterizes a key projection.

Proposition 6.5 *'Domain' is the only permutation-invariant continuous operator obeying*

(i) $\text{dom}(?(P)) = P$
(ii) $\text{dom}(0) = 0$
(iii) $dom(R \circ S) = \text{dom}(R \circ ?(\text{dom}(S)))$

Proof. Cf. van Benthem 1991b. ⊣

6.1.5 Calculus of Preconditions and Postconditions

We now turn to more general uses of the above system. Propositional dynamic logic may be used to analyze systems of dynamic semantics as presented in Chapter 2. We give the broad technique here, elaborating the sketch in Chapter 2. Intuitively, dynamic evaluation of ϕ moves to states where some static content of ϕ holds, while values for its relevant variables have been set. For this static tracing, a simple technique from computer science exists which can be justified in propositional dynamic logic. Imperative programs can be explained in terms of 'preconditions' and 'postconditions'. We recall the main idea from Chapters 1, 2 . Starting from a set of states described by some input predicate A, execution of a program π moves to a new state set (being the standard set-theoretic image of A under the relation π) defined by the *strongest postcondition* of A under π: $\mathrm{SP}(A,\pi)$. Conversely, one may compute *weakest preconditions* for an output predicate under a program, by inverse images $\mathrm{WP}(\pi, A)$. The two notions are related by conversion:

$$\mathrm{SP}(A,\pi) \leftrightarrow \mathrm{WP}(\pi^{\smile}, A).$$

For the simple programs of dynamic predicate logic, these operators may be computed recursively, using the proof calculus of PDL.

Fact 6.6

$$\begin{aligned}
\mathrm{WP}(Pt, A) &= A \wedge Pt \\
\mathrm{WP}(\exists x, A) &= \exists x\, A \\
\mathrm{WP}(\phi_1 \wedge \phi_2, A) &= \mathrm{WP}(\phi_1, \mathrm{WP}(\phi_2, A)) \\
\mathrm{WP}({\sim}\phi, A) &= A \wedge \neg \mathrm{WP}(\phi, \top).
\end{aligned}$$

Proof. Note first that $\mathrm{WP}(\pi, A)$ is defined by the formula $\langle\pi\rangle A$. Here are the calculations for programs induced by DPL-formulas, using the PDL-axioms. Note the different senses for, e.g., 'Pt' and '$\exists x$' to the left and right of the equivalences:

(test axiom)	$\langle Pt\rangle A \leftrightarrow Pt \wedge A$
(direct inspection)	$\langle \exists x\rangle A \leftrightarrow \exists x\, A$
(composition axiom)	$\langle \phi_1 \wedge \phi_2\rangle A \leftrightarrow \langle\phi_1\rangle\langle\phi_2\rangle A$
(cf. Section 3.4.4)	$\langle {\sim}\phi\rangle A \leftrightarrow A \wedge \neg\langle\phi\rangle\top.$ ⊣

This reduction also works for strongest postconditions:

Fact 6.7

$$\begin{aligned}
\mathrm{S}(A, Pt) &= Pt \wedge A \\
\mathrm{S}(A, \exists x) &= \exists x A \\
\mathrm{S}(A, \phi_1 \wedge \phi_2) &= \mathrm{S}(\mathrm{S}(A, \phi_1), \phi_2) \\
\mathrm{S}(A, {\sim}\phi) &= A \wedge \neg \mathrm{WP}(\phi, \top) \quad (!)
\end{aligned}$$

Proof. We show two cases, using the above connection with WP, plus the fact that atomic reassignment is a symmetric relation (equal to its own converse) — as are tests:

$$\begin{aligned}
\mathrm{S}(A, \exists x) &\leftrightarrow \mathrm{WP}(\exists x^{\smile}, A) \leftrightarrow WP(\exists x, A) \leftrightarrow \exists x\, A \\
\mathrm{S}(A, {\sim}\phi) &\leftrightarrow \mathrm{WP}(({\sim}\phi)^{\smile}, A) \leftrightarrow \mathrm{WP}({\sim}\phi, A) \leftrightarrow A \wedge \neg\mathrm{WP}(\phi, \top).
\end{aligned}$$ ⊣

We repeat a sample computation (cf. Chapter 2) on a sequence of dynamic formulas, with trace points indicated in bold-face subscripts — which should now be obvious:

$$\mathbf{0}\ \exists x\ \mathbf{1}\ (Ax\ \mathbf{2}\ \wedge Bx\ \mathbf{3})\ .\ \exists x\ \mathbf{4}\ Cx\ \mathbf{5}\ .\ (\exists x\, Dx)\ \mathbf{6}$$

SP
0 $\top$
1 $\exists x \top = \top$
2 Ax
3 $Ax \wedge Bx$
4 $\exists x\,(Ax \wedge Bx)$ *
5 $\exists x\,(Ax \wedge Bx) \wedge Cx$
6 $\exists x\,(Ax \wedge Bx) \wedge Cx \wedge \neg\exists x\, Dx$, etcetera

At the trace point *, the initial $\exists x$ becomes an ordinary quantifier after all.

In addition to its practical use, the SP/WP reduction, being effective, makes validity in DPL at most as complex as that of PDL, or by translation, of standard predicate logic. This validity meant that, in every model, the range of the sequential composition of the premises is included in the domain of the conclusion. It has the following description:

Fact 6.8

$$\phi_1, \ldots, \phi_n \models_{\text{DPL}} \psi \textit{ iff } \models_{\text{PDL}} \text{S}(\top, \phi_1 \wedge \ldots \wedge \phi_n) \rightarrow \text{WP}(\psi, \top).$$

A similar analysis works for eliminative update semantics (US). The latter calculus is close to propositional logic, or the modal logic S5. Various reductions bring this out. We copy the clauses given in Chapter 2, which are similar to the preceding ones for strongest postconditions. But the T-formulas $T(P, \phi)$ do not describe a set of states, but one single next state, consisting of all propositional valuations satisfying them.

Fact 6.9

$$\begin{aligned}
T(P, q) &= P \wedge q\\
T(P, \phi \wedge \psi) &= T(T(P, \phi), \psi)\\
T(P, \phi \vee \psi) &= T(P, \phi) \vee T(P, \psi)\\
T(P, \neg\phi) &= P \wedge \neg T(P, \phi)\\
T(P, \Diamond\phi) &= P \wedge \Diamond T(P, \phi).
\end{aligned}$$

As an immediate consequence, we have the following reduction of US validity, which said that updates resulting from successive premises are fixed points for the conclusion:

Fact 6.10

$$\phi_1, \ldots, \phi_n \models_{\text{upd}} \psi \textit{ iff}$$
$$\models_{\text{S5}} T(T(P, \phi_1 \wedge \ldots \wedge \phi_n), \psi) \leftrightarrow T(P, \phi_1 \wedge \ldots \wedge \phi_n)).$$

Hence, for our propositional language, dynamic update-test consequence is decidable, with pleasant meta-properties. It is interesting to see how far these reductions can be extended to further dynamic operators, both in DPL and US. Similar static reductions work e.g., for preferential upgrading and other dynamic operations from Chapter 2.

6.2 Dynamic Modal Logic

So far, we considered purely procedural issues, without any special structure on states. But even a minimal enrichment will allow us to raise many further issues of information structure and flow. We start by recalling some basic ideas from Chapter 2.

6.2.1 Base Calculus

Consider transition models with a partial order of inclusion or 'possible development', as found in intuitionistic or relevant logics:

$$\mathbf{M} = (S, \subseteq, \{R_p \mid p \in P\}).$$

Our two-level architecture gives us various new propositional and program operators. In particular, we obtained the following new modes creating dynamic procedures from static standard propositions, both 'forward' and 'backward':

'loose updating'	$\lambda P.\, \lambda xy.\, x \subseteq y \wedge Py$
'strict updating'	$\lambda P.\, \lambda xy.\, x \subseteq y \wedge Py \wedge \neg\exists z\, (x \subseteq z \subset y \wedge Pz)$
'loose downdating'	$\lambda P.\, \lambda xy.\, y \subseteq x \wedge \neg Py$
'strict downdating'	$\lambda P.\, \lambda xy.\, y \subseteq x \wedge \neg Py \wedge \neg\exists z\, (y \subset z \subseteq x \wedge \neg Pz)$.

This formalism was rich enough to perform all cognitive tasks covered in the well-known AGM-theory of epistemic dynamics (cf. Chapter 2). To be more definite, one has to choose a procedural repertoire over these atomic epistemic actions. For a start, we take standard propositional and modal operators

$$\{\neg, \wedge, \vee, \Box_{\text{up}}, \Box_{\text{down}}\},$$

while the program component employs the usual relational repertoire

$$\{-, \cap, \cup, \circ, ^{\smile}, \Delta\}.$$

In addition, we introduce following modes and projections:

{ test, loose and strict updates, loose and strict downdates }
{ fixed point, domain, range }

The resulting dynamic logic validates various connections between these operators.

Example 6.11 (Some Valid Identities)

$$\begin{array}{rclrcl} \subseteq & = & \mathrm{upd}(\top) & \Delta & = & (\top)? \\ \mathrm{upd}(P) & = & \subseteq \circ (P)? & \mathrm{range}(\pi) & = & \mathrm{dom}(\pi^{\smile}) \\ \Diamond_{\mathrm{up}}(P) & = & \mathrm{dom}(\mathrm{upd}(P)) & \mathrm{fix}(\pi) & = & \mathrm{dom}(\Delta \cap \pi). \end{array}$$

Indeed, for a minimal vocabulary, various subsets of these operators suffice.

Fact 6.12 *The usual operations from relational algebra, plus test, update and domain define the whole vocabulary of Dynamic Modal Logic.*

All other notions are definable in terms of these. More generally, this system encodes a theory of cognitive processes, including interactions between updates and downdates, as observed in Chapter 2, Section 3. It also handles various earlier notions of dynamic inference. E.g., inference in dynamic predicate logic $P_1, \ldots, P_n \models^{\mathrm{dyn}} C$ amounts to validity of the universal-existential program modality

$$[P_1 \circ \cdots \circ P_n]\mathrm{dom}(C),$$

and inference in propositional update semantics is the validity of

$$[P_1 \circ \cdots \circ P_n]\mathrm{fix}(C).$$

Modal reductions for dynamic styles of inference return systematically in Chapter 7. Solely on the strength of these, one can derive the earlier-mentioned deviant structural properties of Update-to-Test or Update-to-Domain inference.

Example 6.13 (Modified Cut Rule) DPL-consequence satisfied the following Cut rule (Section 2.2.3):

$$X \models^{\mathrm{dyn}} \phi \text{ and } \phi \models^{\mathrm{dyn}} \psi \text{ imply } X \models^{\mathrm{dyn}} \phi \wedge \psi.$$

Transposed modally, this is the following valid principle of PDL (where the turnstyle $\models$ refers to global consequence in models, cf. Chapter 7):

$$[X_1 \circ \cdots \circ X_n]\langle\phi\rangle\top, [\phi]\langle\psi\rangle\top \models [X_1 \circ \cdots \circ X_n]\langle\phi \wedge \psi\rangle\top. \quad \dashv$$

Here is one more way of thinking about this system. On the one hand, it is a relational algebra *specialized* to special atomic operations of informational update. But through the marriage between its two levels, it is also an *enriched* modal logic over information models (cf. van Benthem 1996d). For instance, in addition to the S4-modalities, one can also define the much more expressive modal operators *Until* and *Since*:

$$\begin{array}{rcl} \subset & = & \mathrm{upd}(\top) \cap -(\top)? \\ \mathit{Until}\,\phi\psi & \leftrightarrow & \langle \subseteq \cap -(\subseteq \circ (\neg\psi)? \circ \subset)\rangle\phi. \end{array}$$

6.2.2 Expressive Power and Definability

Dynamic modal logic raises some obvious technical questions. By general reasoning, its universal validities must be recursively enumerable (due to the embedding into first-order logic presented below). The latter have a complete modal-style axiomatization using a 'difference operator' $D\phi$, whose interpretation runs as follows:

$$\mathbf{M}, x \models D\phi \text{ iff } \phi \text{ is true in at least one state } \textit{different from } x.$$

Its corresponding semantic inequality relation $\neq$ is definable as $-\Delta$. Its deductive principles include explicit modal definitions for the main modes, such as

$$(q \wedge \neg Dq) \rightarrow (\langle \text{strict-upd}(P) \rangle A \leftrightarrow \Box_{\text{up}}(A \wedge P \wedge \Box_{\text{down}}(\Box_{\text{down}} q \rightarrow \neg P)))$$

There is still an 'asymmetry' in our design so far. The static component of dynamic modal logic is not 'in harmony' with the dynamic one. To see why, consider the earlier reduction from dynamic procedures to static propositions, via strongest postconditions (Section 6.1.5). Equivalently, it may be set up with 'weakest preconditions', whose atomic clauses for tests and loose updates or downdates read like this:

$$\begin{aligned}
\text{WP}((P)?, A) &= P \wedge A \\
\text{WP}(\text{upd}(P), A) &= \Diamond_{\text{up}}(P \wedge A) \\
\text{WP}(\text{downd}(P), A) &= \Diamond_{\text{down}}(\neg P \wedge A)
\end{aligned}$$

Now, to deal with the strict versions, more complex 'temporal' operators are needed over information patterns:

$$\begin{aligned}
\text{WP}(\text{strict-upd}(P), A) &= \textit{Until}\,(P \wedge A, \neg P) \\
\text{WP}(\text{strict-downd}(P), A) &= \textit{Since}\,(\neg P \wedge A, P).
\end{aligned}$$

Thus, we can make various selections of primitives to set up systems like this. Clearly then, dynamic modal logics lie on a ladder of ascending expressive strength, all the way up to a complete standard first-order logic over information models.

Translation into Standard Logic. As in modal logic, there is a straightforward translation from dynamic propositions and procedures to unary and binary formulas in a standard first-order predicate language referring to partial orders of states with unary predicates. There are two parts to this. * takes propositions to formulas $\phi(x)$, while $^\#$ takes programs to formulas $\phi(x, y)$. These call each other recursively by modes and projections.

$$\begin{aligned}
(p)^* &= Px \\
(\neg\phi)^* &= \neg(\phi)^* \\
(\phi \wedge \psi)^* &= (\phi)^* \wedge (\psi)^* \\
(\Diamond_{\text{up}}\phi)^* &= \exists y\,(x \subseteq y \wedge [y/x](\phi)^*) \\
(\Diamond_{\text{down}}\phi)^* &= \exists y\,(y \subseteq x \wedge [y/x](\phi)^*)
\end{aligned}$$

$$
\begin{aligned}
(?\phi)^{\#} &= x = y \wedge (\phi)^{*} \\
(\text{upd}(\phi))^{\#} &= x \subseteq y \wedge [y/x](\phi)^{*} \\
(\text{strict-upd}(\phi))^{\#} &= x \subseteq y \wedge [y/x](\phi)^{*} \wedge \neg\exists z\,(x \subseteq z \subset y \wedge [z/x](\phi)^{*}) \\
(\text{downd}(\phi))^{\#} &= y \subseteq x \wedge [y/x](\phi)^{*} \\
(\text{strict-downd}(\phi))^{\#} &= y \subseteq x \wedge \neg[y/x](\phi)^{*} \wedge \\
&\qquad \neg\exists z\,(y \subset z \subseteq x \wedge \neg[z/x](\phi)^{*}) \\
(\Delta)^{\#} &= x = y \\
(-\pi)^{\#} &= \neg(\pi)^{\#} \\
(\pi_1 \cap \pi_2)^{\#} &= (\pi_1)^{\#} \wedge (\pi_2)^{\#} \\
(\pi_1 \circ \pi_2)^{\#} &= \exists z\,([z/y](\pi_1)^{\#} \wedge [z/x](\pi_2)^{\#}) \\
(\pi^{\smile})^{\#} &= [y/x, x/y](\pi)^{\#} \\
(\text{fix}(\pi))^{*} &= [x/y](\pi)^{\#} \\
(\text{dom}(\pi))^{*} &= \exists y\,(\pi)^{\#}
\end{aligned}
$$

Proposition 6.14 *The above is a correct translation from dynamic modal logic into first-order predicate logic over state transition models.*

As the first-order theory of partial orders with extra predicates is clearly recursively enumerable, we get the following application:

Corollary 6.15 *Universal validity in dynamic modal logic is recursively axiomatizable.*

Fine-Structure of Finite-Variable Levels. The previous conclusion holds even for the full first-order language over information models, which is more expressive than DML. The fine-structure of this language may be studied as in Chapter 4. In particular, one can look at finite-variable levels for definitions of update operators and information modalities. At level 2, we have a decidable language, with a standard modal logic in the static component, and a curtailed relational algebra in the dynamic component. At level 3, we have the full relational algebra, and a 'temporal' logic in the static component. (De Rijke 1992 shows functional completeness of DML for this fragment.) A typical notion at level 4 would be a modality referring to *suprema* and *infima* of states in the information order, witness

$$x = \sup(y, z) \quad y \subseteq x \wedge z \subseteq x \wedge \forall u\,((y \subseteq u \wedge z \subseteq u) \rightarrow x \subseteq u)).$$

6.2.3 Axiomatizability

A complete system of dynamic modal logic is presented in De Rijke 1992. It employs the earlier-mentioned sparse vocabulary. In addition to DML-equivalents for the core axioms of propositional dynamic logic, it has the following more specific principles. Let $\underline{D}\phi$ be the universal dual of the difference modality $D\phi$ ('at all different worlds'), and let $E\phi$ abbreviate

$\phi \vee D\phi$ ('somewhere'). These difference operators are a mere convenience here, as we may define them away:

$$D\phi = \langle -(\top)?\rangle\phi$$

Difference
$$\underline{D}(\phi \to \psi) \to (\underline{D}\phi \to \underline{D}\psi)$$
$$\phi \to \underline{D}D\phi$$
$$DD\phi \to (\phi \vee D\phi)$$
$$\langle\pi\rangle\phi \to E\phi$$

Booleans and Converse
$$\langle\pi_1 \cap \pi_2\rangle\phi \to \langle\pi_1\rangle\phi \wedge \langle\pi_2\rangle\phi$$
$$E(\phi \wedge \neg D\phi) \to (\langle\pi_1\rangle\phi \wedge \langle\pi_2\rangle\phi \to \langle\pi_1 \cap \pi_2\rangle\phi)$$
$$E(\phi \wedge \neg D\phi) \to (\langle -\pi\rangle\phi \leftrightarrow \neg\langle\pi\rangle\phi)$$
$$\phi \to [\pi]\langle\pi^{\smallsmile}\rangle\phi$$
$$\phi \to [\pi^{\smallsmile}]\langle\pi\rangle\phi$$

Update
$$\phi \to \langle \mathrm{upd}(\phi)\rangle\top$$
$$\langle \mathrm{upd}\rangle\langle \mathrm{upd}(\phi)\rangle\top \to \langle \mathrm{upd}(\phi)\rangle\top.$$

These axioms may be analyzed by modal frame correspondences, as in Chapter 3. Then they express obvious relations between accessibility relations for various programs. For instance, the clauses for the relational operations # enforce the right equalities between $R_a \# R_b$ and R_a, R_b. In particular, although no obvious simple reduction axiom works for intersection or complement, the uniqueness prefix '$E(\phi \wedge \neg D\phi)$' makes sure that we are dealing with singleton sets, where reduction *is* possible. Likewise, the S4-style update axioms express that inclusion is a pre-order. If inclusion should be a partial order, then this fact, too, can be expressed using a difference operator, by an axiom

$$(\phi \wedge \neg D\phi) \to [\mathrm{upd}(\top)](\langle \mathrm{upd}(\phi)\rangle\top \to \phi).$$

As for proof rules, we take Modus Ponens and Necessitation for all universal modalities (including $\underline{D}$), plus a new 'Difference Rule':

> if $(p \wedge \neg Dp) \to \phi$ is derivable (where p does not occur in ϕ),
> then ϕ is derivable.

This says that if ϕ is valid at all singletons, then it is valid everywhere. The reference to singletons is significant for the following result.

Theorem 6.16 *The above axiomatic calculus is complete for universal validity in DML.*

Proof. The completeness proof employs a standard Henkin argument (cf. Goldblatt 1987), but with the following twist. One defines worlds as maximally consistent sets Δ, using the standard accessibility relation. This will not quite work to prove the usual Truth Lemma (stating that membership of Δ and truth at Δ coincide), as we need to make sure that all accessibility relations in the model have their standard interpretation. This can be enforced (by mimicking the above semantic correspondence observations), if the difference operator has its standard interpretation on the Henkin model. Now, the latter may be achieved by making sure that every world has some *unique propositional constant* true at it and false at all others. That this may be achieved consistently, through a more careful step-wise construction, is precisely the import of the additional modal derivation rule. (This technique is due to Venema 1991.) $\dashv$

Can one do without the additional Difference Rule, axiomatizing DML over a standard modal base calculus? Venema 1991 shows that such rules are essential for axiomatizing full relational algebra (whose non-finite-axiomatizability in standard terms was noted in Chapter 3). In DML, however, one has an atomic repertoire of an inclusion order plus propositional updates, and hence this system might be simpler. In any case, the above completeness theorem is merely a pilot result. In particular, to reduce complexity, one can restrict the relation-algebraic procedural repertoire to regular PDL operations. Then, earlier derivabilities will disappear — so that one must reinstate, for instance, minimal updates and downdates as primitive operators. Thus, the earlier semantic options in defining expressively different dynamic modal logics return as combinatorial questions about deductive strength and complexity.

6.2.4 Undecidability

Next, there is an issue of complexity. Is dynamic modal logic *decidable*? Its static modal relative S4 has the latter property, but its dynamic relative of full relational algebra over *arbitrary* relations does not. De Rijke 1992 (following Spaan 1993) has established undecidability. The proof technique is of general interest by itself, and hence we reproduce it here. Roughly speaking, modal logics become undecidable when they can encode *two-dimensional grids*, which provide enough room to encode computations of Turing machines.

Theorem 6.17 *The satisfiability problem for DML is* Π^0_1*-hard.*

Proof. We outline an effective reduction of the so-called *Unbounded Tiling Problem* (UTP; cf. van Emde Boas 1990, Johnson 1990) to satisfiability in DML. Consider a set T of square tiles $\{t_1, \ldots, t_m\}$ with four sides whose colours are in $C = \{c_1, \ldots, c_k\}$. UTP is the task of putting one tile on each point in the grid $\mathbb{N} \times \mathbb{N}$ such that adjacent edges have the same color. This problem is known to be Π^0_1-complete. Its reduction to DML proceeds by defining a formula ϕ_T whose models look like grids, covered with tiles, with matching colours enforced for neighbours to the right end above. The special relation R defining the grid is upd($\top$). The construction to follow ensures that

> ϕ_T = GRID $\wedge$ COVER $\wedge$ MATCH is satisfiable in a DML-model iff the set T can tile the $\mathbb{N} \times \mathbb{N}$ plane.

We start with GRID. For purposes of orientation, fix two proposition letters p, q. The following explanations can all be reproduced formally within the formalism of DML.

(1) The UP relation says that we make an immediate successor step in the inclusion order, shifting only the polarity (true/false) of q. The RIGHT relation makes a similar step, but shifting the polarity of p.

(2) EQUAL(π_1, π_2) says that the relations π_1, π_2 are equal throughout the model.

(3) CR (for 'Church Rosser') says that UP ; RIGHT = RIGHT ; UP.

(4) Let $A\phi$ be the 'universal modality' defined as $\neg E \neg \phi$.

(5) PARTIAL says that R is a partial order.

(6) GRID says: $p \wedge q \wedge A\langle\text{UP}\rangle\top \wedge \text{A}\langle\text{RIGHT}\rangle\top \wedge \text{CR} \wedge \text{PARTIAL}$.

We proceed with COVER. For each color c_i, introduce four (self-explanatory) color atoms up_i, down_i, right_i, left_i.

(7) Each tile t_j has a describing formula t_j stating its four colours, and denying that other color propositions would hold about it.

(8) COVER says (using the operator A from (4)) that everywhere at least one tile description holds, and no two tile descriptions hold at the same point.

Finally,

(9) MATCH says that whenever some color atom up_i holds, every UP successor has down_i, and whenever right_i holds, every RIGHT successor has left_i. $\dashv$

6.3 Notes

Inverse Logic. Various analogies between definability questions in Generalized Quantifier Theory and in Dynamic Modal Logic invite exploration. See van Benthem 1986, Westerståhl 1989, Keenan and Westerståhl 1996 on GQT, and van Benthem and Cepparello 1994 on setting up the theory of DML in a similar fashion. On connections between dynamic modalities and generalized quantifiers, see Alechina 1995a, Ben-Shalom 1994 — as well as the general survey van Benthem and Westerståhl 1995.

Hoare Logic and Dynamic Semantics. This link between computer science and natural language semantics has been studied systematically in van Eijck and de Vries 1991, 1995, which deal with dynamic predicate logic and eliminative update semantics. We give a simplified calculus here.

Similar analyses may be given for upgrading and other less conventional dynamic procedures (van Benthem et al. 1993).

Dynamic Modal Logic. Key references for the background of this system are Gärdenfors 1988, van Benthem 1989c, de Rijke 1992, 1994. The main idea in analyzing the AGM Theory is to identify deductively closed databases (theories) with 'epistemic types' for states s — that is, the set of all statements A whose 'knowledge version' $\Box_{\text{up}}A$ is true at s. Unlike DML statements in general, the latter are upward preserved along informational inclusion, and hence epistemic types grow upward, like theories. DML has been used by Jaspars 1994, Jaspars and Krahmer 1995 as a tool for *comparing* systems of dynamic semantics, and making them compatible.

Modal Rules of Inference. The new technique sketched here changes the rules of the completeness game, by allowing both new inference rules and new axioms in a description of modal validity for some given semantic model class. See Åqvist 1979 (where it was still a flaw), Gabbay 1981b (who makes it a virtue), Venema 1991 (for an elegant general Sahlqvist Theorem allowing extra rules, plus applications to relation algebra where the rule-based strategy is essential), de Rijke 1993b, Mikulas 1995. Blackburn et al. 1996 give details, and further applications.

Complexity of Modal Logics. This area is still under construction, and one usually has to find results ad-hoc. Spaan 1993 is the first source for general techniques, refining modal filtration methods, and demonstrating thresholds of modal expressive power at which complexity may typically jump up.

Richer States. More realistic modal information theories must complicate their notion of 'state'. Thus, Segerberg 1995 uses *partitions* of the space of possible worlds to do revisions (keeping a route open into the 'cognitive past'). Similar enrichments of states in eliminative update semantics have been proposed by Veltman. Interestingly, partitions of logical space have also become the norm in the semantics of *questions* (cf. Groenendijk and Stokhof 1996). Richer notions of state may also provide the proper level to formulate intuitive 'minimality principles' of *least effort* in interpretation.

Cognitive and Physical Action. Dynamic Modal Logic may be combined with components describing real actions. Of the vast literature, we mention the work of the Utrecht Group in computer science (Van der Hoek et al. 1995), starting from Meyer's 1988 paper on resolving paradoxes of deontic logic via dynamic logics, the Amsterdam Free University AI Group (cf. Treur and Engelfriet 1994 on epistemic temporal logic for default reasoning), and of course, the San Jose IBM Group who initiated the biannual TARK Conferences on pure and applied epistemic logic (cf. Fagin et al. 1995, Friedman and Halpern 1994a, 1994b).

6.4 Questions

Computing WPs and SPs. The same program operators seem to support simple recursion rules for calculating their WPs, and to enjoy safety for bisimulation. What is the underlying connection?

Modal Logic of Information. Consider partial orders and a modal language with the usual modalities plus *Since*, *Until*. Provide a perspicuous axiomatization. Is this system *decidable*? Same questions for S4 expanded with modal operators for *suprema* and *infima* modeling 'merges' in the information order — as in the clause

$$\mathbf{M}, x \models \phi + \psi \text{ iff}$$
$$\text{there exist } y,\ z \text{ s.t. } x = \sup(y, z) \text{ and } \mathbf{M}, y \models \phi,\ \mathbf{M}, z \models \psi.$$

Meta-Properties of DML and Its Fragments. One key conjecture is still open: decidability of the original DML with both loose and strict updates and downdates, but with only the regular relation-algebraic operations of PDL. (Recall from Chapter 2 that the DPL-fragment of DML with only $\{\sim, \circ\}$ is decidable.) Given that *conditional logics* also lie at level 3 in the finite-variable hierarchy, how do DML systems relate to the former? Many DML-*fragments* still have open questions of completeness and decidability. Also, what is the *complexity* of the decidable logics in the DML family? As for the full system, can we axiomatize it without additional modal inference rules?

Arrow Versions. Retaining the whole vocabulary of DML (including all relation-algebraic operations), one can generalize the semantics given here in the 'arrow style' presented in Chapter 8, and obtain decidable versions after all. What becomes of the precise complexity then? What will explicit axiomatizations look like?

7

Dynamic Styles of Inference

Current dynamic systems show a proliferation of notions of semantic consequence, or 'coherence'. This reflects the coexistence of different styles of reasoning in human cognition. We classify these by charting their structural rules (not involving specific process operators), and then prove some representation and completeness theorems. There are general methods for this purpose — which may be seen as modal completeness arguments, now modified to deal with non-Boolean fragments. Also, we show how styles can be investigated set-theoretically in the earlier semantic perspective of 'inverse logic'. Finally, we present a recent architecture for cooperation between different styles in one logical system via the addition of 'structural modalities.'

7.1 Styles and Structural Rules

7.1.1 Botany of Styles

The standard explication of valid inference demands 'automatic transmission of truth':

"in every situation where all premises are true, so is the conclusion".

To approximate this Tarskian standard style in a dynamic setting, the following seems appropriate. Each procedure may have its 'fixed points', being those states at which it loops (the state already 'satisfies' the goal of the procedure). Thus, we can formulate

Test-to-Test Consequence

In all models, each state which is a fixed point for all premises is also a fixed point for the conclusion

In the notation of Relational Algebra (Chapter 5), this says the following:

$$\text{fix}(P_1) \cap \cdots \cap \text{fix}(P_n) \subseteq \text{fix}(C)$$

In a diagram:

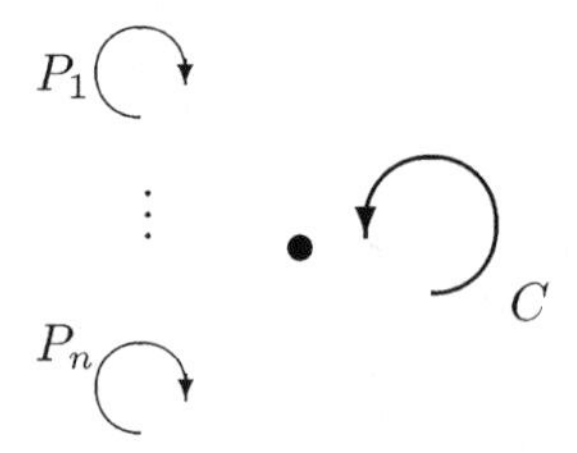

But there are also genuine dynamic styles, starting from the idea that inferences invite us to update our current information state with their premises, and then see about the conclusion. The latter process admits of variation. We start with a notion from van Benthem 1991b (inspired by valid reasoning in Categorial Grammar: cf. Chapter 12), which makes the premises a recipe for achieving the conclusion:

Update-to-Update Consequence $P_1 \circ \cdots \circ P_n \subseteq C$

in all models, each transition for the sequential composition of the premises must be admissible for the conclusion:

P_1 ... P_n C

Here, Veltman 1991 proposes a shift in the verification procedure for the conclusion:

Update-to-Test Consequence $\text{range}(P_1 \circ \cdots \circ P_n) \subseteq \text{fix}(C)$

first process all premises consecutively, then test if the conclusion is satisfied by the resulting state:

P_1 ... P_n C

A related notion is that of Groenendijk and Stokhof 1991:

Update-to-Domain Consequence $\text{range}(P_1 \circ \cdots \circ P_n) \subseteq \text{dom}(C)$

first process all premises consecutively, then test if the conclusion can be processed from the resulting state:

P_1 ... P_n C

Thus, there appears to be a genuine variety of dynamic styles of inference, reflecting different intuitions about reasoning, and suggesting possibly different applications. These differ in what they declare valid.

Example 7.1 (Dynamic Inferences) The following table shows how the above notions differ (in an ad-hoc notation, with $\circ$ read as composition and ? as fixed point):

	$A, B \models A \circ B$	$A \models (A)?$
Test-to-Test	+	+
Update-to-Update	+	− (1)
Update-to-Test	− (2)	+
Update-to-Domain	− (3)	− (4)

Here are some counter-examples. (1) Update-to-Update obviously does not claim that ranges consist of fixed points. (2) US does not validate $\Diamond p$, $\neg p \models \Diamond p \circ \neg p$. (3), (4) DPL has neither $Ax, \exists x \neg Ax \models Ax \circ \exists x \neg Ax$ nor $Ax \circ \exists x \neg Ax \models (Ax \circ \exists x, \neg Ax)?$.

The list is not even complete, because there are 'backward' counterparts to the above, focusing on initial states rather than final ones. For instance, van Eijck and de Vries 1991 propose this variant of Update-to-Domain, as 'implication between preconditions':

Domain-to-Domain Consequence $\mathrm{dom}(P_1 \circ \cdots \circ P_n) \subseteq \mathrm{dom}(C)$

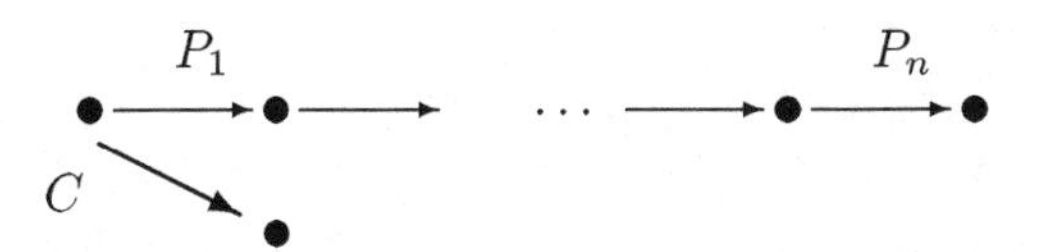

Finally, Beaver 1995 proposes dynamic *presupposition*, a converse to Update-to-Test deriving preconditions (rather than postconditions) for succesful discourse processing:

Domain-to-Test $\mathrm{dom}(P_1 \circ \cdots \circ P_n) \subseteq \mathrm{fix}(C)$

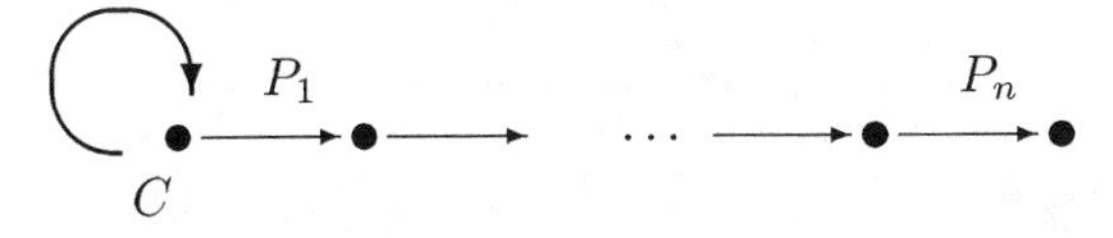

7.1.2 Capturing Styles via Structural Rules

One way of defining a style of inference is through its general properties, expressed in the usual 'structural rules'. For instance, the classical style

Test-to-Test has all the usual properties of standard inference:

$/\ C \Rightarrow C$	Reflexivity
$X \Rightarrow D$ and $Y, D, Z \Rightarrow C\ /\ Y, X, Z \Rightarrow C$	Cut Rule
$X, P_1, P_2, Y \Rightarrow C\ /\ X, P_2, P_1, Y \Rightarrow C$	Permutation
$X, P, Y, P, Z \Rightarrow C\ /\ X, P, Y, Z \Rightarrow C$	Contraction
$X, P, Y, P, Z \Rightarrow C\ /\ X, Y, P, Z \Rightarrow C$	Contraction
$X, Y \Rightarrow C\ /\ X, P, Y \Rightarrow C$	Monotonicity.

These say that only the bare set of premises matters to a classical conclusion: details of presentation, such as ordering or multiplicity are irrelevant. Moreover, Monotonicity says 'the more the better': further information can never invalidate earlier conclusions.

By contrast, the dynamic style Update-to-Update satisfies only Reflexivity and Cut. For instance, from $S \subseteq S$, Monotonicity would derive the invalid principle $R \circ S \subseteq S$. And likewise, from $S \circ S \subseteq S \circ S$, contraction would derive the invalid $S \subseteq S \circ S$. This deviation is natural, as order and multiplicity of instructions may be crucial to the outcome of a recipe. But new religions need not be defined by the old dogmas which they accept or reject. Their point may be precisely that these old dogmas are too crude. Inferential styles may modify standard structural rules, reflecting more delicate handling of dynamic premises. For instance, Update-to-Test Consequence has none of the above structural properties (counter-examples are easy to produce), but we do have this

Fact 7.2 *Update-to-Test Consequence satisfies*

Left Monotonicity	$X \Rightarrow C\ /\ P, X \Rightarrow C$
Left Cut	$X \Rightarrow D\ \&\ X, D, Z \Rightarrow C\ /\ X, Z \Rightarrow C$

And here are some modified structural properties for Domain-to-Domain Consequence.

Fact 7.3 *Domain-to-Domain Consequence satisfies*

Reflexivity	$/A \Rightarrow A$
Right Monotonicity	$X \Rightarrow C\ /\ X, P \Rightarrow C$
Right Cut	$X \Rightarrow D\ \&\ Y, D \Rightarrow C\ /\ Y, X \Rightarrow C$

For a general defense of this way of classifying styles of inference, cf. Chapter 13, establishing a connection with the early logical research program of Bernard Bolzano.

7.2 Representation Theorems

The preceding observations characterize our dynamic styles of reasoning completely. Indeed, there exist precise mathematical representation results to this effect.

Proposition 7.4 *{Monotonicity, Contraction, Reflexivity, Cut} completely determine the structural properties of classical inference, defined by* $P_1 \cap \cdots \cap P_k \subseteq C$.

Proof. Let R be any abstract relation between finite sequences of objects and single objects satisfying the classical structural rules. Now, define

$$a^* = \{A \mid A \text{ is a finite sequence of objects such that } ARa\}.$$

Then, it is easy to show the following two assertions:

1. if $a_1, \ldots, a_k Rb$, then $a_1^* \cap \cdots \cap a_k^* \subseteq b^*$, using Cut and Contraction
2. if $a_1^* \cap \cdots \cap a_k^* \subseteq b^*$, then $a_1, \ldots, a_k Rb$, by Reflexivity and Monotonicity. ⊣

This fact explains why non-classical logics in the literature tend to be 'non-monotonic' — or otherwise deviant. Structural rules are a sensitive logical measuring instrument. We prove a second result in more detail. Let 'propositions' denote binary relations. A sequent is 'true' if the composition of its premises is contained in its conclusion. Valid consequence $\Delta \models \sigma$ for sequents says that truth of all sequents in Δ implies that of σ. Our representations actually show stronger results than the preceding one.

Proposition 7.5 *{Reflexivity, Cut} completely axiomatize Update-to-Update sequent consequence.*

Proof. Evidently, Reflexivity and Cut are valid on the above semantic interpretation. Conversely, suppose that some sequent σ cannot be derived from a set Δ using these two principles. Let all finite sequences of basic syntactic items occurring in sequents be our underlying state set, and define this map * from basic items to binary relations:

$$C^* = \{(X, XY) \mid Y \Rightarrow C \text{ is derivable from } \Delta \text{ using Reflexivity and Cut}\}.$$

Then we have that, for sequents $X = X_1, \ldots, X_n$,

Claim. $X \Rightarrow C$ *is derivable from* Δ *iff* $X_1^* \circ \cdots \circ X_n^* \subseteq C^*$.

Proof. 'If'. By Reflexivity, $X_1 \Rightarrow X_1, \ldots, X_n \Rightarrow X_n$ are derivable from Δ. Hence the pairs $(\langle\ \rangle, X_1)$, (X_1, X_1X_2), $\ldots$, $(X_1 \ldots X_{n-1}, X_1 \ldots X_{n-1}X_n)$ belong to $X_1^*, \ldots, X_n^*$, respectively. So $(\langle\ \rangle, X)$ is in the composition of the consecutive premises, whence it belongs to C^*. But then by definition, $X \Rightarrow C$ is derivable from Δ.

'Only if'. Consider any sequence of consecutive state transitions for the successive premises:

$$(X, XY_1)(Y_1 \Rightarrow X_1 \text{ is derivable}), (XY_1, XY_1Y_2)(Y_2 \Rightarrow X_2 \text{ derivable}), \text{etc.},$$

up to $Y_n \Rightarrow X_n$. Then, n successive applications of Cut to the derivable sequent $X \Rightarrow C$ will derive $Y_1 \ldots Y_n \Rightarrow C$, and hence $(X, XY_1 \ldots Y_n)$ is in C^*, by the definition of *. ⊣

The required counter-example now arises by observing that every sequent in Δ is derivable from it and hence true under the intended relational interpretation, whereas the original nonderivable sequent σ has become false in the latter model. $\dashv$

These arguments can be extended to deal with less standard structural properties.

Proposition 7.6 *{Left Monotonicity, Left Cut} determine Update-to-Test consequence.*

Proof. It suffices to give the main representation involved. This time, the following map $^{\#}$ from syntactic items to binary relations will work:

$$C^{\#} = \{(X, X) \mid X \Rightarrow C \text{ is derivable}\} \cup \{(X, XC) \mid \text{all sequences } X\}.$$

What may be shown now is the following equivalence:

Claim. *$X \Rightarrow C$ is derivable iff it is valid under this interpretation.*

Proof. 'If'. The pairs $(\langle\ \rangle, X_1)$, ..., $(X_1 \ldots X_{n-1}, X_1 \ldots X_{n-1}X_n)$ belong to the successive premise relations. Because of mixed validity then, (X, X) must be in $C^{\#}$, which can only mean that $X \Rightarrow C$ is derivable.

'Only if'. Here is an example, with $n = 4$. Consider the following sequence of 'mixed' transitions for the premises:

$$(U, UX_1), (UX_1, UX_1X_2), (UX_1X_2, UX_1X_2) \text{ (with } UX_1X_2 \Rightarrow X_3\text{)}, (UX_1X_2, UX_1X_2X_4).$$

Then we have $X_1 \ldots X_4 \Rightarrow C$ (by assumption), $UX_1 \ldots X_4 \Rightarrow C$ (Left Monotonicity), $UX_1X_2X_4 \Rightarrow C$ (Left Cut, using $UX_1X_2 \Rightarrow X_3$). That is, the final pair of objects $(UX_1X_2X_4, UX_1X_2X_4)$ is in $C^{\#}$. $\dashv$

Finally, we analyse one of the above 'backward-looking' notions of dynamic consequence.

Proposition 7.7 *{Reflexivity, Right Monotonicity, Right Cut} completely axiomatize Domain-to-Domain Consequence.*

Proof. Again, we merely state the key representation over a state set of finite sequences (here, '$-$' is the empty sequence):

$$A^{\bullet} = \{(AX, X) \mid \text{ all sequences } X\} \cup \{(X, -) \mid X \Rightarrow A\}.$$

That this is adequate, may be shown as follows. (i) Suppose that P_1, ..., $P_n \Rightarrow C$. Let X be in the domain of $P_1^{\bullet} \circ \ldots \circ P_n^{\bullet}$. Case 1. $X = (P_1, \ldots, P_n, Y)$. By Right Monotonicity, $X \Rightarrow C$, whence $(X, -) \in C^{\bullet}$ and X is in the domain of $C^{\bullet}$. Case 2. $X = (P_1, \ldots, P_i, Y)$, with $i < n$ and $Y \Rightarrow P_{i+1}$, $- \Rightarrow P_{i+2}$, ..., $- \Rightarrow P_n$. Applying Right Cut as often as needed on P_1, ..., $P_n \Rightarrow C$ with the empty sequences, we see that P_1, ..., P_i, $P_{i+1} \Rightarrow C$. Then Right Cut with $Y \Rightarrow P_{i+1}$ gives $X \Rightarrow C$.

(ii) Conversely, suppose that $\text{dom}(P_1^{\bullet} \circ \cdots \circ P_n^{\bullet}) \subseteq \text{dom}(C^{\bullet})$. Evidently, the sequence $X = (P_1, \ldots, P_n)$ is in $\text{dom}(P_1^{\bullet} \circ \cdots \circ P_n^{\bullet})$. Hence it is in dom($C^\bullet$). By the above definition, this means either that $C = P_1 -$ and $X \Rightarrow C$ follows by Reflexivity and Right Monotonicity, or that $X \Rightarrow C$ straightaway. ⊣

Practical experience has shown that styles of dynamic consequence can usually be characterized via simple representations. But is there a *general method* behind this? For this purpose, Groeneveld 1995 employs a familiar perspective from modal logic relating representation arguments to Henkin-style completeness proofs. This move has further uses. Structural representation theorems can be uninformative. E.g., Update-to-Domain consequence is characterized by Left Monotonicity only. Its truly interesting properties emerge only in a richer vocabulary. E.g., structural forms of Cut fail for this dynamic style — but it does satisfy a variant of Cut involving sequential composition (as was noted already in Section 2.2.3):

$$X \Rightarrow A \ \&\ A \Rightarrow B \,/\, X \Rightarrow A \circ B.$$

Kanazawa 1993a extends the above results to deal with composition of propositions. His ingenious arguments move closer to full-fledged modal completeness proofs.

7.3 A Modal Perspective

7.3.1 Structural Rules as Modal Consequences

All the above notions of dynamic consequence can be represented inside poly-modal logic or PDL (cf. Chapters 3, 4, 6), even without propositional variables (a constant atom $\top$ suffices). Some of them involve merely the standard modalities $\Box$, $\Diamond$:

Domain-to-Domain Consequence $[a_1] \dots [a_k]\langle b\rangle\top$
Domain-to-Domain Consequence $\langle a_1\rangle \dots \langle a_k\rangle\top \to \langle b\rangle\top$.

Others need a fixed-point operator ( ):

Update-to-Test Consequence $[a_1] \dots [a_k](b)\top$
Domain-to-Test Consequence $\langle a_1\rangle \dots \langle a_k\rangle\top \to (b)\top$.

Adding fixed points yields an axiomatizable decidable extension of PDL (Section 3.4; the result is due to the Sofia modal logicians), as we shall prove later on. In principle, these observations reduce the theory of dynamic consequence to decidable versions of PDL. Here, one must take one decision as to modal consequence. Deriving sequents from sequents amounts to inferences from modal formulas $\phi_1, \dots, \phi_k$ to ψ. These can be taken either *locally*, from truth at a world to truth at that world, or *globally*, from and to truth in a model. The former implies the latter, but not vice versa. The difference shows with the rule of *Necessitation*: ϕ implies $[a]\phi$ globally, but not locally. Our previous notion of valid consequence for sequents (cf. Proposition 7.5) was global.

Example 7.8 (Modal Derivation of Structural Properties)

(i) *Domain-to-Domain Consequence.* Reflexivity is the modal axiom $\langle a\rangle\top \to \langle a\rangle\top$. Right Monotonicity is the (even locally valid) inference from $\langle a_1\rangle \ldots \langle a_k\rangle\top \to \langle b\rangle\top$ to $\langle a_1\rangle \ldots \langle a_k\rangle\langle c\rangle\top \to \langle b\rangle\top$. This uses the minimal modal validity $\langle a_1\rangle \ldots \langle a_k\rangle\langle c\rangle\top \to \langle a_1\rangle \ldots \langle a_k\rangle\top$. Finally, Right Cut is the globally valid modal transition from $\langle a\rangle\top \to \langle b\rangle\top$ and $\langle c\rangle\langle b\rangle\top \to \langle d\rangle\top$ to $\langle c\rangle\langle a\rangle\top \to \langle d\rangle\top$.

(ii) *Update-to-Test Consequence.* Left Monotonicity is just the globally valid modal transition from ϕ to $[a]\phi$. Left Cut is the locally valid transition from $[a](b)\top$ and $[a][b]\phi$ to $[a]\phi$, which rests on that from $(b)\top$ and $[b]\phi$ to ϕ.

Remark 7.9 (Distinguishing Two Levels of Structural Rules) Single sequents $\phi_1, \ldots, \phi_k \vdash \psi$ represent transitions between dynamic inferences (here: modal formulas), possibly encoding non-standard structural rules. But the modal calculus of these sequents themselves is classical, using all standard structural rules.

The general background formalism employed here is a very simple modal system.

Theorem 7.10 *Global consequence for polymodal logic with fixed points is decidable.*

Proof. Recall the argument for standard modal logic. Suppose that $\phi_1, \ldots, \phi_k$ do not globally imply ψ. Then there is a model in which $\phi_1, \ldots, \phi_k$ hold, while ψ fails somewhere. Filtrating this model with respect to the set Φ consisting of $\phi_1, \ldots, \phi_k, \psi$ plus all their subformulas gives a finite counter-model where the ϕ_i hold throughout, while ψ fails — of a size computable from the input formulas (cf. Goldblatt 1987). To deal with a modal language with fixed-points (or atoms for loops), one has to filtrate and then 'unravel' the model obtained, to avoid the emergence of a-loops at states which fail to validate the formula $(a)\top$. More precisely, consider all equivalence classes E of states with respect to the formulas in the filtration set Φ. Choose one representative if no two members of E stand in a relevant relation R_a, while the formula $(a)\top$ is denied by the states in E. Otherwise, choose two representatives, without R_a-loops, and make these behave similarly in all further relational behaviour. By a simple induction, one shows that both single and duplicated representatives agree with all members of their equivalence class on all filtration formulas. ⊣

7.3.2 Modal Axiomatizations for Dynamic Fragments

In practice, however, one wants explicit axiomatizations for the relevant *fragments*, in order to extract genuine dynamic information.

Definition 7.11 (Fragments) We consider the following two sublanguages. The fragment *UT* has only formulas consisting of a sequence of $[a]$, (b) ending in $\top$. The fragment *UDD* has only similar sequences with $[a]$, $\langle a\rangle$ plus their outer Boolean combinations. The latter defines both Update-to-Domain and Domain-to-Domain Consequence.

Consider the usual Henkin completeness construction for modal logic (standard in all good textbooks, and hence presupposed here). There is no guarantee that this will work for our fragments, as these lack the pervasive Boolean structure that occurs throughout standard completeness proofs. On the other hand, our experience in categorial logics (cf. Chapter 12) shows that completeness arguments may indeed simplify considerably for small fragments — witness, e.g., cheap formula-based canonical models for (non-) associative Lambek calculi. In particular, our first fragment can be axiomatized by a simple modification of modal completeness proofs. One can also axiomatize Update-to-test Consequence for the special case where all update relations involved are *functions* (Groeneveld and Veltman 1996). This additional restriction shows in the validity of a further modified structural rule which makes life simpler, namely

Cautious Monotonicity $X \Rightarrow A,\ X \Rightarrow B \,/\, X, A \Rightarrow B$.

They also have a smooth representation argument when the relations involved are reflexive, satisfying $A \Rightarrow A$. In our modal calculus, this amounts to having a special axiom $[a](a)\top$. We can understand the resulting situation more generally.

Theorem 7.12 *Global validity for sequents in UT with a functional interpretation is axiomatized by the following modal principles:*

(o) $\phi \vdash \Box\phi$

(i) *from* $\phi_1, \ldots, \phi_k \vdash \psi$ *infer* $\Box\phi_1, \ldots, \Box\phi_k \vdash \Box\psi$

(ii) $(a)\phi \vdash \phi$

(iii) $(a)\phi, [a]\psi \vdash \psi$

(iv) $(a)\phi, \psi \vdash (a)\psi$

(v) $\vdash [a](a)\top$.

Proof. Consider the following Henkin model $\mathbf{M}$. Its states are sets of formulas closed under derivability in sequent form. Moreover, it has the standard accessibility relations

$$R_a\Sigma\Delta \text{ iff for all formulas } \alpha, \text{ if } \Sigma \vdash \Box\alpha, \text{ then } \Delta \vdash \alpha.$$

Truth Lemma. *For all formulas* ϕ*,* $\Sigma \vdash \phi$ *iff* $\mathbf{M}, \Sigma \models \phi$.

Proof. Induction on ϕ. Case 1. $\phi = [a]\psi$. If $\Sigma \vdash \Box\psi$, then, for all sets Δ with $R_a\Sigma\Delta$: $\Delta \vdash \psi$. By the inductive hypothesis, $\mathbf{M}, \Delta \models \psi$ and hence $\mathbf{M}, \Sigma \models \Box\psi$. Conversely, suppose that $\mathbf{M}, \Sigma \models \Box\psi$. The set $\Box^{-1}(\Sigma)$ is an

R_a-successor of Σ, where ψ holds. By the inductive hypothesis, $\Box^{-1}(\Sigma) \vdash \psi$ and hence $\Sigma \vdash \Box\psi$ by the Distribution Rule (i).

Case 2. $\phi = (a)\psi$. Suppose that $\Sigma \vdash (a)\psi$. Then, by (ii), also $\Sigma \vdash \psi$ — and by the inductive hypothesis, ψ is true at Σ. Moreover, for any Σ-derivable $[a]\beta$, β is derivable too, by axiom (iii). It follows that R_a loops at Σ, whence $\mathbf{M}, \Sigma \models (a)\psi$. Conversely, suppose that $(a)\psi$ is true at Σ. Then R_a loops at Σ, whence ψ is true, and hence derivable at Σ (by the inductive hypothesis). Moreover, since $[a](a)\top$ is derivable at Σ, the definition of R_a says that $(a)\top$ is derivable there, too. Together with axiom (iv), this gives Σ-derivability of $(a)\psi$. $\dashv$

Now, consider any non-derivable sequent $\phi_1, \ldots, \phi_k \vdash \psi$. Take the deductive closure Δ of its antecedents. By the Truth Lemma, in the Henkin Model $\mathbf{M}$, the consequent will fail there. Moreover, we can restrict attention to the *generated submodel* consisting of Δ plus all its hereditary R_a-successors. By the Necessitation Rule (i), all antecedents will be derivable (and hence true) throughout the latter. So, we have found a global counter-example to $\phi_1, \ldots, \phi_k \vdash \psi$. $\dashv$

The above principles derive Left Monotonicity and Left Cut. This argument extends to the general case without Reflexivity, by the earlier semantic argument unraveling undesired fixed points. Next, we consider UDD. Outer Boolean compounds may be written as conjunctions of sequents $\phi_1, \ldots, \phi_k \vdash \psi_1, \ldots, \psi_m$, read in the usual $\bigwedge\bigvee$ fashion. We shall manipulate the latter, using a construction from Jaspars 1994.

Theorem 7.13 *Global validity for sequents in UDD is axiomatized by the usual Boolean principles, plus Necessitation, and the following modal rules of inference*

(i) from $\phi_1, \ldots, \phi_k \vdash \psi_1, \ldots, \psi_m, \alpha$ *infer* $\Box\phi_1, \ldots, \Box\phi_k \vdash \Diamond\psi_1, \ldots, \Diamond\psi_m, \Box\alpha$.

(ii) from $\phi_1, \ldots, \phi_k, \alpha \vdash \psi_1, \ldots, \psi_m$ *infer* $\Box\phi_1, \ldots, \Box\phi_k, \Diamond\alpha \vdash \Diamond\psi_1, \ldots, \Diamond\psi_m$.

Proof. This time, worlds are deductively closed *splitting* sets of formulas Δ such that, if $\Delta \vdash \psi_1, \ldots, \psi_m$, then at least one ψ_i belongs to Δ. We say that Δ 'splits into' Σ if the latter ψ_i can always be found in Σ. The following result is very useful.

Lindenbaum Lemma. *If* Δ *splits into* Σ, *then there exists a splitting set* Δ^+ *extending* Δ *but contained in* Σ.

Proof. A straightforward argument using an enumeration of all disjunctions plus the propositional disjunction rules (plus the structural Cut Rule). $\dashv$

Next, we define the alternative relations R_a in the Henkin model:

$$R_a\Delta\Sigma \quad \text{iff} \quad \text{for all } \phi, \text{ if } \Delta \vdash [a]\phi, \text{ then } \Sigma \vdash \phi \\ \text{and if } \Sigma \vdash \phi, \text{ then } \Sigma \vdash \langle a\rangle\phi.$$

The valuation V is as usual: atoms p hold in the worlds Δ where they are derivable.

Truth Lemma. *For all formulas ϕ, $\Delta \vdash \phi$ iff $\mathbf{M}, \Delta \models \phi$.*

Proof. Induction on ϕ. Two cases are routine, by the definition of R_a: those starting from $\Delta \vdash [a]\phi$ and from $\mathbf{M}, \Delta \models \langle a\rangle\phi$. Case 3. $\Delta \vdash \langle a\rangle\phi$. Consider the set $\Sigma = \Box^{-1}(\Delta) \cup \{\phi]\}$. It splits into $\{\beta \mid \Delta \vdash \Diamond\beta\}$, because of rule (ii) plus splittingness of Δ. Now take any splitting completion Δ^+ of the former set, contained in the latter. This is an R_a-successor of Δ containing ϕ. Clearly, then, $\mathbf{M}, \Delta^+ \models \phi$, and $\mathbf{M}, \Delta \models \langle a\rangle\phi$. Case 4. $\mathbf{M}, \Delta \models [a]\phi$. Suppose that not $\Delta \vdash [a]\phi$. Then the set $\Box^{-1}(\Delta)$ splits into the set $\{\beta \mid \Delta \vdash \Diamond\beta\} - \{\phi\}$. This is easy to see, using the above rule (i). So, by the Lindenbaum Lemma, some R_a-successor Δ^+ exists in between the two. In the latter, ϕ will hold, and hence it will be derivable (by the inductive hypothesis). But this contradicts the choice of the upper bound for the preceding splitting. $\dashv$

7.3.3 Extended Modal Logics

Modal formalisms beyond these are needed to render the earlier notion of Update-to-Update Consequence. Indeed, Kanazawa 1993a observed that sequent derivability under the latter regime (with Reflexivity and Cut only) is equivalent to unrestricted rewriting in type-0 grammars. This analysis proves the following negative result.

Proposition 7.14 *Sequent entailment for Update-to-Update Consequence is undecidable.*

The increase in complexity can also be understood semantically. To formalize the latter notion of consequence, one can no longer use purely 'local' statements about states. Whole transitions are to be recorded, using Boolean relational operators to exclude cases $a \cap -b$. This requires richer dynamic languages with relation-algebraic operators. As in previous chapters, the result is a *modal hierarchy* of increasingly complex definitions for dynamic consequence. The *bounded fragments* of Chapter 4 are relevant here. The definition for Update-to-Test reads as follows

$$\forall x \forall y_1 (P_1 x y_1 \rightarrow \forall y_2 (P_2 y_1 y_2 \rightarrow \ldots \rightarrow \forall y_k (P_k y_{k-1} y_k \rightarrow C y_k y_k).$$

This lies inside the Guarded Fragment, and hence structural rules for this notion are decidable. By contrast, the transcription for Update-to-Update is of a non-guarded form

$$\forall x \forall y_1 (P_1 x y_1 \rightarrow \forall y_2 (P_2 y_1 y_2 \rightarrow \ldots \rightarrow \forall y_k (P_k y_{k-1} y_k \rightarrow C x y_k).$$

7.4 Inverse Logic

7.4.1 Structural Properties as Defining Constraints

Another interplay between structural rules and logical constants arises as follows. One may license certain structural behaviour not for *all* propositions, but for *special* kinds only (cf. Girard 1987 for linear logic, van Benthem 1989c for non-monotonic logic). For instance, let O be some operator that is to admit of arbitrary monotonic insertion:

$$X, Y \Rightarrow C \,/\, X\,, O(P), Y \Rightarrow C.$$

It is easy to show that this can be the case in the dynamic style if and only if $O(P)$ is a 'test' contained in the diagonal relation. Thus, only test actions support unrestricted Monotonicity. Here is a slightly less trivial result characterizing tests.

Proposition 7.15 *An operator O allows unlimited contraction iff for all P, $O(P)$ is either empty or it contains the diagonal relation.*

Proof. 'Only if'. If $O(P)$ is empty, then compositions including it are empty, and hence Contraction holds vacuously. If $O(P)$ includes the identity, then any relation Y dynamically implies both $Y \circ O(P)$ and $O(P) \circ Y$, whence Contraction holds too.

'If'. Suppose that $O(P)$ allows unlimited contraction under the dynamic interpretation of sequents. If $O(P)$ is not empty, then there exist x, y with $xO(P)y$. Consider any state z. Let R be $\{(y, z)\}$. The sequent $O(P)$, R, $O(P) \Rightarrow O(P) \circ R \circ O(P)$ is dynamically valid, and hence by Contraction, so is $O(P)$, $R \Rightarrow O(P) \circ R \circ O(P)$. Hence (x, z) must be in $O(P) \circ R \circ O(P)$, which can only be the case if $zO(P)z$. ⊣

7.4.2 Digression: Dynamic Negation

Inverse logic also characterizes important operators inside the procedural repertoire. To show there is more to life than completeness theorems, we serve some delightful intricacies from van Benthem and Cepparello 1994. (This digression may be skipped.)

Theorem 7.16 *DPL negation is the only permutation-invariant operator in Relational Algebra satisfying the following conditions:*

(i) $\sim 0 = \mathrm{Id}$

(ii) $\sim(\bigcup_i \pi_i) = \bigcap_i \sim\pi_i$

(iii) $\sim\sim\pi \leq \pi \circ 1$

(iv) $\sim\pi \circ \pi = 0$

Proof. We start with an auxiliary observation.

Lemma. *Conditions (i), (ii) imply that* $\sim\pi \leq \mathrm{Id}$.

Proof. We know from Relational Algebra that $\pi = \pi \cup 0$. From this it follows by (ii) that $\sim\pi = \sim(\pi \cup 0) = \sim\pi \cap \sim 0$, whence $\sim\pi \leq \sim 0 = \mathrm{Id}$. ⊣

Now comes the main argument. Given any relation π, the relation $\sim\pi$ can be retrieved from the values $\sim(\{(x,y)\})$, for $(x,y) \in \pi$. This follows from (ii) again, because $\pi = \bigcup_{\pi xy}\{(x,y)\}$. Now we have seen that $\sim\pi \leq \text{Id}$. Hence, by the permutation invariance of $\sim$,

$$\sim(\{(x,y)\}) = \{(z,z) \mid \phi(x,y,z)\} \text{ for some condition } \phi,$$

can only refer to the 'Venn zones' consisting of $\{x\}$, $\{y\}$ and their Boolean combinations. The argument then proceeds case by case.

Case 1: $x = y$. Here are the options for the object z:

- $z = 1 - \{x\}$. This is what we want.
- $z = 0$. If the domain has one object, this is our previous case. If the domain has more objects, we derive a contradiction. Let $\sim(\{(x,x)\}) = 0$ while the domain contains some $y \neq x$. From the assumption, by (i), $\sim\sim(\{(x,x)\}) = \text{Id}$. But by (iii),

  $$\sim\sim(\{(x,x)\}) \leq (\{(x,x)\}) \circ 1,$$

  whence $\text{Id} \leq (\{(x,x)\}) \circ 1$. But this cannot be true, as the domain of Id is larger than $\{x\}$.
- $z = \{x\}$. This conflicts with (iv) — as we would have $\sim\{(x,x)\} \circ \{(x,x)\}) \neq 0$.
- $z = 1$. This is again in conflict with (iv), for the same reason.

Case 2 $x \leq y$. Here are the options for the object z:

- $z = \{x\}$. This is in conflict with (iv).
- $z = 0$. This can be disposed of by the same argument as before.
- $z = 1 - \{x\}$. This is our intended choice.
- $z = \{y\}$. If x, y are the only objects, this falls under the last case. Otherwise, by (iv), x cannot occur in the outcome set. Now suppose $\sim(\{(x,y)\}) = \{(y,y)\}$. Then

  $$\sim\sim(\{(x,y)\}) = \sim(\{(y,y)\}) = 1 - \{y\}$$

  (by Case 1). By (iii) then, $1 - \{y\} \leq (\{(x,y)\}) \circ 1$ — *quod non*, because the domain contains at least one object $z \neq x$, y.
- $z = 1 - \{x,y\}$. If x, y are the only objects, then this falls under the case $z = 0$. Otherwise, $\sim(\{(x,y)\}) = \text{Id} - \{(x,x),(y,y)\}$. Then

  $$\sim(\{(x,y)\}) = \bigcup_{z \neq x,y} \{(z,z)\}.$$

  Hence, by (ii),

  $$\sim\sim(\{(x,y)\}) = \sim \bigcup_{z\neq x,y} \{(z,z)\} = \bigcap_{z \neq x,y} \sim(\{(z,z)\}) = \{(x,x),(y,y)\}.$$

  But this is again in conflict with (iii). $\dashv$

The result seems the best possible, as all stated conditions are necessary. For instance, one can satisfy all other conditions (i), (ii), (iii) via the non-standard negation $\mathrm{Id} - \pi$. Unique definability does not guarantee complete axiomatization. Nevertheless, the above principles do derive some other basic properties of strong test negation.

Fact 7.17 *Conditions (ii) and (iii) imply that* $\sim\pi \cap \pi = 0$.

Proof. By the previous Lemma, $\sim\pi \leq \mathrm{Id}$. From this we have, using some valid (non-negation!) principles of Relational Algebra, that

$$\begin{aligned} \sim\pi \cap \pi &= \sim\pi \cap \pi \cap \mathrm{Id} \\ &= (\sim\pi \cap \pi) \cap (\sim\pi \cap \mathrm{Id}) \\ &= (\sim\pi \cap \pi) \circ (\sim\pi \cap \mathrm{Id}) \leq \sim\pi \circ \pi \\ &= 0. \quad \dashv \end{aligned}$$

To understand this behaviour, recall the fact (Chapters 2, 12) that DPL semantics is complete for the full relational set algebra of its special vocabulary $\{\sim, \circ\}$.

7.5 Switching Between Styles

7.5.1 Varieties of Translation

An abundance of inferential styles raises a new issue. How are they going to co-exist? Also, it is natural to ask whether reasoning in one style can be reduced to another. Here logical constants become relevant. Often, one inferential style can be *simulated* inside another by adding suitable logical operators. One example is the above classical Test-to-Test style. Let us introduce a relational fixed point operator Φ sending relations R to their diagonal $\lambda xy.\,(Rxy \wedge y = x)$. Then we have the evident equivalence

Fact 7.18 $P_1, \ldots, P_n$ *imply* C *classically iff* $\Phi(P_1), \ldots, \Phi(P_n)$ *imply* $\Phi(C)$ *dynamically.*

In the opposite direction, there is no similar formula-wise faithful embedding from the dynamic style into the classical one. For, such an embedding would import classical Monotonicity into the dynamic style (since adding translations of dynamic premises would not disturb the classical translation of a dynamic inference): quod non. Still there may be more global embeddings that do the trick, translating whole sequents at once.

7.5.2 A Modal Perspective on Translations

Let us consider three dynamic styles and their possible reductions, in a triangle with some obvious ad-hoc notation:

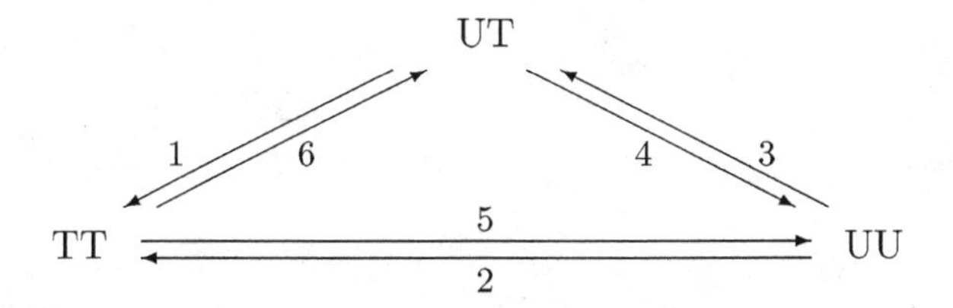

The following relational operations will achieve a number of embeddings:

$$\begin{aligned} \mathrm{fix}(R) &:= \lambda xy.\, x = y \wedge Rxx \\ \mathrm{range}(R) &:= \lambda xy.\, \exists z\, Rzy. \end{aligned}$$

For simplicity, we consider single premises. (One of the following four facts is redundant.)

Proposition 7.19

$$\begin{aligned} P \models_{\mathrm{TT}} C \quad &\textit{iff} \quad \mathrm{fix}(P) \models_{\mathrm{UU}} \mathrm{fix}(C) \\ P \models_{\mathrm{TT}} C \quad &\textit{iff} \quad \mathrm{fix}(P) \models_{\mathrm{UT}} C \\ P \models_{\mathrm{UT}} C \quad &\textit{iff} \quad \mathrm{range}(P) \models_{\mathrm{TT}} C \\ P \models_{\mathrm{UT}} C \quad &\textit{iff} \quad \mathrm{fix}(\mathrm{range}(P)) \models_{\mathrm{UU}} \mathrm{fix}(C) \end{aligned}$$

These results involve modal definitions in the spirit of the preceding section, using a propositional modal language with standard modalities plus fixed points. E.g., what the second line says is that global truth in any labeled transition model of an implication $\langle P\rangle\top \to \langle C\rangle\top$ is equivalent to global truth of the modal formula $[\mathrm{fix}(P)]\langle C\rangle\top$. This equivalence is a sinmple property of the fixed point modality. Likewise, the third line expresses another validity of this modal logic: global truth of $[P]\langle C\rangle\top$ is equivalent to that of the formula $\langle \mathrm{range}(P)\rangle\top \to \langle C\rangle\top$. By contrast, no obvious modal definition of this sort exists for Update-to-Update consequence. This is as it should be:

Proposition 7.20

- *No modal embedding exists for UU in terms of UT .*
- *No modal embedding exists for UU in terms of TT .*

Proof. The second assertion follows from the first, given the above positive facts. To refute the existence of an embedding, we appeal to our standard technique, namely bisimulations for the modal language (this time) with *fixed points*. Consider an earlier counter-example (cf. Chapters 3, 4):

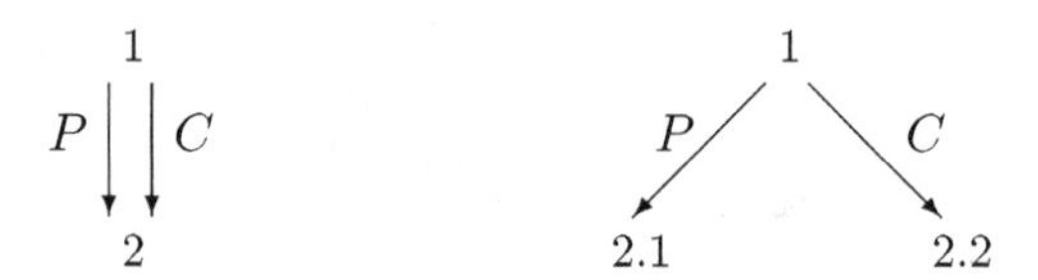

$P \models_{UU} C$ holds on the left, but not on the right. But connecting the obvious points on both sides is a bisimulation for the modal language with loops. ⊣

Bisimulations for a modal language with loops do raise interesting issues of their own (e.g., concerning their safe repertoire), which we forego here.

7.5.3 Modal Operators for Structural Control

Powerful mechanisms for switching styles by introduction of new modalities have emerged in linear and categorial logic. To demonstrate this technique, we transpose some recent results from Kurtonina and Moortgat 1996 to more dynamic inference. Categorial logics with different sets of structural rules can embed each other by adding modalities allowing special behaviour. We generalize this to a modal logic with binary 'product modalities' $\phi \bullet \psi$ and arbitrary Booleans. Intuitively, the product modality is true at a state x if x is a combination of states y, z where ϕ, ψ hold. This was the pattern of ternary relational semantics for relevant logic (Section 2.3.1). But ternary combinations like this will also be a central feature in the 'arrow frames' of Chapter 8. Now we show how adding vocabulary may allow us to manipulate structural rules.

Theorem 7.21 *The minimal modal logic of three conjugate existential binary modalities over ternary relational frames can be faithfully embedded into its* associative *extension with additional unary tense operators F, P.*

Proof. We begin by defining the relevant translation:

$$\begin{aligned}
\underline{\phi \bullet_1 \psi} &= F(\underline{\phi} \bullet_1 \underline{\psi}) \\
\underline{\phi \bullet_2 \psi} &= \underline{\phi} \bullet_2 P\underline{\psi} \\
\underline{\phi \bullet_3 \psi} &= \underline{\phi} \bullet_3 P\underline{\psi}
\end{aligned}$$

By induction on derivations, all translated laws of the minimal logic are easily shown derivable in its associative extension. Conversely, if a modal principle is underivable, we perform a simple model transformation. Consider any counter-example $\mathbf{M}$ on a ternary frame (W, R). Here, the above translation shows its mettle. One replaces all 'triangles' Rx, yz by situations xSx', Rx', yz, where x' is some unique new point for this particular triangle, and S is a new binary accessibility relation for the additional tense operators. Call the resulting model $\mathbf{M}'$. An easy induction shows

$\mathbf{M}, u \models \phi$ iff $\mathbf{M}', u \models \underline{\phi}$, for all formulas ϕ in the original language.

Therefore, the translated sequent has a counter-example on $\mathbf{M}'$. Now observe that the new ternary relation R on the latter model is *associative*: whence we have a counter-example to derivability in the richer associative modal calculus. $\dashv$

This technique sometimes requires further twists. For instance, arguments similar to the preceding one establish a faithful embedding from the associative modal calculus into its 'temporalized' extension having a symmetric ternary accessibility relation, reflecting the structural rule of Permutation. The relevant modalized translation is as follows:

$$
\begin{aligned}
\underline{\phi \bullet_1 \psi} &= F\underline{\phi} \bullet_1 \underline{\psi} \\
\underline{\phi \bullet_2 \psi} &= F\underline{\phi} \bullet_2 \underline{\psi} \\
\underline{\phi \bullet_3 \psi} &= P\underline{\phi} \bullet_3 \underline{\psi}.
\end{aligned}
$$

The key semantic move again transforms counter-examples, replacing triangles Rx, yz by situations $R, x, y'z$, Rx, zy' for some new world y' with $Sy'y$. But now, there is a new phenomenon in the result obtained. To make sure that all original principles remain valid under this translation, one needs to add all *translations of associativity axioms* to the target calculus. The same strategy can be used in the opposite direction, descending from more permissive styles to stricter ones. For instance, one can embed the above associative modal logic into its (unary temporalized) *minimal* counterpart, by our first translation. But again, translations of associativity axioms must be postulated.

The results of this technique raise an interesting issue of 'packaging'. In describing a certain style of reasoning, one chooses a representation in some formal language plus an appropriate logical calculus. But under translation, one gets a different representation with a different calculus manipulating the latter. Thus, one would like to compare the advantages of different packages — rather than languages or calculi by themselves. One can investigate the whole landscape of the earlier structural rules from this new modal viewpoint. The resulting architecture is as in Chapter 6, but with a new feature: dynamic styles of inference turn out to co-exist through a web of translations in enriched modal languages. We still know very little about its structure.

7.6 Notes

Styles of Inference. Natural modifications of the above styles occur in Artificial Intelligence. The non-monotonic style of Shoham 1988 considers only most *preferred* models of the premises. Makinson 1994 is a general survey. Specifically dynamic styles of inference occur in areas like *planning* (Moore 1984, Morreau 1992).

Representation and Completeness Theorems. Many further results are in Groeneveld 1995. Groeneveld, Veltman, and van der Does 1996 axiomatize Update-to-test Consequence for a full propositional language with update *functions*. The border-line between representation and completeness arguments is very delicate in Kanazawa 1993a. Blackburn and Venema 1995 bring representation techniques to bear from relational set algebra. Other contributions concern completeness theorems for propositional update logic: cf. Veltman 1991, van der Does 1994.

Related Themes in Categorial Logics. Models for categorial logics, and some completeness results are in van Benthem 1991b (continuing earlier work of Došen and Buszkowski). Kurtonina 1995 is a congenial study addressing such issues as: completeness for non-Boolean fragments, representation arguments and 'cheap' completeness arguments via hereditary sets or filters, semantic analysis of non-Boolean languages via non-symmetric 'directed bisimulations'. (Incidentally, non-Boolean fragments also occur in *feature logics*: cf. Rounds 1996.)

Modal Architecture. Modal languages extended with relation-algebraic operators for analyzing styles of inference are found in Blackburn and Venema 1995. A good recent survey of substructural logics (categorial, relevant, linear) is the anthology Došen and Schröder-Heister 1994. The strategy of adding modalities for 'structural control' is a key contribution from Girard 1987. Kurtonina and Moortgat 1996 study modal interpretations switching between categorial logics with different structural rules, i.e., different ways of handling linguistic resources. As for 'upward' results of this kind, the *non-associative Lambek Calculus* NL (cf. Chapter 12) can be embedded into its associative counterpart L, and the latter again into its *undirected extension* LP allowing Permutation. 'Downward' results exist, too, going from liberal calculi to more austere ones, using modalities to encapsulate exceptions. Versmissen 1996 shows how one can achieve the result of extra modal operators in translation also by adding new *propositional constants*, which are then substituted into binary products.

Alternative Architectures. Other architectures for combining different styles of inference have been proposed in linear and relevant logic. Instead of translating styles, one can also *combine* them in formats performing different types of reasoning at the same time. For instance, sequents may be heterogeneous, with 'classical' and 'non-classical' premise parts:

$$X \mid Y \Rightarrow C,$$

where X, Y are subject to different structural rules. (Cf. R. Meyer's work reported in Dunn 1985, Girard 1993, or Gabbay 1992).

7.7 Questions

Representation and Completeness. Find a *general* representation method generalizing our specific cases in this chapter (where each trick was found ad-hoc). Unify these representation arguments with standard modal completeness arguments. Unify the different modal completeness strategies that have been used to axiomatize DPL-style and US-style dynamic inference (Groeneveld 1995 leaves this issue open). Prove a completeness theorem for a modal logic with an added operator needed to encode the Update-to-Update style. (This will be a fragment of DML in Chapter 6.)

More on Modal Logics. Find appropriate bisimulations for the extended modal logics that arise here: without Booleans, with loops, with additional operators. What about *safe* operational repertoires for these bisimulations in a relational set algebra?

Complexity. Determine the complexity of inference in our various dynamic styles. Also there are matters of comparison between calculi: in particular, what is the computational cost of performing deduction modalized inside another system?

Translations with Added Modalities. Are new modalities needed at each switch, or does 'equilibrium' occur occasionally? More specifically, when are there embeddings between different logics as they stand, without adding modalities? Also, when are modal expansions *conservative* over the original calculi? (In the above cases, they are.) Find *general* interpretation results for dynamic styles, allowing us to make predictions. Apply the strategy of additional modalities to non-monotonic logics in AI.

Alternatives. Redo our representations in the Arrow Logic of Chapter 8.

8

Decidable Remodelling: Arrow Logic

8.1 Core Content Versus Formal Wrappings

Dynamic logics over transition relations between computational states may have high complexity (witness the undecidability of relational algebra or dynamic modal logic). But dynamic semantics was motivated by a wish to model *simple* cognitive procedures. We seek a 'computational core', avoiding spurious complexity induced by mathematical tools. To this end, we develop a new modal logic of 'arrows' which treats transitions as dynamic objects in their own right. This chapter presents a decidable system of Arrow Logic with first-order relational operations and infinitary Kleene iteration, as a dynamic core calculus. We prove completeness for its minimal version, and establish its connections with dynamic and categorial logics. In the process, we find a new landscape of possible arrow logics, whose semantic content can be analyzed using modal frame correspondences. One can do such a reanalysis for other standard logics. Our general concern is therefore: *what is genuine 'computation' and what is 'extraneous mathematics' in the logical analysis of (cognitive) processes?* Thus, this chapter is also a case study for a more general program of lowering complexity.

8.2 Arrow Logic in a Nutshell

8.2.1 Language and Models

Arrow Logic is a research line initiated in van Benthem 1991b, Venema 1991, 1994. Binary relations may be thought of as denoting sets of arrows. Key examples are 'arcs' in multi-graphs, or 'transitions' for dynamic procedures in computer science, but one can also think of 'preferences' with ranking relations (as in theories of reasoning in Artificial Intelligence, or social choice and economics). Arrows may have internal structure, whence they need not be identified with pairs ⟨source, target⟩: several arrows may share the same input-output pair, but also certain pairs may not be instantiated by an arrow. This motivates the following definition:

Definition 8.1 (Arrow Frames) *Arrow frames* are tuples (A, C^3, R^2, I^1) with

1. A: a set of objects ('arrows') carrying three predicates:
2. C^3x, yz: x is a 'composition' of y and z
3. R^2x, y: y is a 'reversal' of x
4. I^1x: x is an 'identity' arrow.

Given the abstractness of this modelling, other mathematical models are possible too. Arrow Logic says that dynamic transitions need not be identified with the ordered pairs over some underlying state set. This idea has really two different aspects. Distinct arrows may correspond to the same pair of ⟨input, output⟩, but also, not every such pair need correspond to an available arrow. This shows in the following example:

> Let arrows be functions $f : A \to B$ giving rise to, but not identifiable with, ordered pairs $\langle A, B\rangle$ of 'source' and 'target'. Then, the relation C expresses the partial function of composition of mappings, while the relation R will hold between a function and its inverse, if available.

Here, one can think of categories, with arrows as morphisms — where reversals do not always exist. There is also a 'conservative' variant, where arrows remain ordered pairs, which gives up the idea that all ordered pairs are available as arrows. This yields the universally first-order definable class of arrow frames which can be represented via sets of ordered pairs (not necessarily full Cartesian products of an underlying state space).

Next, we introduce a formal language over arrow frames. This is a modal formalism, originally arising from Relational Algebra (Chapters 3, 5) in an abstract perspective. This provides an entirely new view of earlier systems. In particular, it will allow us to 'deconstruct' the mathematical intricacies of Relational Algebra. *Arrow models* **M** add a propositional valuation V here, and one can then interpret an appropriate modal propositional language expressing properties of (sets of) arrows using two modalities reflecting the basic 'ordering operations' of relational algebra:

$$\begin{array}{rll}
\mathbf{M}, x \models p & \text{iff} & x \in V(p) \\
\mathbf{M}, x \models \neg\phi & \text{iff} & \text{not } \mathbf{M}, x \models \phi \\
\mathbf{M}, x \models \phi \wedge \psi & \text{iff} & \mathbf{M}, x \models \phi \text{ and } \mathbf{M}, x \models \psi \\
\mathbf{M}, x \models \phi \bullet \psi & \text{iff} & \text{there exist } y,\ z \text{ with } Cx, yz \text{ and } \mathbf{M}, y \models \phi,\ \mathbf{M}, z \models \psi \\
\mathbf{M}, x \models \phi^{\smallsmile} & \text{iff} & \text{there exists } y \text{ with } Rx, y \text{ and } \mathbf{M}, y \models \phi \\
\mathbf{M}, x \models \mathrm{Id} & \text{iff} & Ix.
\end{array}$$

Eventually, one can introduce more expressive modal operators into this vocabulary.

8.2.2 Frame Correspondence: Landscape of Arrow Logics

The minimal modal logic of this system is an obvious counterpart of its mono-modal predecessor, whose key principles are the following axioms of Modal Distribution:

$$
\begin{array}{rcl}
(\phi_1 \vee \phi_2) \bullet \psi & \leftrightarrow & (\phi_1 \bullet \psi) \vee (\phi_2 \bullet \psi) \\
\phi \bullet (\psi_1 \vee \psi_2) & \leftrightarrow & (\phi \bullet \psi_1) \vee (\phi \bullet \psi_2) \\
(\phi_1 \vee \phi_2)^{\smile} & \leftrightarrow & {\phi_1}^{\smile} \vee {\phi_2}^{\smile}.
\end{array}
$$

A completeness theorem is provable here along standard lines, using Henkin models. Next, one can consider further axiomatic principles (taking cues from Relational Algebra) and analyze what constraint these impose on arrow frames via the usual *frame correspondences*. All results stated in what follows are direct applications of modal techniques mentioned in Chapter 3. This semantic analysis reveals a whole landscape of options, as different principles turn out to have quite diverse 'arrow content'.

Example 8.2 (Possible Principles for Reversal)

(1) $\neg(\phi)^{\smile} \rightarrow (\neg\phi)^{\smile}$ iff $\forall x \exists y\, Rx, y$

(2) $(\neg\phi)^{\smile} \rightarrow \neg(\phi)^{\smile}$ iff $\forall xyz\, (Rx, y \wedge Rx, z) \rightarrow y = z$.

Proof. (of (1)) We illustrate the typical kind of argument involved. *From right to left.* Consider any arrow frame whose relation R has successors everywhere. Let $\mathbf{M}$ be any model over it such that $\mathbf{M}, x \models \neg(\phi)^{\smile}$. Let y be any successor of x. Then, by the truth definition, $\mathbf{M}, y \models \neg\phi$. This shows that $\mathbf{M}, x \models (\neg\phi)^{\smile}$.

From left to right. Assume that $\neg(\phi)^{\smile} \rightarrow (\neg\phi)^{\smile}$ holds on our frame for all valuations. Let x be any world in it. Set $V(\phi) := A - \{y \mid Rxy\}$. This valuation makes $\neg(\phi)^{\smile}$ true at x, and hence so is $(\neg\phi)^{\smile}$. But the latter means that x must have at least one R-successor. ⊣

Together, these axioms make the binary relation R a unary *function* r of 'reversal'. On top of this, the 'double conversion' axiom makes the function r *idempotent*:

(3) $(\phi)^{\smile\smile} \leftrightarrow \phi$ iff $\forall x\, r(r(x)) = x$.

Let us assume this much henceforth in our arrow frames. Next, the following principles of Relational Algebra regulate the perspicuous interaction of reversal and composition in the characteristic semantic triangles for arrows.

Example 8.3 (Composition Triangles)

(4) $(\phi \bullet \psi)^{\smile} \rightarrow \psi^{\smile} \bullet \phi^{\smile}$ iff $\forall xyz\, Cx, yz \rightarrow Cr(x), r(z)r(y)$

(5) $\phi \bullet \neg(\phi^{\smile} \bullet \psi) \rightarrow \neg\psi$ iff $\forall xyz\, Cx, yz \rightarrow Cz, r(y)x$.

Proof. (of (5).) We also do this case, as it was the most mysterious relation-algebraic axiom of Section 3.3. *From right to left.* Suppose that $\mathbf{M} =$

$(\mathbf{F}, V), x \models \phi \bullet \neg(\phi^{\smile} \bullet \psi)$, where $\mathbf{F}$ is an arrow frame satisfying the given condition. Assume also $\mathbf{M}, x \models \psi$: we derive a contradiction. By the truth definition, there must be y, z with Cx, yz such that $\mathbf{M}, y \models \phi$, $\mathbf{M}, z \models \neg(\phi^{\smile} \bullet \psi)$. Therefore, $Cz, r(y)x$ (by the condition) and also $\mathbf{M}, r(y) \models \phi^{\smile}$. But then, $\mathbf{M}, z \models \phi^{\smile} \bullet \psi$: quod non.

From left to right. Let frame $\mathbf{F}$ validate the arrow axiom $\phi \bullet \neg(\phi^{\smile} \bullet \psi) \to \neg\psi$. Consider any arrows x, y, z such that Cx, yz. Define a valuation by setting $V(\phi) := \{y\}$, $V(\psi) := \{u \mid \text{not } Cz, r(y)u\}$. Then, $\mathbf{M} = (\mathbf{F}, V), x \models \phi \bullet \neg(\phi^{\smile} \bullet \psi)$, because $\mathbf{M}, y \models \phi$ while $\mathbf{M}, z \models \neg(\phi^{\smile} \bullet \psi)$. (To see the latter, if z verified the formula $\phi^{\smile} \bullet \psi$, we would have Cz, uv for some u verifying $\phi^{\smile}$ (the latter must be $r(y)$, given our assumptions about the function r) and some v verifying ψ. But $Cz, r(y)v$ contradicts the above definition of $V(\psi)$.) We may conclude that $\mathbf{M}, x \models \neg\psi$, which is just the required fact $Cz, r(y)x$, by the definition of V. $\dashv$

Together (2), (4), (5) imply the further interchange law $\forall xyz\, Cx, yz \to Cy, xr(z)$. Moreover, there is actually a more elegant form of axiom (5) without negation:

$$\phi \wedge (\psi \bullet \chi) \to \psi \bullet (\chi \wedge (\psi^{\smile} \bullet \phi)).$$

Finally, the propositional constant Id may constrain 'identity loops'.

Example 8.4 (Identity Arrows)

(6) $\mathrm{Id} \to \mathrm{Id}^{\smile}$ iff $\forall x\, Ix \to Ir(x)$

(7) $\mathrm{Id} \bullet \phi \to \phi$ iff $\forall xyz\, (Iy \wedge Cx, yz) \to x = z$.

This completes our correspondence analysis of the basic BRA axioms (Section 3.3). Obviously, there are many further choices here. 'Arrow Logic' really stands for a family of modal logics, whose selection may depend on intended applications. (Indeed, an emphasis on 'master calculi', rather than general ideas seems an outdated fashion.) Nevertheless, what might be the most natural 'computational core' in this field? Our recommendation would be as follows. Correspondence analysis reveals a natural border line in what is expressed by principles like the above. Some of them are purely universal, making no demands on the supply of arrows, whereas others are existential. The former, rather than the latter, seem to form the true logical core of any field:

Universal frame constraints take only those principles concerning composition, converse and identity on arrow frames which lack existential import: i.e., their corresponding constraints can be formulated by purely *universal* first-order sentences over arrow frames.

The complete logic of this class of arrow frames may also be described differently. It is the set of validities for the above concrete dialect of Arrow Logic, whose arrows are ordered pairs — and the only change from standard models is their limited supply. One potential problem for this proposal is Associativity for composition. Its frame condition requires the existence of representants for different groupings of transitions:

(8a) $(\phi \bullet \psi) \bullet \chi \to \phi \bullet (\psi \bullet \chi)$ iff

$$\forall xyzuv\,(Cx, yz \wedge Cy, uv) \to \exists w\,(Cx, uw \wedge Cw, vz))$$

(8b) and likewise in the opposite direction.

Associativity is often presupposed in dynamic semantics: different orders of sequential computation are taken as equivalent. In contrast, we shall distinguish different ways of 'chunking' parts. (Cf. the non-associative categorial calculi of Chapter 12.)

8.2.3 Arrow Logic over Pair Models

There exists a complete axiomatization for the logic of pair arrow models consisting of families of ordered pairs over some base set (not necessarily the full Cartesian product), with the obvious definitions for composition, reversal and identity. As argued above, this logic is entirely without existential assumptions.

Theorem 8.5 *The following axioms are complete for pair arrow models:*

(1) the above minimal arrow logic

(2) converse is an idempotent function

(3) some evident identity principles

$$\begin{array}{lll} \mathrm{Id} \leftrightarrow \mathrm{Id} \bullet \mathrm{Id} & \mathrm{Id} \leftrightarrow \mathrm{Id}^{\smile} & \phi \wedge \mathrm{Id} \to \phi^{\smile} \\ \mathrm{Id} \bullet \phi \leftrightarrow \phi & \phi \bullet \mathrm{Id} \leftrightarrow \phi & \phi \bullet \neg\phi^{\smile} \to \neg\mathrm{Id}. \end{array}$$

(4) triangle principles

$$\phi \bullet \neg(\phi^{\smile} \bullet \psi) \to \neg\psi \qquad \neg(\phi \bullet \psi^{\smile}) \bullet \psi \to \neg\phi$$

(5) limited associativity

$$\begin{array}{rcl} ((\phi \wedge \mathrm{Id}) \bullet \psi) \bullet \chi & \leftrightarrow & (\phi \wedge \mathrm{Id}) \bullet (\psi \bullet \chi) \\ ((\phi \bullet (\psi \wedge \mathrm{Id})) \bullet \chi & \leftrightarrow & \phi \bullet ((\psi \wedge \mathrm{Id}) \bullet \chi) \\ (\phi \bullet \psi) \bullet (\chi \wedge \mathrm{Id}) & \leftrightarrow & \phi \bullet (\psi \bullet (\chi \wedge \mathrm{Id})) \end{array}$$

Proof. (Cf. Marx 1995.) This is a representation argument over modal Henkin models, replacing arrows stepwise by ordered pairs over a growing base domain of objects. ⊣

Known properties of this system include its decidability (which may be established via modal filtration) and Craig interpolation. Another research line adds expressive power to this system. For instance, the universal pair arrow logic retains all its nice properties when we extend its vocabulary with a universal modality, or with 'graded modalities' counting finite numbers of worlds verifying some formula. These moves exemplify a more general research program of 'trading deductive strength for expressive power'.

Arrow Logic is an active mathematical research program. Further information is found in van Benthem 1991b, Németi 1991, Marx 1995, Mikulas 1995, Vakarelov 1996, Venema 1991, 1994. We mention some salient points that link up with earlier Chapters.

(1) The above frame correspondences can all be computed from the *Sahlqvist Theorem* in Modal Logic (Chapter 3).

(2) Arrow logics are counterparts of algebraic theories arising by *relativizing* classes of relational algebras: Marx 1995 has a joint presentation. Through this connection, much is known about arrow calculi in the above landscape. Most properties of pair arrow logic, including its decidability, can be derived from existing algebraic results.

(3) Andréka, Kurucz, Németi, Sain and Simon 1994 show that in general, adding a pair-frame requirement of transitivity, validating an axiom of associativity in arrow logics leads to *undecidability*. The reason is, one can then faithfully encode undecidable word problems into the logic. Andréka 1991 proves non-finite-axiomatizability in the presence of Associativity.

(4) The pair-based 'Budapest dialect' of Arrow Logic is hardly different in its theory from the arrow-based 'Amsterdam Dialect' (Marx, Mikulas, Németi and Sain 1996).

(5) Finally, the arrow language so far has limited expressive power. It cannot even force composition to be a partial function (thus, arrow logic allows many ways of composing two transitions — as happens in multigraphs). To enforce the latter property, *enriched modal formalisms* are needed, e.g., with the 'difference operator' of Chapters 3, 7, a 'universal modality' (Marx 1995) or 'graded modalities' (Mikulas 1995). This raises issues of *transfer* of earlier results to arrow logics in richer vocabularies. (Marx, Mikulas, Németi and Simon 1996 show how the difference operator considerably improves things, by reinstating a deduction theorem.)

By the time one achieves full representability of arrow frames as full sets of ordered pairs, their modal logic will be just as complex as the standard theory of representable relational algebras (Section 3.3). The art is to know when to stop.

8.3 Dynamic Arrow Logic

One important enrichment of vocabulary is needed for programming purposes. Dynamic Arrow Logic adds one infinitary operator to the basic arrow language:

$$\mathbf{M}, x \models \phi^* \quad \text{iff} \quad x \text{ can be } C\text{-decomposed into some finite sequence of arrows satisfying } \phi \text{ in } \mathbf{M}.$$

This says there is a finite sequence of ϕ-arrows in $\mathbf{M}$ which allows successive composition via intermediate arrows so as to arrive at x. (In general, this does not say that x could be obtained by just every route of combinations from these same arrows.) Thus, ϕ^* describes the transitive closure of ϕ. It satisfies the following simple laws:

(9) axiom $\phi \to \phi^*$
(10) axiom $\phi^* \bullet \phi^* \to \phi^*$
(11) rule if $\vdash \phi \to \alpha$ and $\vdash \alpha \bullet \alpha \to \alpha$, then $\vdash \phi^* \to \alpha$.

These principles may be added to the earlier minimal arrow logic. Our preferred choice consists of this minimal basis plus the earlier principles (1)–(5), to obtain a natural axiomatic Dynamic Arrow Logic DAL. Here is an illustration of how this works.

Example 8.6 (Derivation of Monotonicity for Iteration) If $\vdash \alpha \to \beta$ then $\vdash \alpha \to \beta^*$ (axiom (9)). Also $\vdash \beta^* \bullet \beta^* \to \beta^*$ (axiom (10)) whence $\vdash \alpha^* \to \beta^*$ (rule (11)).

Example 8.7 (Derivation of Interchange for Iteration and Converse)

i	$\phi \to \phi^*$	axiom (9)
ii	$\phi^{\smile} \to \phi^{*\smile}$	i plus monotonicity for converse
iii	$(\phi^{*\smile} \bullet \phi^{*\smile}) \to (\phi^* \bullet \phi^*)^{\smile}$	axioms (3) and (4)
iv	$\phi^* \bullet \phi^* \to \phi^*$	axiom (10)
v	$(\phi^* \bullet \phi^*)^{\smile} \to \phi^{*\smile}$	iv plus monotonicity for converse
vi	$(\phi^{*\smile} \bullet \phi^{*\smile}) \to \phi^{*\smile}$	iii, v
vii	$\phi^{\smile *} \to \phi^{*\smile}$	ii, vi plus rule (11)
viii	$\phi^{\smile\smile *} \to \phi^{\smile *\smile}$	by similar reasoning
ix	$\phi \to \phi^{\smile\smile}$	axiom (3)
x	$\phi^* \to \phi^{\smile\smile *}$	monotonicity for iteration
xi	$\phi^* \to \phi^{\smile *\smile}$	x, viii
xii	$\phi^{*\smile} \to \phi^{\smile *\smile\smile}$	xi plus monotonicity for converse
xiii	$\phi^{\smile *\smile\smile} \to \phi^{\smile *}$	axiom (2)
xiv	$\phi^{*\smile} \to \phi^{\smile *}$	xii, xiii

Completeness may be established for DAL, as well as for several of its variants.

Theorem 8.8 *DAL is complete for its intended interpretation.*

Proof. We present one modal-style completeness argument here, because it shows how infinitary notions may be accommodated in our dynamic logics with relative ease. Take some finite set of relevant formulas which is closed under subformulas and which satisfies the following closure condition:

$$\text{if } \phi^* \text{ is included, then so is } \phi^* \bullet \phi^*.$$

Now consider the usual model of all maximally consistent sets in this restricted universe, setting (for all 'relevant' formulas):

$$\begin{array}{rl} Cx, yz & \text{iff} \quad \forall \phi \in y \forall \psi \in z \ \phi \bullet \psi \in x \\ Rx, y & \text{iff} \quad \forall \phi \in y \ \phi^{\smile} \in x. \end{array}$$

Here we can prove the usual 'decompositions' for maximally consistent sets, such as

$$\phi \bullet \psi \in x \text{ iff there exist } y,\, z \text{ with } Cx, yz \text{ and } \phi \in y,\ \psi \in z$$

using the minimal distribution axioms only. The key new case here is the following observation:

Claim. $\phi^* \in x$ *iff some finite sequence of maximally consistent sets each containing* ϕ *'C-composes' to* x *in the earlier sense.*

Proof. 'If'. By induction on the length of the decomposition, using axioms (9), (10) and the closure condition on relevant formulas. 'Only if'. Describe the finite set of all 'finitely C-decomposable' maximally consistent sets as usual by one formula α , viz. the disjunction of all their conjoined 'complete descriptions' δ. Then we have

$$\vdash \phi \rightarrow \alpha,$$

since ϕ is provably equivalent to $\bigvee_{\phi \in \delta} \delta$ in propositional logic, and α contains all these δ by definition. Next, we have

$$\vdash \alpha \bullet \alpha \rightarrow \alpha.$$

To see this, suppose that $(\alpha \bullet \alpha) \wedge \neg \alpha$ were consistent. By Distributivity with respect to successive relevant formulas, $(\delta_1 \bullet \delta_2) \wedge \neg \alpha$ must be consistent for some maximally consistent δ_1, δ_2. Likewise, $(\delta_1 \bullet \delta_2) \wedge \neg \alpha \wedge \delta_3$ is consistent for some maximally consistent δ_3. Now, δ_1, δ_2 must be in α, and also $C\delta_3, \delta_1\delta_2$ by the definition of C and standard properties of deduction. Hence, δ_3 is in α too (by definition), contradicting the consistency of $\neg \alpha \wedge \delta_3$. So, applying the iteration rule (11), we have

$$\vdash \phi^* \rightarrow \alpha.$$

Therefore, if $\phi^* \in x$, then x belongs to α. $\dashv$

Semantic evaluation in the canonical model will now proceed in harmony with the above syntactic decomposition: any relevant formula is true 'at' a maximally consistent set iff it belongs to that set. This completes our analysis of the basic case. Next, in order to deal with the additional axioms (1)–(5), their frame properties must be enforced in our finite canonical model. This may be done as follows:

i one closes the universe of relevant formulas under Boolean operations and converses: the resulting infinite set of formulas will remain logically finite, given the Boolean laws and the above interchange principles for converse,

ii the definition of the relation C is to be modified by adding suitable clauses, so as to 'build in' the required additional frame properties.

First, the required behaviour of reversal is easy to obtain. One may define $r(x)$ to be the maximally consistent set consisting of (all representatives of) $\{\phi^\smile \mid \phi \in x\}$: the available axioms make this an idempotent function inside the universe of relevant maximally consistent sets. For a more difficult case, consider axiom (5) with corresponding frame condition

$$\forall xyz\, Cx, yz \to Cz, r(y)x.$$

One redefines:

$$\begin{array}{lll} Cx, yz & \text{iff} & \text{both } \forall \phi \in y, \psi \in z \colon \phi \bullet \psi \in x \\ & & \text{and } \forall \phi^\smile \in y, \psi \in x \colon \phi \bullet \psi \in z \end{array}$$

This has been designed to validate the given frame condition. Now, we check that the earlier decompositions for maximally consistent sets are still available, to retain the harmony between membership and truth at such sets. Here are the two key cases:

$$\begin{array}{rll} \phi \bullet \psi \in x & \text{iff} & \text{there exist } y,\, z \text{ with } Cx, yz \text{ and } \phi \in y,\, \psi \in z \\ \phi^* \in x & \text{iff} & \text{some finite sequence of maximally consistent} \\ & & \text{sets containing } \phi \text{ `}C\text{-composes' to } x. \end{array}$$

The crucial direction here is from left to right: can we find maximally consistent sets as required with C satisfying the additional condition? What we need is this. In the earlier proof, the sets y, z were constructed 'globally', by showing how successive selection yields a consistent set of formulas x, $\bigwedge y \bullet \bigwedge z$ with $\phi \in y$, $\psi \in z$. (For then, whenever $\alpha \bullet \beta$ is a relevant formula with $\alpha \in y$, $\beta \in z$, $\alpha \bullet \beta$ must belong to x, on pain of inconsistency.) Now, it suffices to show that in this same situation, the set z, $\bigwedge r(y) \bullet \bigwedge x$ is consistent too. Here, we use a rule derived from axioms (2), (5):

$$\text{if} \vdash \phi \to \neg(\psi \bullet \chi) \text{ then} \vdash \chi \to \neg(\psi^\smile \bullet \phi).$$

Then, if $\vdash z \to \neg(\bigwedge r(y) \bullet \bigwedge x)$, then $\vdash x \to \neg(\bigwedge rr(y) \bullet \bigwedge z)$, and hence also $\vdash x \to \neg(\bigwedge y \bullet \bigwedge z)$: contradicting the consistency of x, $\bigwedge y \bullet \bigwedge z$. The argument for iteration is similar. Moreover, the general case with all frame conditions implanted simultaneously employs the same reasoning. $\dashv$

Corollary 8.9 *DAL is decidable.*

Proof. The preceding argument establishes not just axiomatic completeness but also the Finite Model Property within a bound computable from the formula concerned. $\dashv$

The above filtration method accommodating the relevant additional frame properties in the finite counter-model is that of Roorda 1991. Marx 1995 has a uniform version, which deals with different dynamic arrow logics at once (including both the original abstract ones and those based on ordered pair models).

8.4 Dynamic Logic with Arrows

8.4.1 Dynamic Logic as Two-Sorted Modal Logic

Now what about a Dynamic Logic in arrow style? The usual account in the literature considers the addition of a propositional component referring to truth at states essential, as it allows us Boolean negation at least at the latter level. Since this is no longer true now, a state component becomes less urgent. Nevertheless, we do think the resulting two-level system is a natural one: 'arrow talk' and 'state talk' belong together in an analysis of computation and general action. So as usual, add a Boolean propositional language, plus two mechanisms of interaction between the two resulting components:

a *test* 'mode' ? taking statements to programs
a *domain* 'projection' $\langle\ \rangle$ taking programs to statements.

For notational convenience, we shall reserve ϕ, ψ, ... henceforth for state assertions and π, π_1, π_2, ... for describing programs in this two-tier system. In line with the above general modal analysis, let us view this system with some greater abstraction. What we have is a *two-sorted modal logic*, whose models have both 'states' and 'arrows', and whose formulas are marked for intended interpretation at one of these. Both the arrow and state domains may carry internal structure, reflected in certain modalities, such as the earlier $\bullet$ and $\breve{}$ referring to arrows. (States might be ordered by 'precedence' or 'preference' with appropriate modalities.) Our key point, however, is this. Even the modes and projections themselves may now be viewed as 'non-homogeneous' modalities, reflecting certain structure correlating the two kinds of object in our models. For instance, 'test' is again a *distributive* modality, and so is 'domain':

$$(\phi \vee \psi)? \quad \leftrightarrow \quad \phi? \vee \psi?$$

$$\langle \pi_1 \vee \pi_2 \rangle \quad \leftrightarrow \quad \langle \pi_1 \rangle \vee \langle \pi_2 \rangle$$

whose interpretations run as follows:

$$\mathbf{M}, x \models \phi? \quad \text{iff} \quad \text{there exists some } s \text{ with } Tx, s \text{ and } \mathbf{M}, s \models \phi$$

$$\mathbf{M}, x \models \langle \pi \rangle \quad \text{iff} \quad \text{there exists some } x \text{ with } Ds, x \text{ and } \mathbf{M}, x \models \pi.$$

Intuitively, the first relation Tx, s says that x is an identity arrow for the point s, while the second relation Ds, x says that s is a left end-point of the arrow x. On top of this, via the usual correspondences, further axioms on ?, $\langle\ \rangle$ will then impose additional connections between T and D.

Example 8.10 (Connecting Identity Arrows and End-Points) The principle $\langle\phi?\rangle \leftrightarrow \phi$ (a modal 'Sahlqvist form') expresses the conjunction of

$$\forall s \exists x\, Ds, x \wedge Tx, s \qquad \text{and} \qquad \forall sx\, Ds, x \rightarrow \forall s'\, Tx, s' \rightarrow s = s'.$$

Axiomatic completeness proofs are straightforward here, with two kinds of maximally consistent sets: for arrows and for points. Thus everything about Dynamic Logic becomes Modal Logic: not just its two separate components, but also their connections. This provides the proper abstract setting for the two-level architecture of Chapter 6.

8.4.2 Functional Reformulation

Further elegance may be achieved. In Chapter 7, the diagonal map $\lambda R.\, \lambda x.\, Rxx$ turned out to be the only logical projection which is a *Boolean homomorphism*, while the only two logical homomorphic modes are

$$\lambda P.\, \lambda xy.\, Px \text{ and } \lambda P.\, \lambda xy.\, Py.$$

Thus, we can introduce three matching modalities with corresponding new semantic relations:

$$\begin{array}{lll} \mathbf{M}, s \models D\pi & \text{iff} & \text{for some } x,\ \Delta s, x \text{ and } \mathbf{M}, x \models \pi \\ \mathbf{M}, x \models L\phi & \text{iff} & \text{for some } s,\ Ls, x \text{ and } \mathbf{M}, s \models \phi \\ \mathbf{M}, x \models R\phi & \text{iff} & \text{for some } s,\ Rs, x \text{ and } \mathbf{M}, s \models \phi. \end{array}$$

These modalities satisfy not just the Distribution axioms, but they also commute with Boolean negation (just like relational converse), so that we can take Δ, L, R to be functions. This set-up is more elegant, as well as easy to use. (It may be simplified even further by dropping $R\phi$ in favour of $(L\phi)^{\smile}$.) For instance, one source of axiomatic principles is the interaction of various operators over our dynamic frames:

Proposition 8.11 (Correspondence for Axioms)

$DL\phi \leftrightarrow \phi$	*expresses that*	$\forall s\, L\Delta(s) = s$
$LD\pi \leftrightarrow (\pi \wedge \mathrm{Id}) \bullet \top$	*expresses that*	$\forall x \exists y\, Cx, \Delta L(x) y$
$L\phi \bullet \pi \rightarrow L\phi$	*expresses that*	$\forall xyz\, Cx, yz \rightarrow L(x) = L(y)$.

One gets the expressive power of the standard system with these new primitives under the following simple

Translation from old to new format

$$\begin{array}{lll} \phi? & \mapsto & L\phi \wedge \mathrm{Id} \\ \langle\pi\rangle & \mapsto & D(\pi \bullet \top) \end{array}$$

Analyzing the usual axioms of Propositional Dynamic Logic in this fashion is a straightforward exercise. We list the key principles that turn out to be needed (these allow us to represent statements $\langle\pi\rangle\phi$ faithfully as $D((\pi \wedge R\phi) \bullet \top)$:

1 $D\pi \to D(\pi \wedge \mathrm{Id})$
2 $\mathrm{Id} \to (L\phi \leftrightarrow R\phi)$
3a $DL\phi \leftrightarrow \phi$
3b $DR\phi \leftrightarrow \phi$
4 $\pi_1 \wedge RD\pi_2 \leftrightarrow \pi_1 \bullet (\pi_2 \wedge \mathrm{Id})$
5 $(\pi_1 \bullet \pi_2) \wedge R\phi \leftrightarrow \pi_1 \bullet (\pi_2 \wedge R\phi)$

Their corresponding frame conditions can be computed by hand, or again with a Sahlqvist algorithm, as they are all of the appropriate modal form. These principles suffice for deriving various other useful ones, such as the following reductions

$$(\mathrm{Id} \wedge L\phi) \bullet \pi \quad \leftrightarrow \quad L\phi \wedge \pi$$
$$\pi \bullet (\mathrm{Id} \wedge R\phi) \quad \leftrightarrow \quad \pi \wedge R\phi.$$

Finally, there is also a converse route, via two more translation schemata:

Translation from new to old format

$$L\phi \quad \mapsto \quad \phi? \bullet \top$$
$$D\pi \quad \mapsto \quad \langle \mathrm{Id} \wedge \pi \rangle.$$

The same style of analysis applies to richer systems of dynamic logic, with additional structure in their state domains. An obvious example is the dynamic modal logic DML of Chapters 2, 6, which featured an inclusion order $\subseteq$ between information states. This may be treated by introducing another propositional constant at the arrow level, say, E for 'inclusion' (perhaps with suitable axioms expressing its transitivity and reflexivity). Then, the logic of updating and revision will employ special defined arrows, such as the following update moves

$E \wedge R\phi$	update transition for ϕ
$(E \wedge R\phi) \wedge \neg((E \wedge R\phi) \bullet (E \wedge \neg \mathrm{Id}))$	minimal update transition for ϕ.

This yields a workable alternative circumventing the undecidability of the full system, proved in Chapter 6. Roughly speaking, arrow versions of dynamic logics should stay on the right side of the '2D-boundary' which allows embedding of two-dimensional grids in the models, and hence encoding of full Turing machine computation.

8.5 Notes

Semantic Motivation. Undecidability and expressive complexity abound in the semantics of mathematical, natural and programming languages. E.g., in Hoare Logic, infinitary control structures create complexity. Is this inevitable, or can the system be redesigned? Likewise, in knowledge representation, higher-order data structures may generate complexity, as in 'branching temporal logic', which may be avoided by re-analysis in *many-sorted first-order* theories. Thus, there is a general issue of separating intrinsic content from formal wrapping. If we can isolate the former component, many technical results in the literature might split into some relevant computational content plus a recurring mathematical overhead. The broader aim of this chapter is to advocate some general awareness of this phenomenon. See van Benthem 1996a for a principled defense of general remodelling techniques for this purpose: including both algebraic techniques and Henkin-style 'general models'. These are not as ad-hoc as is often thought — and correspond to doing one's 'geometrical homework'.

Arrow Logic and Geometry. Indeed, dynamic arrow logic is much like a geometrical theory with one sort of 'points' and a second sort of 'lines' or 'paths'. What we have been doing may also be viewed as a kind of modal geometry, which is of independent interest.

Arrow Dialects. The Amsterdam version of Arrow Logic emphasizes abstract modal frames. The Budapest version works mainly with families of arrows as set-theoretic ordered pairs. There is also a Sofia version, which endows arrows with even further structure, so that they become more like geometrical paths. For some concrete examples of languages, models and axioms for the latter line of thought, see Arsov and Marx 1994, de Rijke 1993b, Kühler 1994. General references for the whole enterprise have been given in the main text of this chapter.

Technical Development. The technical development of Arrow Logic was briefly surveyed in this chapter. As in Modal Logic, there is not one optimal system, but a landscape of calculi whose general features are the eventual subject of investigation. For state of the art metamathematics of this research program, cf. the dissertations Marx 1995, Mikulas 1995, the textbook Andréka, Németi and Sain 1996, the monograph Venema and Marx 1996a, as well as the conference volume Marx et al. 1996.

Connections with Algebraic Logic. Arrow Logic derives from earlier work on *relativization* in algebraic logic (Chapter 3). This powerful technique comes originally from the Berkeley School in algebraic logic: cf. Henkin et al. 1985, Jónsson 1987, Maddux 1990, Thompson 1990, Monk 1993, as well as Resek and Thompson 1991, Givant 1994, Andréka and Thompson 1988, Henkin, Monk, Tarski, Andréka and Németi 1981. All these references are equally valid for the generalized semantics for predicate logic to follow in Chapter 9.

Game Techniques. A recent illuminating perspective on completeness proofs and representation employs model construction by games (cf. Hodges 1993 for this modern theme). This is work of a London group in pure and applied algebraic logic, including D. Gabbay, R. Hirsch, I. Hodkinson and M. Reynolds. Cf. Hirsch and Hodkinson 1995.

Other Decidable Action Logics. To some extent, the aim of this chapter is also realized by other theories of programs and actions (cf. Kozen 1994, Pratt 1994a). But their conventional wisdom may be biased: e.g., in their insistence that Boolean negation is the main cause of complexity. (The latter is no longer true in Arrow Logic.)

Fixed-Point Operators. Arrow analysis extends to a more powerful system with a fixed-point operator $\mu p.\, \phi(p)$. This is the μ-calculus mentioned in Chapters 4, 5. Its two key derivation rules are as follows:

I if $\vdash \phi(\alpha) \rightarrow \alpha$ then $\vdash \mu p.\, \phi(p) \rightarrow \alpha$

II if $\vdash \beta \rightarrow \mu p.\, \phi(p)$ then $\vdash \phi(\beta) \rightarrow \mu p.\, \phi(p)$.

This language defines our iterations ϕ^* via the fixed-point formula $\mu p.\, \phi \vee p \bullet p$, whose successive approximations give us all C-combinations that were involved in the earlier semantic definition. The derivation rules for iteration then become derivable from the above two rules: I corresponds to rule (11), while II gives the effect of the axioms (9), (10). In semantic reasoning, one needs the following decomposition:

$$\mu p.\, \phi(p) \in x \quad \text{iff} \quad x \text{ belongs to some finite iteration of the operator } \lambda p.\, \phi(p) \text{ starting from the empty set for } p.$$

From Decidable to Feasible. The next step in the remodeling program for complexity must analyze standard *connectives* in addition to quantificational operators (such as relational composition). Decidable logics may be quite complex because of recurrent branchings in their search space. A major cause for this is classical *disjunction*, whose semantics should therefore be reconsidered. Instead of enforcing one choice, one might say that disjunctions refer to 'mixtures' of options (like linear combinations in quantum logic). One abstract schema for this would be

$$\mathbf{M}, x \models A + B \quad \text{iff} \quad \text{there exist } y, z \text{ such that } Rx, yz \text{ and } \mathbf{M}, y \models A, \mathbf{M}, z \models B$$

Eventually, one might combine, not just single representatives for A, B, but their complete set denotations. Formally, of course, this account of disjunction is the same as that for arrow composition — it becomes an existential binary modality, to which our previous methods apply. For instance, in combination with a standard *conjunction* $\wedge$, we get universal validity for *monotonicity principles*:

$$(A \wedge B) + C \quad \Rightarrow \quad A + C$$
$$A + (B \wedge C) \quad \Rightarrow \quad A + B$$

Other traditional principles for disjunction may now be analyzed via *frame correspondences*, seeing how far they take us on the road toward traditional 'case disjunction'. For instance

$A + A \Rightarrow A$	expresses	$\forall xyz\, Rx, yz \rightarrow (x = y \vee x = z)$
$A \Rightarrow A + A$	expresses	$\forall x\, Rx, xx$
$A + B \Rightarrow B + A$	expresses	$\forall xyz\, Rx, yz \rightarrow Rx, zy$

and, e.g., the earlier ubiquitous Distribution Laws for disjunction now say:

$$A \wedge (B + C) \Rightarrow (A \wedge B) + (A \wedge C) \quad \forall xyz\, Rx, yz \rightarrow (x = y \wedge x = z)$$
$$(A \wedge B) + (A \wedge C) \Rightarrow A \wedge (B + C) \quad \forall xyz\, Rx, yz \rightarrow (x = y \vee x = z)$$

These virtually trivialize the 'mixing relation' R. All these correspondences are easy to compute, as they involve only *Sahlqvist forms.*

Without standard disjunction, the resulting systems are more like the *non-Boolean categorial logics* of Chapter 12 (cf. also the non-Boolean modal logics of Chapter 6). Their complexity does go down, from the usual polynomial-space-completeness (or worse) for modal logics, to the polynomial-time level of the 'non-associative Lambek Calculus'.

With standard conjunction and standard *negation*, our logics will still encode choice disjunction via the De Morgan laws. Therefore, one must also generalize the semantics of negation. Here, the natural pattern is that of *relevant* or *intuitionistic* logic. (Indeed, the present analysis seems close in spirit to relevant semantics, even though the latter did not seem driven by complexity concerns.) For some relevant relation N marking a 'refutation range', we set

$$\mathbf{M}, x \models \neg A \quad \text{iff} \quad \text{for all } y \text{ with } Nxy, \text{ not } \mathbf{M}, y \models A.$$

Again, one can then analyze frame correspondence for De Morgan laws and the like. Also relevant here may be the different varieties of negation investigated by Dunn 1996. This sketch is a first promissory note, which may show at least how the technical apparatus of this book seems useful also in making the next steps in our 'minimalist program'.

8.6 Questions

Metamathematics of Arrow Logic. There are many open technical questions concerning the landscape of arrow logics. In particular, there is a scarcity of general results. For instance, one wants a *classification* of all decidable arrow logics having good meta-properties (such as finite axiomatizability, Łoś-Tarski and interpolation). Another general question concerns the range of *modal techniques*. Marx 1995 has a general filtration method for proving decidability: can one formulate such results in abstract terms which will obviate the need for case-by-case analysis of proposed logics? Finally, arrow logics seem like *categorial logics* in many ways (van

Benthem 1991b). Some translations exist for special cases (cf. Kurtonina 1995). Are there systematic *translations* which couple the two landscapes?

Adding Vocabulary. What happens to Arrow Logic when we add modalities, along the lines sketched in Chapters 4, 5? Mikulás 1995, as well as other references in the main text, provide interesting case studies. Again, are there general *transfer results*? In particular, our own case study of dynamic arrow logic suggests that adding iteration should always be a harmless exercise, which leaves decidable arrow logics decidable. Is this in fact the case?

Complexity. Exactly where is the boundary between decidable and undecidable arrow logics? What is the connection between the '2D analysis' for undecidability (Spaan 1993, de Rijke 1992) encoding tilings or Turing computations, and the analysis via word problems in Andréka, Kurucz, Németi, Sain and Simon 1994? (For general techniques, cf. van Emde Boas 1990, Johnson 1990.) Next, what is the complexity of the above arrow logics that are known to be decidable? More generally, following up on the hints in our final note, can we also have general remodelling strategies that rather take us *from decidable to tractable* logics?

Applying Arrows. One can redo a lot of the preceding chapters in arrow semantics. What will happen to the major results on *invariance* and *safety*? Generally speaking, bisimulations should now connect both states and arrows, and safety seems to become invariance, but at the arrow level. Also, what about further special arrow models corresponding to different kinds of application? For instance, the dynamics of changing preferences (Chapter 2) could also be viewed as a dynamics over arrow models. And the 'modal geometry' of an earlier note points at another possible line of development.

9

Modal Foundations for Predicate Logic

Arrow Logic is just one step along a more radical path. In this chapter, we deconstruct the semantics of first-order logic, the undecidable 'standard tool' of logical analysis. Notably, we design 'lighter' modal versions of this system, by locating implicit choice points in its set up. This yields a whole landscape of decidable systems, each with different computational constraints over universes of states related by variable updates. These constraints express properties of 'dependence' between individual variables. We investigate this landscape by the earlier modal techniques.

9.1 The Modal Core of Tarski Semantics

The well-known standard semantics for predicate logic has the following key clause:

$$\mathbf{M}, \alpha \models \exists x\, \phi \text{ iff for some } d \in |\mathbf{M}|\text{: } \mathbf{M}, \alpha_d^x \models \phi.$$

Tarski's main innovation here was the use of variable assignments, which are essential in decomposing quantified statements, which may contain free variables in their matrix. But much less than this is needed to give a compositional semantics for first-order quantification. The abstract core pattern which would make the latter work is this:

$$\mathbf{M}, \alpha \models \exists x\, \phi \text{ iff for some } \beta\text{: } R_x\alpha\beta \text{ and } \mathbf{M}, \beta \models \phi.$$

Here, 'assignments' α, β become abstract states, and the concrete relation $\alpha =_x \beta$ (which holds between α and α_d^x) has become just any binary update relation R_x. Evidently, this abstract pattern involves standard poly-modal models, of the form

$$\mathbf{M} = (S, \{R_x\}_{x \in \text{VAR}}, I),$$

where S is a set of 'states', R_x a binary relation for each variable x, and I a 'valuation' or 'interpretation function' giving a truth value to each atomic formula Px, Rxy, ... in each state α. Disregarding function symbols, which will be ignored throughout this chapter, a typical atomic clause reads

$$\mathbf{M}, \alpha \models Px \text{ iff } I(\alpha, Px).$$

Boolean clauses are as usual. In particular, interpreted as above, existential quantifiers $\exists x$ are unary existential modalities $\langle x \rangle$. This modal view of predicate logic is close in spirit to the dynamic semantics of Chapter 2: first-order evaluation is an informational process which changes computational states. In this manner, the first-order language becomes a *dynamic logic* with a special choice of atoms and without explicit compound programs. From this modal point of view, conversely, 'standard semantics' arises by insisting on three additional mathematical choices, not enforced by the core semantics:

1. states are identified with variable assignments
2. 'update' must be the specific relation $=_x$, and
3. all assignments in the function space D^{VAR} are available in the course of evaluation.

The former are issues of implementation, the latter is a strong existence assumption. (Actually, standard predicate logic can get by with all 'locally finite' assignments — but even that is a strong existence requirement.) Henceforth, we shall regard these further 'set-theoretic' choices as negotiable. This view lends further support to the abstract modal approach. E.g., it is often felt that the usual set-theoretic tricks making predicates sets of tuples should be orthogonal to the nature of logical validity. Finally, as an alternative to even assumptions 1, 2, the Notes to Chapter 2 contained more than one dynamic semantics for predicate logic manipulating states involving variable stacks whose update relations R_x differ considerably from the standard one.

The universal validities produced by a general modal semantics are well-known. One obtains a *minimal poly-modal logic*, whose principles are

- all classical Boolean propositional laws
- Modal Distribution: $\forall x\, (\phi \to \psi) \to (\forall x\, \phi \to \forall x\, \psi)$
- Modal Necessitation: if $\vdash \phi$, then $\vdash \forall x\, \phi$
- a definition of $\exists x\, \phi$ as $\neg \forall x\, \neg \phi$.

A *completeness theorem* with respect to the above abstract models may be proved via the standard modal Henkin construction with maximally consistent sets for the states in S, and the relations R_x defined via: $\Delta_1 R_x \Delta_2$ iff for all $\phi \in \Delta_2$: $\exists x\, \phi \in \Delta_1$. This logic can be analyzed in a standard modal fashion (cf. Andréka, van Benthem and Németi 1995a), yielding meta-properties such as Craig Interpolation or Łoś-Tarski Preservation. Moreover, it is *decidable* by standard modal techniques (filtration, semantic tableaus). One can now usefully pursue standard first-order model theory in tandem with its modal counterpart. E.g., consider modal *bisimulations* for these models, relating states with the same atomic behaviour, and zigzag conditions for the relations R_x. Specialize these to standard Tarski models. The result is a notion of *partial isomorphism* between models (cf. Chapters 2, 3), related but not equal to the standard one.

The modal perspective reveals a new landscape below standard predicate logic, with a minimal modal logic at the base, ascending up to standard semantics via successive frame constraints. In particular, it contains *decidable* sublogics of predicate logic, sharing its desirable meta-properties. (The minimal modal logic is an example.) Thus, the 'undecidability of predicate logic' largely reflects mathematical accidents of its Tarskian modeling, in particular, encoding set-theoretic facts about function spaces D^{VAR} — rather than the core logic of quantification and variable assignment. We shall explore the resulting view of first-order logic. As with other fine-structure landscapes (e.g., the substructural hierarchy of Chapters 7, 12), we find a family of natural calculi in our original language, but also in richer languages for the more sensitive semantics. In particular, abstract core models support distinctions between various forms of quantification ('monadic' and 'polyadic') that get collapsed in standard predicate logic.

9.2 Dependency Models

In addition to our two choices so far, there are further inhabitants of the landscape between standard logic and its minimal modal core. For instance, one may retain the general implementation of Tarski semantics (the above items 1, 2), while giving up its existence assumption (3). The result is a 'half-way house' where S is some family of assignments (not necessarily the full function space D^{VAR}), and the R_x are the standard relations $=_x$ (cf. Németi 1995). For instance, with two variables $\{x, y\}$, a domain with objects $\{1, 2\}$ supports 2^4 assignment sets. One is the standard model with all four maps from variables to objects. Another has just the two assignments $\{\alpha, \beta\}$ with $\alpha(x) = 1$, $\alpha(y) = 2$ and $\beta(x) = 2$, $\beta(y) = 1$.

Definition 9.1 (Generalized Assignment Models) First-order evaluation is as usual over *generalized assignment models* $(\mathbf{M}, \mathbb{V})$, where $\mathbf{M}$ is a standard model, with a set $\mathbb{V}$ of 'available assignments'. An existential quantifier $\exists x\, \phi$ says some x-variant of the current state *inside* $\mathbb{V}$ satisfies ϕ:

$$(\mathbf{M}, \mathbb{V}), \alpha \models \exists x\, \phi \text{ iff for some } d \in |\mathbf{M}|,\ \alpha_d^x \in \mathbb{V}\text{: } (\mathbf{M}, \mathbb{V}), \alpha_d^x \models \phi.$$

'Assignment gaps' reflect an interesting phenomenon. Intuitively, one often wants to model 'dependencies' between variables: i.e., a situation where changes in value for one variable x may induce, or at least be correlated with, changes in denotation for another variable y. Examples include natural reasoning (Fine 1985b), probabilistic logic (van Lambalgen 1991), but also plural anaphoric discourse (van den Berg 1996). This phenomenon cannot be modeled in standard models, which change values for variables completely independently: Starting from state α, one can move to any α_d^x. But in a model with assignment gaps, the only way to change values for a variable x, starting from an assignment α, may be by incurring a change in y too. Thus, in the above two-assignment model, any shift in value for x also produces one for y. Standard models are then 'degenerate cases' where all dependencies between variables have been suppressed. This shows in the standard validity of the quantifier exchange law $\exists x \exists y\, \phi \leftrightarrow \exists y \exists x\, \phi$, which becomes typically invalid on generalized models. The logic of the latter is called **CRS** (for the algebra behind this acronym, cf. Section 3.3). There are many modelings for dependence phenomena. Thus, Alechina 1995a proposes a semantics where (stated in our framework) the key evaluation clause becomes

$$\mathbf{M}, \alpha \models \exists x\, \phi \text{ iff for some } \beta\text{: } R_{x,\mathbf{y}}\alpha\beta \text{ and } \mathbf{M}, \beta \models \phi,$$

with $\mathbf{y}$ some sequence of 'relevant context variables' — which might consist, e.g., of the free variables in $\exists x\, \phi$. In this case, even Modal Distribution fails. One level up, even assignment sets *themselves* may be made into new states (van den Berg 1996).

9.3 What Do First-Order Axioms Say?

The above three semantic levels have further fine-structure. This may be brought out in two ways. First, one can study natural mathematical constraints on modal frames or generalized assignment models, reflecting various aspects of 'dependence'. But also, one can analyze possible validities expressible in our first-order language. The latter strategy involves modal *frame correspondences* (cf. Chapter 3) Minimal predicate logic retains some basic first-order principles. One half of Modal Distribution expresses the ubiquitous Monotonicity for the existential quantifier: $\forall x\, (\phi \rightarrow \psi) \rightarrow (\exists x\, \phi \rightarrow \exists x\, \psi)$. Let us see, over this base, what is expressed by further principles of predicate logic. Usually, all first-order validities are together in one big bag. In our generalized semantics, however, we can read their different messages in terms of required structure of computational states and accessibility. For a concrete illustration, take the axioms in the well-known textbook Enderton 1972. (Cf. Section 3.1. Other axiomatizations might provide interesting different cuts of the cake.) These consists of all universal closures of arbitrary Boolean propositional laws plus three quantifier axioms

(1) $\forall x\, (\phi \rightarrow \psi) \rightarrow (\forall x\, \phi \rightarrow \forall x\, \psi)$

(2) $\phi \rightarrow \forall x\, \phi$ provided that x do not occur free in ϕ

(3) $\forall x\ \ \phi \rightarrow [t/x]\ \phi$ provided that t be free for x in ϕ.

Moreover, the system has one inference rule, namely Modus Ponens. From the current perspective, the propositional component is base valid (both the axioms and the rule). The first quantifier axiom is base valid, too, using Modal Distribution. Moreover, the use of universal closures for all axioms is equivalent to postulating a Necessitation rule for universal quantifiers. Now, let us analyze the two further quantifier axioms. From our present perspective, the least conspicuous one is already immensely powerful.

The axiom $\phi \to \forall x\, \phi$. Let us analyze this principle inductively, using an equivalent formulation with atoms and their negations, $\wedge$, $\vee$, $\exists$, $\forall$. Our argument will be heuristic, determining the effect of various instances of this principle. The first instance is the atomic pair:

(2.1) $$P\mathbf{y} \to \forall x\, P\mathbf{y} \qquad \neg P\mathbf{y} \to \forall x\, \neg P\mathbf{y}.$$

This expresses a 'local property' of states: truth values for atoms not involving the variable x are not affected by R_x-transitions. When states are assignments, and predicate interpretation works as usual for atoms, this is equivalent to the strong condition that R_x imply $=_x$. In our abstract semantics, however, validity of (2.1) is not naturally translateable into a frame correspondence. It rather suggests a restriction on the range of admissible abstract interpretation functions I for our models. These should satisfy a 'Heredity Principle' stating that

'if $I(\alpha, P\mathbf{y})$, then $I(\beta, P\mathbf{y})$ for all states β with $R_x\alpha\beta$'.

Such restrictions on valuations are also well-known from Kripke semantics for intuitionistic logic. (On the importance of Locality, cf. Section 4.4.3.) Pure frame conditions, however, will emerge eventually with compound cases of axiom (2).

(2.2) Boolean cases $\phi_1 \wedge \phi_2$, $\phi_1 \vee \phi_2$.
There is no new information to be extracted here. Suppose, inductively, that we already know that $\vdash \phi_1 \to \forall x\, \phi_1$ and $\vdash \phi_2 \to \forall x\, \phi_2$. Then, in the base logic, we have automatically (using Distribution) that $\vdash (\phi_1 \wedge \phi_2) \to \forall x\, (\phi_1 \wedge \phi_2)$. The case for disjunction is entirely analogous. The real impact is in the quantified cases:

(2.3) Quantifier cases $\exists y\, \phi$, $\forall y\, \phi$.
Here we must distinguish two subcases.

(2.3.1) the quantified variable y is x itself. Then we have, without any assumptions, that

$$\exists x\, \phi \to \forall x\, \exists x\, \phi \qquad \forall x\, \phi \to \forall x \forall x\, \phi.$$

Here we recognize two modal S5-axioms for $\langle x\rangle$, namely

$$\langle x\rangle\phi \to [x]\langle x\rangle\phi \qquad [x]\phi \to [x][x]\phi.$$

Their frame content is well-known.

Fact.

- *$\forall x\, \phi \rightarrow \forall x \forall x\, \phi$ expresses that R_x is transitive*
- *$\exists x\, \phi \rightarrow \forall x \exists x\, \phi$ expresses that R_x is euclidean.*

Adding the simplest instance of Enderton's quantifier axiom (3), viz. $\forall x\, \phi \rightarrow \phi$, (expressing reflexivity of the R_x), one gets full S5, where all R_x are equivalence relations. (The modal character of first-order deduction is very clear in Henkin et al. 1985.) Henceforth, we shall assume these S5-principles. (Without at least S4, the following analysis becomes messier.) Thus, consider the remaining case where

(2.3.2) the variables x, y are distinct. Inductively, we may assume that $\vdash \phi \rightarrow \forall x\, \phi$, and then we need

$$\vdash \exists y\, \phi \rightarrow \forall x \exists y\, \phi \qquad \vdash \forall y\, \phi \rightarrow \forall x \forall y\, \phi.$$

Modulo S4, these inference rules express two well-known quantifier shifts:

Fact.

- *The rule 'if $\vdash \phi \rightarrow \forall x\, \phi$, then $\vdash \exists y\, \phi \rightarrow \forall x \exists y\, \phi$' is equivalent with the axiom $\exists y \forall x\, \phi \rightarrow \forall x \exists y\, \phi$*
- *The rule 'if $\vdash \phi \rightarrow \forall x\, \phi$, then $\vdash \forall y\, \phi \rightarrow \forall x \forall y\, \phi$' is equivalent with the axiom $\forall y \forall x\, \phi \rightarrow \forall x \forall y\, \phi$, or equivalently $\exists y \exists x\, \phi \rightarrow \exists x \exists y\, \phi$.*

Proof. Here is the first case. 'From axiom to rule'. If $\vdash \phi \rightarrow \forall x\, \phi$, then, by a derived rule of minimal modal logic, $\vdash \exists y\, \phi \rightarrow \exists y \forall x\, \phi$ — and hence, by our axiom, $\vdash \exists y\, \phi \rightarrow \forall x \exists y \phi$. 'From rule to axiom'. (This part uses S4.) Start from the axiom $\vdash \forall x\, \phi \rightarrow \forall x \forall x\, \phi$, and apply the rule. This gives $\vdash \exists y \forall x\, \phi \rightarrow \forall x \exists y \forall x \phi$. Once more in S4, we have $\vdash \forall x \phi \rightarrow \phi$. Using two derived rules in the minimal modal logic, this gives $\vdash \forall x \exists y \forall x \phi \rightarrow \forall x \exists y\, \phi$. Hence, together: $\vdash \exists y \forall x\, \phi \rightarrow \forall x \exists y \phi$. $\dashv$

Both quantifier shifts are Sahlqvist Forms whose frame contents are easily computed:

Fact.

- *$\exists y \exists x\, \phi \rightarrow \exists x \exists y\, \phi$ expresses Path Reversal:*

$$\forall \alpha \beta \gamma\, ((R_x \alpha \beta \wedge R_y \beta \gamma) \rightarrow \exists \delta\, (R_y \alpha \delta \wedge R_x \delta \gamma)).$$

- *$\exists y \forall x \phi \rightarrow \forall x \exists y\, \phi$ expresses Confluence:*

$$\forall \alpha \beta \gamma\, ((R_y \alpha \beta \wedge R_x \alpha \gamma) \rightarrow \exists \delta\, (R_x \beta \delta \wedge R_y \gamma \delta)).$$

In S5-models, Path Reversal and Confluence are semantically equivalent. And indeed, Henkin et al. 1985 has an algebraic proof of this fact. Our general conclusion so far is that first-order predicate-logical axioms express modal Sahlqvist forms, to which standard modal correspondence and completeness techniques may be applied. (Further illustrations may be found in Venema 1991, de Rijke 1993a, Marx 1995.)

One may also use general facts about predicate logic to suggest natural constraints on dependency models. E.g., the Finiteness Lemma says that evaluation of formulas only depends on values for their free variables. This is no longer true in generalized assignment models, where free variables may carry implicit dependencies. But one can study Finiteness as an interesting condition per se. Such conditions may be on models rather than frames. Westerståhl 1995 redoes the above correspondence analysis on modal frames with heredity restrictions on admissible valuations.

It remains to analyze Enderton's final quantifier axiom, stating that $\forall \mathbf{x}\, \phi \rightarrow [\mathbf{t}/\mathbf{x}]\phi$ *provided that* **t** *be free for* **x** *in* ϕ. Here we shall treat the substitution operator $[t/x]$ as a semantic instruction in its own right. It will denote 'controlled value assignment' $x := t$, being the natural semantic companion to the 'random assignment' for $\exists x$ which inspired the dynamic predicate logic of Chapter 2.

9.4 Quantifiers and Substitutions

There is a folklore idea in dynamic logic that syntactic substitutions $[t/x]$ work semantically as program instructions $x := t$. Goldblatt 1987 uses the latter notation to avoid syntactic complexities in Harel's quantified dynamic logic. Another instance of this duality shows up with the well-known Substitution Lemma for predicate logic:

$$\mathbf{M}, \alpha \models [t/x]\phi \text{ iff } \mathbf{M}, \alpha^{x}_{\text{value}(\mathbf{M},\alpha,t)} \models \phi.$$

This expresses a procedural equivalence between 'call by name' and 'call by value'. Finally, random assignment $[\![\exists x]\!]$ naturally invites its specific counterpart $[\![t/x]\!]$. The proper treatment of substitutions is like the earlier modal semantics for quantifiers.

Abstract Assignment Frames. We expand the previous models by adding abstract relations $A_{x,y}$, whose concrete interpretation in standard Tarskian models is as follows:

$$\begin{aligned} \alpha A_{x,y} \beta \quad \text{iff} \quad & \beta(x) = \alpha(y) \text{ and} \\ & \beta(z) = \alpha(z) \text{ for all } z \text{ distinct from } x. \end{aligned}$$

Henceforth, for convenience, we only consider substitutions of variables for variables. The truth definition then treats the substitution operator $[y/x]$ literally as a modality:

$$\mathbf{M}, \alpha \models [y/x]\phi \text{ iff for all } \beta \text{ with } \alpha A_{x,y}\beta\text{: } \mathbf{M}, \beta \models \phi.$$

The result is like that for quantifiers. There will be a universally valid minimal logic, on top of which further principles express special constraints on the relations via frame correspondences. (All principles involved even turn out to have Sahlqvist forms.) Thus, even the usual syntactic 'definition' of substitution acquires semantic import.

Semantic Analysis of Substitution.

Atomic Cases $[y/x]Px \leftrightarrow Py$
$[y/x]Pz \leftrightarrow Pz$ (z distinct from x)

As before, on abstract frames, these principles can be viewed as heredity constraints on admissible valuations. On concrete assignment frames, they express that the relations $A_{x,y}$ are to behave as in the above concrete definition.

Boolean Cases $[y/x](\phi \wedge \psi) \leftrightarrow [y/x]\phi \wedge [y/x]\psi$
$[y/x]\neg\phi \leftrightarrow \neg[y/x]\phi$

The first is universally valid in minimal modal logic. The second modal axiom says that the relation $A_{x,y}$ is a *function*. For convenience, we make this assumption henceforth.

Quantified Cases $[y/x]\exists x\,\phi \leftrightarrow \exists x\,\phi$
$[y/x]\forall x\,\phi \leftrightarrow \forall x\,\phi$
$[y/x]\exists z\,\phi \leftrightarrow \exists z\,[y/x]\phi$ (z distinct from x, y)
$[y/x]\forall z\,\phi \leftrightarrow \forall z\,[y/x]\phi$ (z distinct from x, y)

These all express obvious interactions between compositions of the abstract relations $A_{x,y}$ and R_x. The only remaining quantified cases are $[z/x]\exists z\,\phi$ and $[z/x]\forall z\,\phi$. Here we want to allow only those substitutions which are 'valid' in the sense of our intended semantics. Nothing can be allowed in general when there are free occurrences of x in the matrix ϕ. Without such occurrences, we want the two equivalences

$$[z/x]\exists z\,\phi \leftrightarrow \exists z\,\phi \qquad [z/x]\forall z\,\phi \leftrightarrow \forall z\,\phi.$$

More generally, here, we want a principle not unlike the earlier quantifier axiom (2):

$$[z/x]\phi \leftrightarrow \phi \text{ whenever } x \text{ does not occur freely in } \phi.$$

Analysis. *From right to left.* This is taken care of by quantifier axiom (2) plus axiom (3) (see below). Namely, if x does not occur freely in ϕ, then $\forall x\,\phi$ follows, which again implies $[y/x]\phi$ (since y is free for x in ϕ).

From left to right. If x does not occur freely in ϕ, we already have $\vdash \phi \rightarrow \forall x\,\phi$. In the minimal modal logic then, we have $\vdash [z/x]\phi \rightarrow [z/x]\forall x\,\phi$. Now, by an earlier principle, we get $[z/x]\forall x\,\phi \rightarrow \forall x\,\phi$. Then, with one S4-axiom, we have $[z/x]\forall x\,\phi \rightarrow \phi$. This is all we need.

Finally, we return to the initial quantifier axiom (3), which read:

$$\forall x\,\phi \rightarrow [y/x]\phi \text{ provided that } y \text{ be free for } x \text{ in } \phi.$$

Its business now becomes merely to relate the two modalities $[x]$ and $[y/x]$:

$$A_{x,y} \text{ is contained in } R_x.$$

The proviso is taken care of on the fly by the preceding list of principles for substitution.

This completes our semantic analysis of a complete axiom system for predicate logic. It consists of all accumulated principles on abstract assignment frames

$$(S, \{R_x\}_{x \in \text{VAR}}, \{A_{x,y}\}_{x,y \in \text{VAR}})$$

plus some constraints on admissible valuations. Of course, this is just one pass through predicate logic, via Enderton's particular axiomatic presentation. One can also analyze different presentations (natural deduction, algebraic identities). What comes out in general is the idea that the usual 'predicate-logical validities' form a very diverse bunch, which can be layered in many ways, for different purposes.

9.5 Landscape of Deductive Strength

Let us summarize the main line so far. First-order predicate logic is a dynamic logic of variable assignment, whose atomic processes shift values in registers x, y, z, ... This view opens up a hierarchy of fine-structure underneath standard predicate logic — which becomes the (undecidable) theory of one particular mathematical class of 'rich assignment models'. More generally, we discover a broad semantic landscape (as in Modal Logic or Arrow Logic), with a minimal modal system at the bottom, where intermediate systems arise by imposing some, but not all the usual requirements on assignments and their R_x (and $A_{x,y}$) structure:

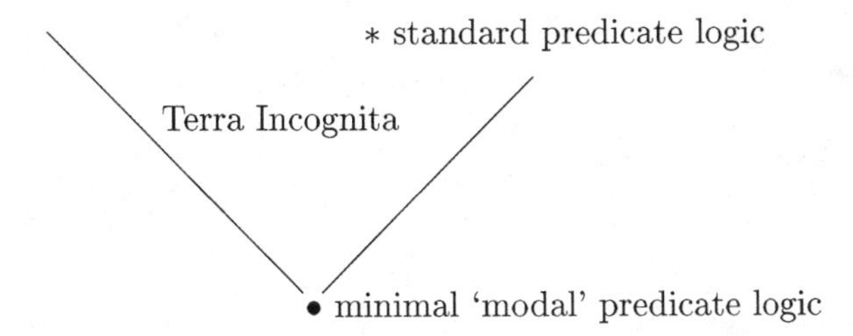

What are natural landmarks in this area? We would like to find logics (1) that are reasonably expressive, (2) that share the important meta-properties of predicate logic (such as Interpolation, Effective Axiomatizability, perhaps even 'Gentzenizability') and (3) that might even *improve* on this, by being decidable. The minimal predicate logic satisfies these three demands — but can we ascend in the above landscape and get more powerful logics with the same behaviour? Fortunately, the area is not unexplored. Research in *Cylindric Algebra* has already identified some very interesting intermediate systems (Henkin et al. 1985, Németi 1985, Venema 1995a, Marx 1995). E.g., Németi 1995 has a 'non-commutative' cylindric algebra, which becomes decidable by giving up the quantifier interchange axioms for $\exists x \exists y$ and $\forall x \forall y$. All this is much like the well-known lattice of *modal logics* (Bull and Segerberg 1984, Blok 1979).

An attractive candidate in this landscape is the logic **CRS** of all generalized assignment models. It may also be described as the set of predicate-logical validities that hold in those abstract state frames for quantification and substitution which satisfy all *universal frame conditions* true in standard assignment models. These obey all general logical properties of assignments, but they do not make any *existential* assumptions about the supply of available assignments. The former conditions seem more truly 'logical', whereas the latter would be more 'mathematical' or 'set-theoretic' (cf. Chapters 8, 13). For instance, in the above correspondences, universal S5-type conditions emerged, but also existential ones for quantifier interchange principles. In later sections, we shall analyze the purely universal kind in more detail, by a representation method turning abstract state frames into assignment frames. Two important known facts about **CRS** are that it is *decidable* and *non-finitely axiomatizable* (Németi 1995, Andréka 1991). Moreover, in Section 9.8 below we shall give a representation method showing that it has a first-order definition by means of universal *Horn* clauses, from which one can derive Craig Interpolation (Marx 1995). **CRS** as defined here is axiomatized by the laws of modal poly-S5 plus the earlier atomic 'locality conditions'. With substitutions added, its axioms become more complex — and we refer to the cited literature. Further modal analysis of this system is found in Section 9.9 on decidability, as well as the notes to this chapter (on interpolation).

On top of **CRS**, one may continue, and add further axioms up to the cliffs of complexity:

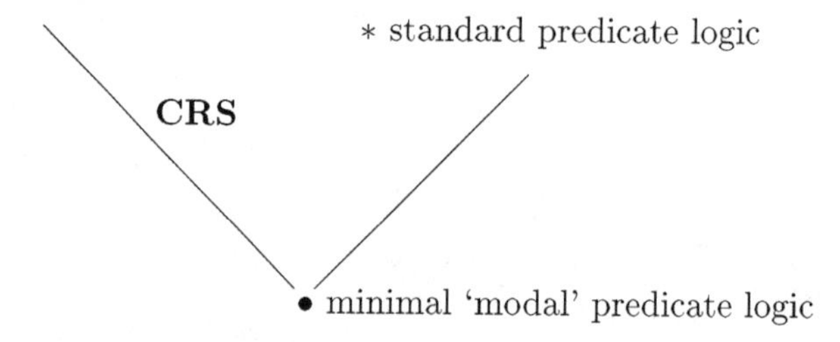

This landscape of dynamic predicate logic can be investigated model-theoretically by modal techniques, just like the Arrrow Logic of Chapter 8. As observed before, basic modal notions like bisimulation generalize model-theoretic counterparts over standard assignment models. There are still interesting discrepancies. E.g., bisimulation relates complete variable assignments, whereas its model-theoretic counterpart of 'partial isomorphism'

relates finite sequences of objects. This reflects another meta-property of standard first-order logic: no variable has a special identity. In the present abstract semantics, this is no longer the preferred option. With possible dependencies present, variables do gain 'individuality'. Other modal themes include axiomatization techniques, decision methods, and model constructions across this whole landscape.

Remark 9.2 (Two First-Order Languages) Do not confuse two uses of 'first-order languages' here. One lives at an object level, as the 'dynamic modal language' of assignment change. Another is at a meta-level, as our 'working language' for stating frame conditions. In particular, one can be a modal minimalist at the object level, and a standard logician meta-wise.

9.6 Rethinking the Language

The modal perspective suggests that first-order predicate logic reflects only part of the expressive resources of abstract state models. There is an obvious first-order (meta-) language over these models, whose variables run over states (again: do not confuse with our first-order *object language*!). This is the language into which poly-modal logic 'translates' in the usual sense (Chapter 3), which contains many items without a modal counterpart. One example is an *unrestricted* existential quantifier $\exists\alpha$ over states. By contrast, modal object quantifiers $\exists x$ induce *restricted* state quantifiers of the form $\exists\beta(R_x\alpha\beta \wedge$. 'Random assignment jumps' seem a natural meaning for isolated quantifier symbols $\exists$ not tagged by any variable. Also, one can have *global predicates* of states, not reducible to assertions about their object values at some set of variables. All this is just one instance of a broad theme mentioned at the beginning. A more general semantics below standard predicate logic usually suggests new notions, that were invisible in the 'classical system'. We list a few directions for such extensions.

Stronger Modalities. Add modal operators, such as the "universal modality", or more complex ones about internal structure of state transitions (cf. the "Since", "Until" of Chapters 3, 4).

Dynamic Operators. Add program constructions, starting from individual variables as atomic programs. E.g., the Path Principles of Section 9.8 suggest addition of both sequential composition and conjunctive intersection. Propositional dynamic logic with these two operations is still decidable: and hence so is our minimal base logic.

Polyadic Quantifiers. A most interesting extension in expressive power is that to polyadic quantifiers. In standard predicate logic, a tuple notation $\exists xy\, \phi$ is just shorthand for either $\exists x \exists y\, \phi$ or $\exists y \exists x\, \phi$. But here, it becomes a notion sui generis. On generalized assignment models, $\exists xy\, \phi$ says that there exists some assignment agreeing with the current one *up to* $\{x, y\}$ *values* where ϕ holds. The corresponding transitions encode a form of concurrency vis-a-vis the single transition relations R_x and R_y. This is not reducible to either iterated version, which require the existence of 'intermediate states'. More generally, abstract state models admit natural definitions of polyadic quantifiers $\exists x_1 \ldots x_k\, \phi$ stating the existence of some $R_{\langle x_1,\ldots,x_k\rangle}$-accessible state where ϕ holds. In standard logic, this assertion is equivalent to any of its linearized versions $\exists x_1 \ldots \exists x_k\, \phi$. But with possible 'gaps' in our models, it is not so reducible. Polyadic quantification has independent linguistic interest. Thus, in formalizing natural reasoning, one may now treat sequences of variables as either 'dependent' or 'independent'. Moreover, adding the latter expressive resource leaves the basic predicate logic **CRS** *decidable* (Mikulás 1995; Section 9.10 presents a new proof via the Guarded Fragment of Chapter 4). There is a more general issue, as to how adding vocabulary affects meta-properties of a logic in our landscape. Adding too much expressive power might reinstate standard first-order logic. Polyadic language extensions also make sense in the presence of explicit *substitutions*. For instance, the latter needs both sequential composition and 'concurrent conjunction' (to deal with irreducibly polyadic *multiple substitutions* of the form $[t_1/x_1, \ldots, t_k/x_k]$).

Partial-State Frames. Some newer versions of dynamic semantics employ partial assignments (cf. Chapter 2). These account for the intuitive difference between two processes: 're-assignment' R_x, which changes an old value for x, and 'new assignment' R_x^+, which gives x a value for the first time. Such actions have corresponding first-order quantifiers $\exists x$ and $\exists^+ x$. In partial-state frames, R_x will remain transitive and Euclidean, but not reflexive (x-values are not always defined). We only have the weaker axiom $\exists x\, \top \wedge \phi \rightarrow \exists x\, \phi$. By contrast, R_x^+ is asymmetric, and it satisfies, e.g., $\forall \alpha\beta\, (R_x^+ \alpha\beta \rightarrow \neg\exists\gamma\, R_x^+ \beta\gamma)$. The connection between the two variable update relations is the valid principle $\exists x\, \top \leftrightarrow \neg\exists^+ x\, \top$. A new notion in these frames is *extension* of partial states. It can be expressed by a corresponding existential modality:

$$\mathbf{M}, \alpha \models \Diamond\phi \text{ iff } \exists\beta \supseteq \alpha\, \mathbf{M}, \beta \models \phi.$$

Using it, one can also define substitutes for $\exists^+ x\, \phi$ such as $\Diamond\exists x\, \phi$. The complete modal logic of partial assignment frames is an interesting extension of predicate logic.

9.7 Applications and Repercussions

Extended first-order logics suggest interesting applications. It now becomes of practical interest just *how much* of predicate logic is used in natural language, common sense reasoning in AI or even mathematical proof. Can our decidable subcalculi of predicate logic be applied to the 'natural logic' underlying natural language? The latter revolves around Monotonicity and Conservativity (van Benthem 1986, Sanchez Valencia 1991), which are derivable in weak calculi in our landscape. Likewise, could there be useful decidable systems of arithmetic or other parts of mathematics using these ideas? E.g., what is the theory of the natural numbers with all possible families of variable assignments? Perhaps, the usual predicate-logical base for applied theories is much too strong for its purpose. The thrust of this line of research is broad. It does not just apply to dynamics of changing variable assignments, but also to *updating information states*. Abstract models can carry further structure, such as 'composition' of information states, which supports new dynamic connectives (cf. Chapters 2, 5, 6). Similar issues to those above will arise, affecting the *propositional* component of first-order predicate logic — which was inviolate so far. Finally, there is a broader philosophical significance to the above technical work (cf. also Chapter 13). If one views generalized models as the natural semantics for first-order predicate logic, rather than some tinkering device, then many received views of the field must be challenged. In standard text books, one learns that 'predicate-logical validity' is one unique notion, specified once and for all by Tarski, and canonized by Gödel's Completeness Theorem. Moreover, it is essentially complex, being undecidable by Church's Theorem: the old ideal of Leibniz's 'Calculus Ratiocinator' just will not work. On the present view, however, 'standard predicate logic' has arisen historically by making several semantic decisions that could have gone differently. The genuine core of first-order predicate logic may well be decidable — and the real issue is not one 'completeness theorem', but rather the model-theoretic and proof-theoretic analysis of the above rich family of languages and axiomatic calculi.

The rest of this Chapter is a technical exploration of the above framework (especially **CRS**) with techniques from modal logic. These give a feel for how it really 'works'.

9.8 Representation

A systematic view on semantic options in our landscape arises as follows. We analyze what it takes to represent any abstract modal frame as a family of assignments with the standard variable update relation $=_x$. The following method is very straightforward. How can abstract states become assignments? The obvious idea is to create 'objects' (α, x) for each state α and variable x, and then set

$$\alpha^*(x) := (\alpha, x).$$

This stipulation will indeed turn states into assignments, and represent abstract state frames as assignments frames with arbitrary abstract update relations R_x. (Thus, the latter option is not really different from the most general one.) But if the latter relations are to become the standard updates $=_x$, then some refinement is necessary.

Representing State Frames. For a start, we assume all universal properties of standard assignment frames. What is needed on the way will eventually be collected in the statement of our results. Let Z denote some sequence of variables. Extend the notion of accessibility as follows:

$$\begin{aligned} \alpha R_{\emptyset} \beta &:= \alpha = \beta \\ \alpha R_{Z \bullet y} \beta &:= \exists \gamma \; \alpha R_Z \gamma \wedge \gamma R_y \beta. \end{aligned}$$

We use sequences here rather than sets, because we do not assume the usual existential quantifier interchange principles suppressing the ordering. Now define

$$(\alpha, x) \approx (\beta, y) \text{ if } x = y \wedge \exists Z \, x \notin Z \wedge \alpha R_Z \beta.$$

It is easy to check that $\approx$ is an equivalence relation, using the symmetry of the update relations R_z. This observation allows us to use equivalence classes for values:

$$\alpha^*(x) := (\alpha, x)^{\approx}.$$

Now, let us analyze what it takes to prove the following key equivalence:

$$\text{Adequacy of Representation: } \alpha R_x \beta \text{ iff } \alpha^* =_x \beta^*.$$

The direction from left to right is immediate. Let y be any variable distinct from x. Set $Z = \langle x \rangle$. Then, since $\alpha R_x \beta$, by the above definition, $(\alpha, y) \approx (\beta, y)$, and hence $\alpha^*(y) = \beta^*(y)$. From right to left, suppose that $\alpha^* =_x \beta^*$. By definition, this means that $\forall y \neq x \, \exists Z \, y \notin Z \wedge \alpha R_Z \beta$. What we want from this bunch of facts is $\alpha R_x \beta$. Here is a special case. With only *two* variables, the latter information applied to the variable y says that α, β are related via some finite sequence (possibly empty) of R_x-steps . Using only reflexivity and transitivity, then, we get the desired conclusion. Thus, we have found (as more often in the algebraic literature) that the two-variable fragment of predicate logic is particularly simple:

Proposition 9.3 *With only two variables, an abstract state frame is representable as an assignment frame iff all its relations R_x are equivalence relations.*

The general situation is more complex. E.g., with *three* variables, we may have the following two paths:

$$\begin{array}{ccccccc} & & \gamma & & & & \\ & x & & z & & & Z_y = \langle x, z \rangle \\ \alpha & & & & \beta & & \\ & y & & x & & & Z_z = \langle y, x \rangle \\ & & \delta & & & & \end{array}$$

In the standard assignment model, this implies that α, β agree on both y and z, whence the arrows for y and z must be identity transitions, and we have $\alpha R_x \beta$. More generally, all *Path Principles* of the following form are valid under the standard Tarskian interpretation (notice that there are infinitely many of these):

- If $\alpha R_{Z_1}\beta$, ..., $\alpha R_{Z_k}\beta$, and the only variable occurring in all of Z_1, ..., Z_k is x, then $\alpha R_x \beta$.
- If no variable occurs in all connecting sequents, then $\alpha = \beta$.

Proposition 9.4 *An abstract frame is representable as an assignment frame iff its relations R_x are equivalence relations satisfying, in addition, all Path Principles.*

Proof. Continue the above argument. Suppose that

$$\forall y \neq x \ \exists Z \ y \notin Z \wedge \alpha R_Z \beta.$$

Let y be any specific variable distinct from x, and select a connecting path Z_y. For any of the (finitely many!) variables u occurring on this path distinct from x, select some connecting path Z_u on which u fails to occur. Then x is the only variable in the intersection, and the path principle for Z_y and the Z_u's will say that $\alpha R_x \beta$. ⊣

Three points may clarify this. (1) Transitivity for relations R_x follows from the Path Principles. (2) Reflexivity is needed when the intersection of all occurrence sets of variables on the paths is empty. (3) One should take care with the first Path Principle. For instance, it does *not* say that the two R_y-transitions displayed must be identical ones:

$$\begin{array}{ccccccc} & & & y \quad z & & & \\ & & x & & y & & \\ \alpha & & & & & & \beta \\ & & & z \quad x & & & \end{array}$$

Next, the second Path Principle also implies that our representation is one-to-one.

Fact 9.5 *In the above case, the map from states α to assignments α^* is injective.*

Proof. (We need at least two variables x, y.) Suppose that $\alpha^* = \beta^*$. Then we have, in particular, that $\alpha^* =_x \beta^*$ and $\alpha^* =_y \beta^*$. By the above observation, this implies that $\alpha R_x \beta$ and $\alpha R_y \beta$. But then, by the second Path Principle, $\alpha = \beta$. ⊣

Analyzing this simple representation from a logical point of view — especially, the crucial family of Path Principles — we see the following:

- The class of representable abstract frames is definable by a set of first-order sentences which are all *universal Horn*.
- This definition employs infinitely many frame conditions.

The former property has pleasant consequences, including Interpolation for predicate-logical validity over this frame class (by general results in modal logic; cf. our notes). The second property hints at a certain complexity (cf. the non-finite axiomatizability result by Andréka). Finally, it is easy to see that few path principles correspond to a modal formula in the predicate-logical language. This completes our analysis of **CRS**.

Remark 9.6 (Sets Instead of Sequences) In the full standard case, with the two quantifier exchange axioms, it suffices to define a relation $\alpha R_X \beta$, where X is a *set* of variables, postulating some connecting sequence of transitions indexed by variables in X. The Path Principles then reduce to

$$\text{if } \alpha R_X \beta \text{ and } \alpha R_Y \beta, \text{ then } \alpha R_{X \cap Y} \beta.$$

Representing State Models. Our representation extends to models $\mathbf{M}$ that interpret structured atomic formulas. These are abstract frames $(S, \{R_x\}_{x \in \text{VAR}})$ plus an interpretation function I interpreting atoms over states (a 'modal valuation'), needed to interpret the predicate-logical object language. Define the following standard interpretation function over the represented frame (with one binary predicate letter Q, for convenience):

$$I^*(Q) = \{((\alpha, x)^{\approx}, (\alpha, y)^{\approx}) \mid I(\alpha, Qxy)\}.$$

We need to show the following assertion of adequacy:

Claim. $\mathbf{M}, \alpha \models \phi$ *iff* $\mathbf{M}^*, \alpha^* \models \phi$, *for all predicate-logical* ϕ.

Unfortunately, we do not quite succeed. The following is as far as we get.

Proof Attempt. The assertion is automatic for Booleans, and it holds for quantifiers by the above proof. The atomic case presents a difficulty, though. From left to right, its assertion is trivial. If $\mathbf{M}, \alpha \models Qxy$, then $I(\alpha, Qxy)$ holds, and hence $I^*(Q)$ holds of $(\alpha, x)^{\approx}$, $(\alpha, y)^{\approx}$ (i.e., $\alpha^*(x)$, $\alpha^*(y)$) by definition. But from right to left, we encounter an obstacle. Let $I^*(Q)$ hold in $\mathbf{M}^*$ of $\alpha^*(x)$, $\alpha^*(y)$, that is, of $(\alpha, x)^{\approx}$, $(a, y)^{\approx}$. Thus, there exists γ with $(\alpha, x) \approx (\gamma, x)$, $(\alpha, y) \approx (\gamma, y)$ such that $I(\gamma, Qxy)$. By the definition of $\approx$ then, there are two finite sequences Z_x (not containing x) and Z_y (not containing y) with $\alpha R_{Z_x} \gamma$, $\alpha R_{Z_y} \gamma$. Now, we need to show that $I(\alpha, Qxy)$. Here, the earlier atomic invariance principles $P\mathbf{y} \rightarrow \forall x\, P\mathbf{y}$ and $\neg P\mathbf{y} \rightarrow \forall x\, \neg P\mathbf{y}$ should help. But these are not strong enough. We need a more complex path principle stating that $Qxy \rightarrow [Z_x \cap Z_y]Qxy$. This is beyond our modal predicate-logical language, however — as it involves essentially a further operation of 'program intersection'. $\dashv$

One way of overcoming this difficulty uses an extension of our representation to a richer predicate-logical language. Two options are presented in digressions below. Westerståhl 1995 presents the most elegant solution so far, combining our previous representation with ideas from Section 9.9 below. One can extend these representation arguments to abstract frames with transition relations $A_{x,y}$ reflecting the earlier *substitution*. We forego this extension here (some relevant details are in the notes).

Option 1. (Pointwise Equality of States) The following useful relation turns up implicitly in the above arguments:

$$\alpha R_x \beta \text{ iff } \alpha(x) = \beta(x).$$

This suggests the use of enriched state models

$$(S, \{R_x\}_{x \in \text{VAR}}, \{R^x\}_{x \in \text{VAR}}, I).$$

The new relations R^x are easier to handle than the old R_x , being equivalence relations. They can be used to define the latter, via the following equivalence:

$$\alpha R_x \beta \leftrightarrow \bigwedge_{y \neq x} \alpha R^y \beta.$$

Option 2. (Polyadic Quantifiers) Another natural extension of our framework (cf. Section 6) uses indifference relations between states involving finite sets of variables X:

$$\alpha R_X \beta \text{ iff } \alpha(y) = \beta(y) \text{ for all variables } y \text{ outside of } X.$$

Now we can use the old representation, setting $(\alpha, x) \approx (\beta, x)$ iff there is some finite set Y not containing x with $\alpha R_Y \beta$. The earlier Path Principles are replaced by these:

- $\alpha R_X \beta$ and $\alpha R_Y \beta$ imply $\alpha R_{X \cap Y} \beta$
- $\alpha R_X \beta$ and $\beta R_Y \gamma$ imply $\alpha R_{X \cup Y} \gamma$.

9.9 Decidability

This Section presents a new proof for decidability and finite model property of **CRS** (without substitutions) — using modal filtration and unwinding, plus representation. Our secondary aim is to show modal logic at work in an unconventional environment.

Filtration of Generalized Assignment Models. Consider generalized models $(\mathbf{M}, \mathbb{V})$ for predicate logic, where $\mathbb{V}$ is the range of 'available assignments'. Here are two relevant extra constraints. 'Atomic Locality' says that assignments α, $\beta \in \mathbb{V}$ which agree on all free variables $\text{FV}(\phi)$ of an atomic formula ϕ must give ϕ the same truth value. 'Locality' says the same for all formulas. In what follows, we fix some formula ϕ with variables VAR_ϕ (free or bound) and subformulas SUB_ϕ. Everything will be restricted to such *finite* syntax sets. First, we define a multi-S5 finite filtration over generalized assignment models.

Definition 9.7 For $\alpha, \beta \in \mathbb{V}$, set $\alpha \sim \beta$ if α, β give the same truth values to all formulas in SUB_ϕ. The *ϕ-filtration* of $(\mathbf{M}, \mathbb{V})$ is the Kripke model $\text{FILT}_\phi(\mathbf{M}, \mathbb{V}) = (S, \{R_x\}_{x \in \text{VAR}}, V)$ obtained as follows. State Universe: S consists of all $\sim$-equivalence classes $\widetilde{\alpha}$. Accessibility: $\widetilde{\alpha} R_x \widetilde{\beta}$ holds if α, β give the same truth value to all relevant formulas $\exists x\, \psi$ and to all relevant atomic formulas not containing x. Valuation: $V(\widetilde{\alpha}, \psi) = 1$ for relevant atoms ψ iff ψ is true at α in $(\mathbf{M}, \mathbb{V})$.

Filtration Lemma. *For all relevant formulas ψ and all assignments α, $(\mathbf{M}, \mathbb{V}), \alpha \models \psi$ iff $\mathrm{FILT}_\phi(\mathbf{M}, \mathbb{V}), \widetilde{\alpha} \models \psi$.*

Proof. Induction on ψ. Atoms: by definition. Booleans: use routine. Existentials. By the truth definition in generalized models, $(\mathbf{M}, \mathbb{V}), \alpha \models \exists x\, \psi$ implies that there exists some $\beta \in \mathbb{V}$ with $\alpha =_x \beta$ and $(\mathbf{M}, \mathbb{V}), \beta \models \psi$. Hence (by the inductive hypothesis) $\mathrm{FILT}_\phi(\mathbf{M}, \mathbb{V}), \widetilde{\beta} \models \psi$, and also $\widetilde{\alpha} R_x \widetilde{\beta}$ (mapping assignments to their equivalence classes is a homomorphism). Therefore, $\mathrm{FILT}_\phi(\mathbf{M}, \mathbb{V}), \widetilde{\alpha} \models \exists x\, \psi$. Conversely, $\mathrm{FILT}_\phi(\mathbf{M}, \mathbb{V}), \widetilde{\alpha} \models \exists x\, \psi$ implies the existence of $\widetilde{\beta}$ such that $\widetilde{\alpha} R_x \widetilde{\beta}$ and $\mathrm{FILT}_\phi(\mathbf{M}, \mathbb{V}), \widetilde{\beta} \models \psi$. Then, by the inductive hypothesis, $(\mathbf{M}, \mathbb{V}), \beta \models \psi$ — and so $(\mathbf{M}, \mathbb{V}), \beta \models \exists x\, \psi$, whence $(\mathbf{M}, \mathbb{V}), \alpha \models \exists x\, \psi$ by definition of the relations R_x. ⊣

Filtration also works for generalized models with Locality (for all relevant formulas), to yield a finite model with that property. One makes two equivalence classes R_z-accessible when they agree on all formulas in which variable z does not occur free.

Unwinding Kripke Models. The above (filtrated) Kripke models are abstract. They may lack some key properties of generalized assignment models. Notably, the earlier 'Path Principles' may fail. (E.g., there may be two different links R_x, R_y between two distinct states.) We can improve this behaviour by path unraveling, to get a basis for concrete representation.

Definition 9.8 The *unwinding* $\mathrm{UNW}(\mathbf{M})$ of a rooted Kripke model $(\mathbf{M}, s)$ consists of all finite sequences $(s, x_1, \ldots, s_{k-1}, x_{k-1}, s_k)$ where all s_i are worlds in $\mathbf{M}$, and always $s_i R_{x_i} s_{i+1}$. The relations R_{x_i} are the reflexive symmetric transitive closures of the relations consisting of all pairs $(X, X^\frown\langle z, w\rangle)$ with $\mathrm{last}(X) R_z w$ in $\mathbf{M}$. Finally, the valuation V for sequences X is copied from that for $\mathrm{last}(X)$ in $\mathbf{M}$.

Unwinding Lemma. *For all formulas ψ, and all sequences X,*

$$\mathrm{UNW}(\mathbf{M}), X \models \psi \text{ iff } \mathbf{M}, \mathrm{last}(X) \models \psi.$$

Proof. The function sending X to $\mathrm{last}(X)$ is a bisimulation. ⊣

The only non-routine fact here is that the map 'last' is a homomorphism with respect to the relations R_z in the unwinding. (This part of the argument will work as long as our frame conditions are universal Horn.) One further observation may be made.

Corollary 9.9 *Formulas satisfiable in finite Kripke models are also satisfiable in finite unwound Kripke models.*

Proof. This is the multi-S5 version of the well-known modal Finite Depth Lemma. Evaluating a formula from the root involves only finitely many alternations in depth across different relations R_z — as may be seen through normal forms for multi-S5. $\dashv$

This 'cut-off' at the modal depth of ϕ preserves Atomic Locality (in its abstract Kripke version, as a constraint on the valuation V) — though not necessarily full Locality. Finally, we note that unwound Kripke models do satisfy all Path Principles.

Representing Unwound Kripke Models. Unwound multi-S5 models can be represented as generalized assignment models. (This is a more concrete version of the representation in Section 9.8.) The idea is easily explained. Take an arbitrary assignment (x_i, d_i) $(1 \leq i \leq k)$ of different objects to the relevant variables at the root. Then, follow longer sequences X upward. If an assignment ass(X) has already been defined, then choose a supply of *new objects*, and change values at z only for steps from X to $X^\frown\langle z, w\rangle$. This is well-defined.

Definition 9.10 The *object representation* OBJ(**M**) of an unwound Kripke model **M** has just been described. Its admissible assignments are those produced in the process. Its interpretation function $I(Q)$ for predicates Q collects all tuples d contributed by those ass(X) where some atom $Q\mathbf{z}$ was true in **M** at X.

Representation Lemma. *For all formulas* ψ, *and all sequences* X, $\mathbf{M}, X \models \psi$ *iff* $\mathrm{OBJ}(\mathbf{M}), \mathrm{ass}(X) \models \psi$.

Proof. The map from X to ass(X) is a bisimulation. Atomic clause. If $\mathbf{M}, X \models Q\mathbf{z}$, then $\mathrm{OBJ}(\mathbf{M}), \mathrm{ass}(X) \models Q\mathbf{z}$, by the definition of $I(Q)$. Next, if $\mathrm{OBJ}(\mathbf{M}), \mathrm{ass}(X) \models Q\mathbf{z}$, then by that same definition, $\mathbf{M}, Y \models Q\mathbf{z}$ for some sequence Y whose ass(Y) agrees with ass(X) on all the variables $z \in \mathbf{z}$. Hence, by construction, X, Y are equal or connected by a chain of relations R_u with u outside of $\mathbf{z}$. Atomic Locality, i.e., the truth of all formulas $Q\mathbf{z} \to \forall u\, Q\mathbf{z}$ in **M**, then yields $\mathbf{M}, X \models Q\mathbf{z}$. (Unicity of the relevant atoms is guaranteed by our 'free' assignment of objects.) Zigzag clauses. Inspection of the above construction shows that, if XR_zY, then $\mathrm{ass}(X) =_z \mathrm{ass}(Y)$ — and also vice versa. From left to right, this is easy. From right to left, suppose that not XR_zY. Then, in the most general case, they must lie in some tree situation

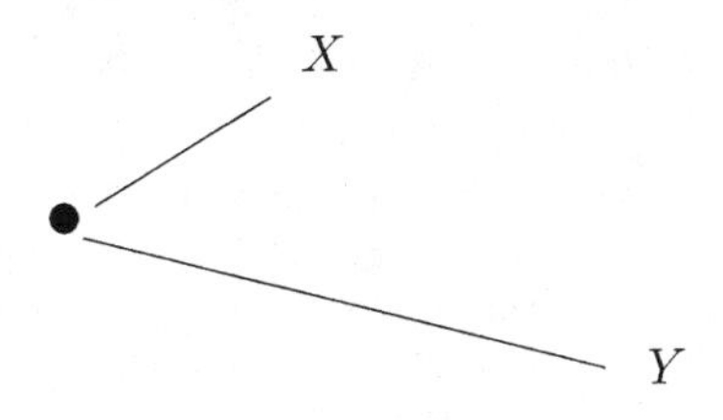

where on the minimal connecting path shown, some value has changed for a variable u distinct from $\mathbf{z}$. As our representation chooses different objects all the time, upwards and sideways in branches, this difference will show up in the pair ass(X), ass(Y): which then lacks the relationship $=_z$. ⊣

We can now apply these results to obtain

Theorem 9.11 *Validity in generalized assignment semantics is decidable.*

Proof. Combining the previous facts, a formula ϕ is satisfiable in a generalized assignment model iff it has a finite abstract Kripke model satisfying Atomic Locality whose size is bounded by $2^{|\mathrm{SUB}_\phi|}$. The latter property is decidable. ⊣

Theorem 9.12 *Generalized assignment semantics has the Finite Model Property.*

Proof. By the Cut-Off Property established above, finite generalized models will suffice. ⊣

9.10 Translations

This chapter and Chapter 4 present two approaches for 'taming' first-order logic: i.e., localizing a well-behaved decidable 'core part'. One can either use standard semantics over non-standard language fragments, or non-standard semantics over the full first-order language. The former approach is more 'syntactical', the latter more 'semantical'. As so often in logic, the distinction is relative. One can translate 'semantic' modal discourse about the above modal first-order models into a restricted syntactic fragment of a *two-sorted* first-order logic, with direct reference to both 'individuals' and 'states'. Specifically, evident analogies exist between generalized semantics and reasoning with 'bounded fragments' of first-order logic. There is a mathematical duality behind this. In particular, our earlier analysis of bounded fragments derives facts about generalized assignment semantics, or equivalently, about relativized cylindric algebras (i.e., **CRS**).

From Bounded Fragments to Generalized Models. Consider any k-variable language $L\{x_1, \ldots, x_k\}$. Let R be a new k-ary predicate. We define a straightforward translation tr_g from k-variable formulas to bounded first-order ones:

Global Relativization $\mathrm{tr}_g(\phi)$ arises from ϕ by relativizating all its quantifiers to the atom $Rx_1 \ldots x_k$.

The relativized images $\mathrm{tr}_g(\phi)$ will even lie inside our Guarded Fragment (Chapter 4). Next, we define a corresponding operation on models. Let $\mathbf{M}$ be any generalized assignment model for $L\{x_1, \ldots, x_k\}$ (as yet without the new predicate symbol R).

Restricted Standard Models The standard model $\mathbf{M}_{\text{rest}}$ is $\mathbf{M}$, viewed as a standard model, and expanded with the following interpretation for the new predicate: $R(d_1, \ldots, d_k)$ iff the variable assignment $x_i := d_i$ $(1 \leq i \leq k)$ is available in $\mathbf{M}$.

The purpose of this construction shows in the following fact.

Proposition 9.13 *For all available assignments α in $\mathbf{M}$, and all formulas ϕ, $\mathbf{M}, \alpha \models \phi$ iff $\mathbf{M}_{\text{rest}}, \alpha \models \text{tr}_g(\phi)$.*

Proof. Induction on first-order formulas. The crucial case is the existential quantifier. In particular, suppose that

$$\mathbf{M}_{\text{rest}}, \alpha \models \text{tr}_g(\exists x_i\, \phi) = \exists x_i\, (Rx_1 \ldots x_k \wedge \text{tr}_g(\phi)).$$

Then, there exists a satisfying k-tuple of objects in R for $\text{tr}_g(\phi)$, which corresponds to an available assignment in $\mathbf{M}$ which is an i-variant of α. I.e., $\mathbf{M}, \alpha \models \exists x_i \phi$. $\dashv$

As a consequence, one can effectively reduce universal validity over all generalized assignment models (i.e., in **CRS**) to standard validity for guarded formulas:

Corollary 9.14 $\models_{\mathbf{CRS}} \phi$ *iff* $\models_{\text{standard}} Rx_1 \ldots x_k \rightarrow \text{tr}_g(\phi)$.

Proof. 'Only if'. If ϕ has a generalized counter-example $\mathbf{M}, \alpha$, then the above model $\mathbf{M}_{\text{rest}}$ falsifies $Rx_1 \ldots x_k \rightarrow \text{tr}_g(\phi)$.

'If'. Suppose, conversely, that the latter formula has a standard counter-example $\mathbf{M}, \alpha$. Now define a corresponding generalized model $\mathbf{M}_g$ by retaining only those assignments whose values for x_1, ..., x_k stand in the relation $R^{\mathbf{M}}$ (in particular, the falsifying assignment α itself remains available). Then ϕ is falsified in $\mathbf{M}_g$ by α as above. $\dashv$

Thus, we have a second proof for the decidability of **CRS** (cf. Section 9.9), this time by reducing it to the Guarded Fragment of Chapter 4. There is more to this analysis. Special classes of generalized models put more constraints on admissible assignments. The first-order theory of such classes, too, will be decidable, if these frame conditions can be stated as *guarded first-order forms*. E.g., this applies to the *locally square* generalized assignment models (needed to set up the full substitution version of **CRS**), in which every permutation or identification of values in an admissible assignment yields another admissible one. Here is an example of the relevant frame condition:

$$\forall xy\, (Rxy \rightarrow (Ryx \wedge Rxx \wedge Ryy)).$$

By contrast, validity is undecidable in the class of generalized assignment models satisfying the so-called 'Patchwork Property' (Németi 1995). Again, this checks out. In first-order form, the latter constraint involves statements for patching admissible assignments like

$$\forall xyzuv\,((Rxyz \wedge Ruyv) \rightarrow (Rxyv \wedge Ruyz)).$$

These are not guarded: variable inclusion holds from matrix to restriction, but the latter is not one single atom. So, one cannot allow arbitrary Boolean combinations of guards.

Translations also help in comparing different models for dependency. Recall the alternative dependency semantics for quantifiers mentioned in Section 9.2. The latter arises from first-order logic through a 'local translation' tr_l like the 'global translation' tr_g, but with a delicate difference. At subformulas $\exists x_i\,\psi$, one only relativizes to an atom R_x where x enumerates all free variables of the local context ψ. This difference explains all deviant behaviour. E.g., tr_g makes Modal Distribution a valid bounded principle, whereas tr_l does not:

$$\forall y\,(\forall x\,(Ax \rightarrow Bxy) \rightarrow (\exists x\,Ax \rightarrow \exists x\,Bxy))$$

$$\mathrm{tr}_g \quad Rxy \rightarrow \forall y\,(Rxy \rightarrow (\forall x\,(Rxy \rightarrow (Ax \rightarrow Bxy)) \rightarrow (\exists x\,(Rxy \wedge Ax) \rightarrow \exists x\,(Rxy \wedge Bxy)))$$

$$\mathrm{tr}_l \quad \forall y\,(Ry \rightarrow (\forall x\,(Rxy \rightarrow (Ax \rightarrow Bxy)) \rightarrow (\exists x\,(Rx \wedge Ax) \rightarrow \exists x\,(Rxy \wedge Bxy)))$$

9.11 Notes

Sources. There are many sources for our modal analysis of predicate logic. Among these, we mention *dynamic logic* (Pratt 1976, Harel 1984), *dynamic semantics* for both natural and programming languages (Chapter 2, but also Janssen 1983), *probabilistic logic* (van Lambalgen 1991, Alechina and van Benthem 1996), and *algebraic logic* (as cited in the many references throughout our text). Especially, one should mention the modal approach to algebraic logic initiated by Venema 1991, 1995a, which shows how modal techniques can inform cylindric algebra, and vice versa. We follow this with a radical twist, as the modal frames have become the real semantics. For recent relevant work, one can take the references of Chapter 8, such as Marx 1995, Mikulás 1995, Venema and Marx 1996a, but also Alechina 1995a, Westerståhl 1995.

Cylindric Algebra. The existing body of research in Cylindric Algebra has many connections with the themes of this chapter. We refer to Henkin et al. 1985, Andréka 1991, Németi 1985, 1995, Thompson 1981, Venema 1994, Marx 1995. For a most recent approach to relativization, linking up with various issues in this chapter, cf. Andréka, Goldblatt and Németi 1996.

Partiality. Partial versions of dynamic semantics are mentioned in Chapters 2, 12. Recent proposals to this effect are Beaver 1995, van den Berg

1996, Vermeulen 1994. Fernando 1992 defends partiality as a necessary condition for joining forces between dynamic semantics for natural language and recursion theory. A concrete stack semantics supporting partiality is in Hollenberg and Vermeulen 1994, Visser 1995.

Further Dynamization. The above analysis extends to other semantic parameters. Not just assignments can change in Tarski semantics, but also *interpretation functions*. (Cf. the general Tarskian Variations of Chapter 2, or van Benthem and Cepparello 1994). Also, one can tackle the propositional connectives, and do an *intuitionistic version*, more along the lines of dynamic modal logic (Chapters 3, 6).

Computational Parallels. There are various parallels between our discussion of predicate logic and issues in the computational literature. For instance, abstract multi-S5 models for **CRS** look like the multi-S5 models employed in Fagin, Halpern, Moses and Vardi 1995 for modeling group knowledge. Then, our representation methods of Section 9.8 show, amongst others, that this logic does not change over more concrete models where states are vectors, with coordinatewise accessibility relations

$$s =_x t \quad \text{iff} \quad s(x) = t(x)$$
$$s =^x t \quad \text{iff} \quad s, t \text{ differ at most in their value for } x.$$

Both options occur in the work of these authors. The first is their standard model for epistemic indistinguishability of distributed agents, the second is their rendering of the perceptional state that the notorious 'muddy children' find themselves in.

More on Polyadics. Polyadic quantification in its own right is a well-known phenomenon in linguistics (van Benthem and Westerståhl 1995, Keenan and Westerståhl 1996). But this notion also provides a further computational parallel for our logics. The translation method of Section 9.10 is easily extended to show that one can embed the complete polyadic version of **CRS** into the Guarded Fragment. (Use the same atomic guard as above, and put in the right finite sequence of quantifiers simultaneously. This construction is allowed in the Guarded Fragment, which is already polyadic itself ...) Hence, polyadic **CRS** may be viewed as a *decidable process logic* for parallelism.

Interpolation Properties. Meta-properties of predicate logic return in our modal landscape, but sometimes with new twists. Modal interpolation gets different forms. The usual Craig version concerns just shared proposition letters (*Weak Interpolation*):

If $\phi \models \psi$, then there exists some modal formula χ over the shared propositional vocabulary of ϕ, ψ (including $\top$) such that $\phi \models \chi \models \psi$.

Strong Interpolation requires in addition that the interpolant χ contain only modal operators that occur in both ϕ and ψ. Both versions hold for a poly-modal base logic. Proofs use bisimulation and saturated models (cf. Chapter 4), plus direct products (cf. Marx 1995, Andréka, van Benthem and Németi 1995a, Madarasz 1996). Weak Interpolation holds for any modal logic with frames defined by *universal Horn conditions*. Strong Interpolation is scarcer — but it does hold, e.g., for multi-S5 (van Benthem 1996c). What we get is (using the fact that the complete modal logic of substitution-free **CRS** is just multi-S5):

Theorem 9.15 *The minimal predicate logic and substitution-free* **CRS** *both have the Strong Interpolation property.*

This may be shown by the representation of Section 9.8. Over models, it axiomatizes **CRS** as *multi-S5 plus Atomic Locality*. (It also shows that virtually no Path Principles have modal definitions.) Axiomatizing **CRS** gets more complicated with accessibility relations for substitutions, and other paraphernalia of a full first-order language (cf. Németi 1995, Mikulás 1995). Standard predicate logic *lacks* Strong Interpolation. E.g., the valid consequence $\exists x\, Px \models \exists y\, Py$ has no interpolant in *shared variables*. This negative outcome is compatible with the preceding positive result. ($\exists x Px \models \exists y Py$ fails in **CRS**.) Translating modal formulas into a first-order state language, the latter is like standard Interpolation, but then with respect to the *accessibility relations* for shared modalities.

Substitutions Revisited. Explicit substitution calculi occur in Németi 1985, Thompson 1981, Venema 1995b, Sagi and Németi 1996. Valid principles look like this, with the relations $A_{x,y}$ as total functions

(i) $x := y\,;\, u := v \leftrightarrow u := v\,;\, x := y$

(ii) $x := y\,;\, x := v \leftrightarrow x := v$

(iii) $x := y\,;\, u := x \leftrightarrow x := y\,;\, u := y$.

To represent abstract frames $(S, \{A_{x,y}\}_{x,y \in \mathrm{VAR}})$, one can use:

$$(\alpha, x) \approx (\beta, x) \quad : \quad \exists y \neq x \exists z \;\; \alpha A_{y,z} \beta$$
$$(\alpha, x) \approx (\beta, y) \quad : \quad \alpha A_{y,x} \beta$$

with a reflexive symmetric transitive closure. Path Principles arise from the equivalence

$$\alpha A_{x,y} \beta \text{ iff } \beta^* = {\alpha^*}^{x}_{\alpha^*(y)}.$$

When this representation is combined with the earlier one over modal state frames $(S, \{R_x\}_{x \in \mathrm{VAR}}, \{A_{x,y}\}_{x,y \in \mathrm{VAR}})$, adjustments are needed for the other equivalence:

$$\alpha R_x \beta \text{ iff } \alpha^* =_x \beta^*.$$

These involve earlier modal interaction principles between substitutions and quantifiers. The functions $A_{x,y}$ are not very complex. In **CRS**, an 'existential principle' like $[y/x]\exists z\, \phi \leftrightarrow \exists z\, [y/x]\phi$ can be treated quasi-universally (cf. Marx 1995). Again, these models suggest richer languages. The relations $A_{x,y}$ are not symmetric, and hence one can look *backwards*. This gives a 'temporal logic' with two directions for substitution:

Future $\quad \mathbf{M}, \alpha \models F_{y/x}\phi$ iff $\exists\beta\ \alpha A_{x,y}\beta \wedge \mathbf{M}, \beta \models \phi$

Past $\quad \mathbf{M}, \alpha \models P_{y/x}\phi$ iff $\exists\beta\ \beta A_{x,y}\alpha \wedge \mathbf{M}, \beta \models \phi$.

E.g., a Hoare assignment axiom $\{[t/x]\phi\}x := t\{\phi\}$ is the temporal conversion axiom $\phi \to H_{t/x}F_{t/x}\phi$. The backward substitution modality combines identity and ordinary quantification: $P_{y/x}\phi(x,y) \leftrightarrow x = y \wedge \exists x\, \phi(x,y)$. (Cf. Andréka and Thompson 1988, Andréka 1991.) These substitution calculi may be extended with multiple substitutions, comparable to polyadic quantifiers.

Other Dependence Models. There are other models for dependence than ours. Alechina 1995a contains an interesting proposal with 'tagged objects' $\langle x, d\rangle$ mixing dependencies for variables and objects. More spectacularly, generalized assignment semantics can be *lifted*. The linguistic literature on plurals uses assignments mapping variables to sets of objects. But one can also work with sets of standard assignments. This allows for finer discrimination, with possible dependencies between values for individual variables. Thus, from a repeated power

$$(2^{\mathrm{DOM}})^{\mathrm{VAR}}$$

we go to

$$2^{(\mathrm{DOM}^{\mathrm{VAR}})}.$$

The following map sends plural assignments A to sets $S(A)$ of individual assignments:

$$S(A) = \{f \mid \text{ for all } x,\ f(x) \text{ is in } A(x)\}.$$

Conversely, another map sends sets S of individual assignments to plural assignments $A(S)$:

$$A(S) = \lambda x.\, \{f(x) \mid \text{ all } f \text{ in } S\}.$$

The map S delivers 'full' sets of assignments. It is 1-1, unlike A. The impact of these two levels depends on the language. (With standard first-order logic for plurality, nothing changes.) Generalized assignment models, too, interpret formulas in a traditional way:

$$\mathbf{M}, S, \alpha \models \phi \text{ (singular state } \alpha \text{ verifies } \phi \text{ in 'plural context' } S).$$

But we can also interpret in the following format:

$$\mathbf{M}, S \models \phi \text{ (plural state } S \textit{ itself} \text{ verifies formula } \phi).$$

where formulas ϕ involve new logical operators, exploiting the richer structure of collective states. We have at least two versions of existential quantification then:

- $\exists_{\text{coll}}x\,\phi$ is true at S iff ϕ is true at some S' with $S =^x S'$, i.e., S, S' have the same assignments up to values for the variable x
- $\exists_{\text{ind}}x\,\phi$ is true at S iff ϕ is true at some S' with $S =_x S'$, i.e., S, S' have the same assignments but all x-values in S' are set to one object.

The resulting modal logic encodes interactions of individual and collective quantifiers, via frame correspondences for the two accessibility relations $R_{\text{coll},x}$ and $R_{\text{ind},x}$:

$$\exists_{\text{coll}}x\exists_{\text{ind}}x\,\phi \leftrightarrow \exists_{\text{ind}}x\,\phi$$
$$\exists_{\text{coll}}x\exists_{\text{ind}}y\,\phi \leftrightarrow \exists_{\text{ind}}y\exists_{\text{coll}}x\,\phi$$
$$\exists_{\text{coll}}x\forall_{\text{ind}}y\,\phi \rightarrow \forall_{\text{ind}}y\exists_{\text{coll}}x\,\phi.$$

Finally, sets of assignments S encode several kinds of dependence between variables. No single intuition need exist covering all kinds of 'correlation' among value ranges.

9.12 Questions

Exploring the Landscape. We have the same metamathematical issues as for the landscape of arrow logic (cf. the questions for Chapter 8). In particular, one wants *general* results and techniques covering large families of (decidable) predicate logics. For instance, one would like one optimal generalization for our modal filtration method, the unraveling method of Chapter 4, and the 'mosaic method' of Németi 1995. One obvious modal source for this are the *subframe logics* of Fine 1985a, being extensions of K4 without existential frame assumptions. But, we need not have transitivity — and also, we may want to impose new constraints on valuations, such as Atomic Locality. Can Fine's theory be applied to our setting? Here is a quite different general issue: how does the modal model theory for **CRS**-type logics relate to *standard first-order model theory*? One might investigate classical model theory anew in this light. In particular, does one get refined versions of the classical results (compactness, preservation theorems, etcetera)?

Mathematical Theories. Axiomatize the complete theories of classical structures, now turned into families of generalized structures with all possible sets of assignments. For instance, what happens to True Arithmetic in this way? How is complexity of standard theories affected in general?

Translations. Are there backward translations from bounded fragments into generalized predicate logics? Moving in both directions, what is the exact mathematical duality that one senses between the two areas? Can one derive behaviour of the Guarded Fragment from known properties of **CRS**, such as the Finite Model Property?

Complexity of Taming. What is the complexity of *decidable* predicate logics? What is the precise complexity of **CRS**? That of multi-S5 is known to be PSPACE-complete: what is the effect (if any) of adding substitutions? Combinatorial questions like these may be raised for about every paper in this incipient research tradition.

Enriched Languages. Axiomatize the complete temporal logic of substitutions. Explore polyadic quantifiers as operators for parallelism, in analogy with parallel programming constructs in computer science. Develop the model theory of these extensions, including their appropriate notions of bisimulation. Where is the border line where expressive power runs into undecidability?

Alternative Semantics. Axiomatize the complete modal logic of standard partial assignment frames. Axiomatize the complete logic of higher dependence models, and develop their model theory. For instance, again, what are the proper bisimulations?

Applications. Analyze existing computational logics from the present viewpoint. E.g., can one get *quantified first-order* dynamic logic (QDL, cf. Harel 1984) to become decidable after all? Analyze existing arguments and linguistic constructions in generalized semantics, to see how much of standard predicate logic is really needed. Formalizations may use finer distinctions, e.g., that between iterated single and polyadic quantifiers.

Part III

Implications

In this third and final part, we show how the logical theory developed so far provides new perspectives in various fields at the intersection of logic, language and computation. Of course, these connections will also bring in some specifics of the intended applications. Our claim here is not that one gets immediate practical benefits — but rather, that dynamic viewpoints enrich and unify our view of these various disciplines sharing a cognitive slant. We start with semantic process theories in computer science, continuing and specializing our studies of bisimulation and modal languages (Chapter 10). Next, we show how dynamic inference arises even inside so-called declarative programming styles, e.g., in proof calculi for logic programming (Chapter 11). Returning to our earlier motivations in linguistics, Chapter 12 discusses dynamic phenomena in syntax, semantics and pragmatics, emphasizing earlier techniques like Tarskian variations, modal fragments and generalized assignment semantics for linguistic representations. The final Chapter 13 summarizes some general philosophical features of the dynamic perspective, as well as some of its potential broader repercussions, that have chimed through occasionally in the book so far.

10

Computational Process Theories

We continue the study of process representation and equivalence, begun in Chapter 4. We start with the first-order model theory of process equivalences over labeled transition systems, including definability and axiomatization over general and special model classes. Next, we move toward connections with the literature on process algebras, via generalizations of modal techniques in Parts I, II, such as invariance and safety for bisimulation, and the use of extended modal formalisms. Except for the later sections on general safety, most results in this chapter come from van Benthem and Bergstra 1994.

10.1 Definition of Process Equivalences

What is the logical complexity of process equivalences? Consider single transition models with simulations $\equiv$ between pairs of states. Let **K** be all those models where $\equiv$ is one of the notions of Chapter 4. We call a process equivalence *elementary* (EC) if **K** is definable by a first-order sentence, and EC_Δ if some infinite set of first-order sentences does so. For instance, bisimulation is EC, as is immediate from its definition.

10.1.1 Finite Trace Equivalence

Proposition 10.1 *Finite Trace Equivalence is* EC_Δ.

Proof. The definition had the first-order form

$$\forall xy\,(x \equiv y \rightarrow (\phi(x) \leftrightarrow \phi(y))),$$

where the ϕ are arbitrary 'successful path formulas' of the form

$$\exists x_1(R_{a_1}xx_1 \wedge \ldots \exists x_n(R_{a_n}x_{n-1}x_n \wedge \surd(x_n))\ldots). \quad \dashv$$

Proposition 10.2 *Finite Trace Equivalence is not* EC.

Proof. We refute EC-ness, even with one atomic label a. Let $\phi(\equiv)$ be any putative first-order definition, of quantifier depth N. Consider two models. One has disjoint chains of length $2N$ ending with success, starting at states which are $\equiv$-related (and this match is the only pair in $\equiv$), the other has two such chains of lengths $2N$, $2N + 1$:

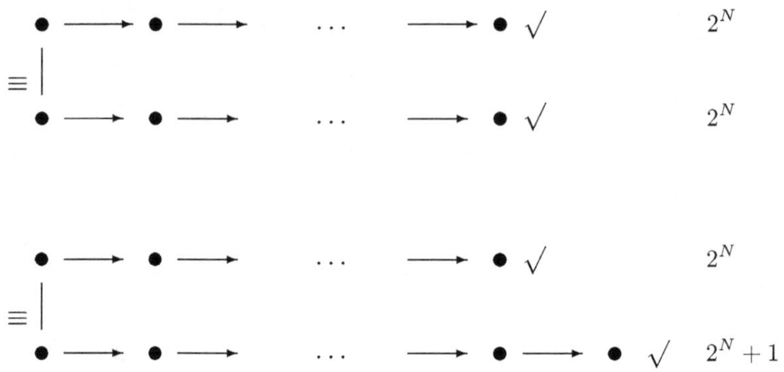

$$\begin{array}{lllllll} \bullet \longrightarrow & \bullet \longrightarrow & \cdots & \longrightarrow \bullet \; \surd & & & 2^N \\ \equiv \mid & & & & & & \\ \bullet \longrightarrow & \bullet \longrightarrow & \cdots & \longrightarrow \bullet \; \surd & & & 2^N \end{array}$$

$$\begin{array}{lllllll} \bullet \longrightarrow & \bullet \longrightarrow & \cdots & \longrightarrow \bullet \; \surd & & & 2^N \\ \equiv \mid & & & & & & \\ \bullet \longrightarrow & \bullet \longrightarrow & \cdots & \longrightarrow \bullet & \longrightarrow \bullet & \surd & 2^N + 1 \end{array}$$

An Ehrenfeucht game for linear orders (cf. Chapter 3 or Doets 1996) shows that these two models are indistinguishable by first-order sentences up to quantifier depth N. (One must match initial and terminal points in the above LTSs, maintaining a judicious invariant throughout with i more moves to go: on both sides, keeping relative distances $< 2^i - 1$ equal between all points chosen so far). But this is a contradiction, as the relation $\equiv$ is a finite trace equivalence in the first model, but not in the second, whereas the purported definition $\phi(\equiv)$ cannot see any difference between the two. $\dashv$

Proposition 10.3 *Complete Finite Trace Equivalence is not* EC_Δ.

Proof. Suppose that $T(\equiv)$ defined this notion. In particular, then,

$$T, \{\phi(x) \leftrightarrow \phi(y) \mid \text{ all succesful path formulas } \phi\} \models x \equiv y.$$

By Compactness, some *finite* set of ϕ-equivalences implies $x \equiv y$ from T. But then complete finite trace equivalence holds when we have it up to some fixed finite depth: *quod non.* $\dashv$

10.1.2 Full Trace Equivalence

Proposition 10.4 *Full Trace Equivalence is not* EC_Δ.

Proof. Again, take one atomic transition only. Consider the LTS with one root from which finite branches fan out of arbitrary length, all ending in success. By a standard Compactness argument, this LTS has an elementary extension with at least one infinite branch. Our first model $\mathbf{M}$ consists of two disjoint copies of the former structure (denoted in our language by unary predicates A, B) — and the second model $\mathbf{M}'$ of the same A-copy but now with the elementary extension in its B-part. Moreover, in both cases, the relation $\equiv$ consists of just the pair connecting the roots in the two components. In a picture, we have the following two tree-like structures:

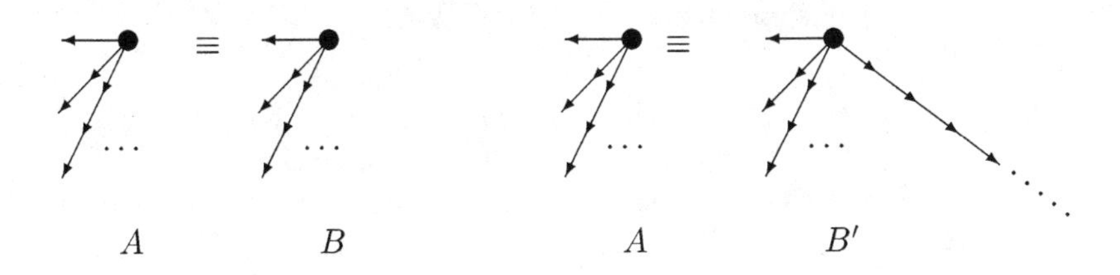

Claim. $\mathbf{M}'$ *is an elementary extension of* $\mathbf{M}$.

Proof. By a simple inductive argument, in these models, each first-order formula is equivalent to a Boolean compound of formulas whose quantifiers are either completely A-relativized or B-relativized. In particular, an atom '$x \equiv y$' says that one of x, y is in A and the other in B, and that both are roots. To prove the Claim, use the fact that both components of $\mathbf{M}'$ are elementary extensions of their counterparts in $\mathbf{M}$. $\dashv$

To complete the main argument, it is clear that $\equiv$ is a full trace equivalence in $\mathbf{M}$, but not in $\mathbf{M}'$: whence it cannot be first-order definable. $\dashv$

Proposition 10.5 *Complete Full Trace Equivalence is not* EC_Δ.

Proof. On the LTS $(\mathbb{N}, >)$ with transitions (i, j) with $i > j$ and success only at the initial state 0, identity of full trace sets coincides with real identity. Thus, $\equiv$ must be equal to $=$: i.e., the first-order formula

$$\forall xy\, (x \equiv y \leftrightarrow x = y)$$

holds. Now take any elementary extension. It has at least one copy of the integers in its tail, while $\equiv$ remains real identity. But all states in integer tails have the same trace set, and hence they share complete full trace equivalence. Thus, our class is not closed under elementary equivalence, whence it is not EC_Δ. $\dashv$

10.1.3 Bisimulation

As observed above, Bisimulation is EC by its definition. This works only if we have *finitely many* atomic actions (which we shall assume henceforth). By contrast:

Proposition 10.6 *Maximal Bisimulation is not* EC_Δ.

Proof. Consider the natural numbers $\mathbb{N}$ with predecessor as the single atomic action:

$$0 \longleftarrow 1 \longleftarrow 2 \longleftarrow \cdots \longleftarrow n \longleftarrow \cdots$$

The identity relation is the maximal bisimulation in this case (non-trivial identifications eventually clash at zero). Now, take any proper elementary extension of the LTS $\mathbb{N}$. It contains states with infinitely many predecessors (and successors) lying in copies of the integers, whose objects can be identified via a bisimulation. So the maximal bisimulation will not be real identity in the latter case. But, if the latter were definable by a set of first-order sentences $T(\equiv)$, then T would hold in both LTSs described: which is a contradiction as before. $\dashv$

10.1.4 Generated Graph Isomorphism

Similar reasoning establishes the following outcome.

Proposition 10.7 *Generated Graph Isomorphism and Complete GGI are not* EC_Δ.

These negative results do not exclude definitions for non-EC_Δ simulations in higher-order logics. For instance, Complete Trace Equivalence is definable in $L_{\omega_1\omega}$ by an infinite conjunction of path equivalences for each finite length. Maximal Bisimulation is rather Δ^1_2, having a form

$$\forall xy\,(x \equiv y \leftrightarrow \exists R\,(\text{“}R \text{ is a bisimulation”} \wedge Rxy)),$$

which reduces to a conjunction of a Σ^1_1 and a Π^1_1 sentence.

10.2 Axiomatization of Process Equivalences

Process equivalences also generate systems of reasoning. When does the set $\mathbf{K}$ of models for a process equivalence have a *recursively enumerable* set of first-order consequences? (By a classical result of Craig's, this implies recursive axiomatizability.)

Proposition 10.8 *The following process equivalences have a recursive axiomatization: Finite Trace Equivalence, Bisimulation, Generated Graph Isomorphism.*

Proof. (i) Finite Trace Equivalence was defined by an RE set of first-order sentences. Hence the latter is also an RE axiomatization for its first-order consequences. The same reasoning applies a fortiori to Bisimulation, with its finite first-order definition.

(ii) To axiomatize the first-order consequences of Generated Graph Isomorphism, we rely on the preservation result for GGI in Chapter 4. We proved (i) GGI preserves all 'restricted first-order formulas' $\phi(x)$ constructed from atoms by Boolean operations and restricted existential quantifiers $\exists y\,(R_a xy \wedge$, and (ii) in any countable recursively saturated model, the relation of agreeing on all restricted formulas is a generated graph isomorphism between states. Now, the relevant first-order theory is axiomatized by all first-order formulas

$$\forall xy\,(x \equiv y \rightarrow (\phi(x) \leftrightarrow \phi(y)), \text{where } \phi \text{ is a restricted first-order formula.}$$

These formulas all follow from GGI, by observation (i). Next, suppose some first-order ψ does not follow from these axioms. Then there is a countable model for the latter where ψ fails (by Completeness and Löwenheim-Skolem), which has a countable *recursively saturated* elementary extension (cf. Keisler 1977). But then, as before, $\equiv$ will be a generated graph isomorphism in the latter model, whence the non-consequence ψ fails in an LTS of the original class. $\dashv$

Proposition 10.9 *The first-order consequences of Maximal Bisimulation and Complete Trace Equivalence are not RE, and not even arithmetically definable.*

Proof. We do Maximal Bisimulation only (CTE is similar). On the natural numbers with predecessor as an atomic action, maximal bisimulation coincided with identity. Consider a first-order language whose vocabulary includes predicates that allow one to express the Peano Axioms for zero, successor, predecessor, addition and multiplication (but without Induction). Call this theory PA*. Now, on LTSs satisfying PA* where the binary relation $\equiv$ coincides with the maximal bisimulation, the following formula defines the initial segment of the natural numbers (the reason is that all numbers in the 'integer tails' admit of non-trivial bisimulations):

$$\phi(x) = \forall y\, (x \equiv y \to x = y).$$

This implies the following effective reduction of standard truth for arithmetical sentences α in terms of their syntactically relativized versions $(\alpha)^{\phi}$:

$\mathbb{N} \models \alpha$ iff

$PA^* \to (\alpha)^{\phi}$ holds in all LTSs where $\equiv$ is the maximal bisimulation.

Thus, the first-order consequences of Maximal Bisimulation are at least as complex as True Arithmetic — which is non-arithmetical by Tarski's Theorem. (Under suitable encoding, this argument will work for non-arithmetical vocabularies too.) $\dashv$

Remark 10.10 (Concrete Axioms) Completed or maximal process equivalences satisfy the theory of their 'compatible notion', while adding certain 'induction principles'. E.g., both Complete Finite Trace Equivalence and Maximal Bisimulation satisfy the following principle:

$$\forall xy\, ((\forall z\, (R_a xz \to \exists u\, (R_a yu \wedge z \equiv u)) \wedge \\ \forall z\, (R_a yz \to \exists u\, (R_a xu \wedge z \equiv u))) \to x \equiv y).$$

10.3 Infinitary Theory of Transition Systems

Many computational notions have a natural formalization in infinitary first-order logic.

10.3.1 Definability of Process Equivalences

Complete finite trace equivalence is easily defined by an infinite conjunction of successful path equivalences. Here are some negative results.

Proposition 10.11 *Full Trace Equivalence, Complete Full Trace Equivalence, Maximal Bisimulation, Generated Graph Isomorphism and Complete Generated Graph Isomorphism are not $L_{\omega_1\omega}$-definable.*

Proof. (i) Maximal Bisimulation. On *well-orders* with downward arrows for transitions, maximal bisimulation is identity. But on non-well-orders, any downward infinite chain yields a non-trivial bisimulation. Suppose that Maximal Bisimulation is $L_{\omega_1\omega}$-definable. Then saying that some relation $<$ is a linear order and $\equiv$ is its maximal bisimulation while

$$\forall xy\,(x \equiv y \leftrightarrow x = y),$$

defines the well-orderings with their maximal bisimulation in $L_{\omega_1\omega}$. This contradicts the non $L_{\omega_1\omega}$-ness of well-order.

(ii) The other assertions follow as in earlier sections, using the Löwenheim-Skolem Theorem for our countably infinitary language. ⊣

10.3.2 Scott Isomorphism Theorem

A corollary of the Invariance Theorem for bisimulation was that two finite LTSs satisfy the same modal formulas in states x, y iff there is a bisimulation $\equiv$ between them with $x \equiv y$. This folklore fact may be extended to the infinitary language $L_{\omega_1\omega}$ via the following modification of Scott's Theorem. The modal fragment is defined as in Chapter 4, with infinite conjunctions and disjunctions.

Theorem 10.12 *For each countable LTS* $\mathbf{M}$ *with state* a*, there exists a modal* $L_{\omega_1\omega}$*-formula* $\phi(x)$ *such that, for each countable LTS* $\mathbf{N}$ *and state* b*,* $\mathbf{N}, b \models \phi$ *iff there exists a bisimulation* $\equiv$ *between* $\mathbf{M}$*,* $\mathbf{N}$ *with* $a \equiv b$*.*

Corollary 10.13 *In countable LTSs* $\mathbf{M}$*,* $\mathbf{N}$*, states* a*,* b *verify the same modal* L_{ω_1}*-theory iff there exists a bisimulation connecting them.*

Proof. An easy adaptation of the usual argument suffices, via an infinitary description up to some countable ordinal. Cf. van Benthem and Bergstra 1994, or de Rijke 1995. ⊣

Here is another piece of folklore. Bisimulation occurs between any two models $\mathbf{M}, x, \mathbf{N}, y$ iff they have the same modal theories in the full infinitary set language $L_{\infty\omega}$.

10.3.3 Invariance under Bisimulation

Earlier preservation results have infinitary versions. Here is a key example.

Theorem 10.14 *The $L_{\infty\omega}$-formulas invariant for bisimulation are precisely the infinitary modal ones, constructed using arbitrary conjunctions and disjunctions.*

Proof. We adapt the proof sketch for the Invariance Theorem for the infinitary language $L_{\omega_1\omega}$ in van Benthem and Bergstra 1994, avoiding compactness or saturation. Of course, infinitary modal formulas (defined in the obvious way) are all invariant for bisimulation. Consider the converse. Let $\phi(x)$ be any formula of $L_{\infty\omega}$ without a modal equivalent. We construct two models $\mathbf{M}$, $\mathbf{N}$ with a bisimulation E between them which has a link aEb such that $\mathbf{M} \models \phi(a)$, $\mathbf{N} \models \neg\phi(b)$. This refutes invariance for bisimulation. The models are constructed from *good triples* A, Σ, Δ, where A defines the relation E, Σ the model $\mathbf{M}$, and Δ the model $\mathbf{N}$. But first, some auxiliary definitions.

(i) By *extended modal formulas* over a set of variables X we mean all formulas of $L_{\infty\omega}$ that are constructed using unary atoms Px ($x \in X$), Boolean operations (finite or infinite), and existential modal quantifiers $\exists y\,(Rxy \wedge [y/u]\psi)$ (where x, $u \in X$). (One can find a more perspicuous normal form for such formulas, but we shall not need this.) If $|X| = 1$, these are just ordinary infinitary modal formulas.

(ii) For convenience, we assume that all formulas have negations pushed inside to atoms, leaving only operators $\bigwedge$, $\vee$, $\forall$, $\exists$.

(iii) Let $\mu = \max(\aleph_0, |\mathrm{TC}(\neg\phi)|)$, where $\mathrm{TC}(\neg\phi)$ is the transitive closure of the formula $\neg\phi$, which contains (amongst others) all its infinitary subformulas.

(iv) Choose two disjoint sets C, D of *new* individual constants of cardinality μ^+ (a regular cardinal). Henceforth, all *formulas* will be C- or D-substitution instances of subformulas of $\neg\phi$. The total cardinality of this set is again μ^+. (In counting this size, recall that formulas of $L_{\infty\omega}$ have only finitely many free variables.)

(v) Call two sets of formulas Σ, Δ *modally inseparable* in A if there is no extended modal formula α over a set of variables X (of cardinality $< \mu^+$) with $\Sigma \models \alpha(\mathbf{c})$ and $\Delta \models \neg\alpha(\mathbf{d})$ — for subsets $\mathbf{c}$, $\mathbf{d}$ of C, D where corresponding pairs c/x, d/x have the atom cEd in A.

(vi) A *good* triple A, Σ, Δ satisfies four conditions: A, Σ, Δ have cardinality $< \mu^+$, A consists of bisimulation atoms cEd, Σ consists of formulas with constants only from C, and likewise for Δ and D — while Σ, Δ are modally inseparable with respect to A. Note that the good triples form a set.

Next, we state some closure conditions on good triples. The first are familiar from infinitary 'consistency properties' describing consistent diagrams of models (cf. Keisler 1971), whence we omit their detailed proofs. Let A, Σ, Δ be a good triple.

- adding to Σ all conjuncts of an infinitary conjunction in Σ again gives a good triple, and the same holds for infinitary conjunctions in Δ. (This addition does not affect modal separation, and it keeps the cardinality of Σ below μ^+.)
- adding to Σ all substitution instances of a universally quantified formula in Σ, with respect to all constants from C already occurring in Σ, again gives a good triple — and the same holds for universal quantifiers in Δ.
- each infinitary disjunction in Σ has at least one disjunct that can be added to Σ to produce a good triple, and the same holds for Δ. (If each disjunct leads to a modal separation for the extended triple, their disjunction will separate the original triple. Here, we need extended modal formulas, as the constants involved may be different. Also, some care is needed, by choosing disjoint sets of variables for each disjunct.)
- each existentially quantified formula in Σ has a substitution instance with some individual constant c that is new to Σ and A which can be added to Σ to form a good triple — and the same holds for Δ. (Notice that the new constant cannot trigger new modal separations, since it has no available bisimulation atoms.)

The final observations guarantee *atomic harmony* and *zigzags* for the bisimulation E:

- if $cEd \in A$ and $Pc \in \Sigma$, then A, Σ, $\Delta \cup \{Pd\}$ is good. (If modal α separates Σ and $\Delta \cup \{Pd\}$, then $Px \wedge \alpha$ separates the original Σ, Δ.) Vice versa for Δ.
- if $cEd \in A$ and $Rcc' \in \Sigma$, there is an atom d' new to A, Σ, Δ such that $A \cup \{c'Ed'\}, \Sigma, \Delta \cup \{Rdd'\}$ is a good triple — and vice versa. (If some modal α separates Σ and $\Delta \cup \{Rdd'\}$, then $\Diamond_x \alpha$ separates the original Σ, Δ.)

Finally, we construct our models. First, we enumerate all 'tasks' at hand, ensuring fair scheduling. Tasks are either C- (D-)formulas, to be verified in the models $\mathbf{M}$ ($\mathbf{N}$), or bisimulation atoms plus R-successor atoms, to be matched in the opposite model. The total number of tasks is μ^+ (by a simple calculation) — whence we can enumerate them in a sequence T_α ($\alpha < \mu^+$), so that each task occurs cofinally often. Now, we construct a corresponding sequence of good triples, starting with an initial triple

$$\{cEd\}, \{\phi(c)\}, \{\neg\phi(d)\}.$$

The latter is good. Modal separation would imply modal definability for ϕ: *quod non*. At each stage α, we take the component-wise union of all previous triples, and add formulas according to the scheduled task (if relevant), via the above closure properties. The final result is a triple A^*, Σ^*, Δ^*.

This defines two obvious models **M**, **N** (over domains consisting of those constants from C, D, respectively, which occur in Σ^*, Δ^*) plus a binary relation E between them in the obvious way. An easy induction then shows that

- C-formulas are in Σ iff they are true in **M**, and likewise for D-formulas and Δ
- E is a bisimulation between **M** and **N**. $\dashv$

This method of proof also works for the finitary first-order language.

10.3.4 Safety for Bisimulation

The main theorem of Chapter 5 has not found its counterpart yet. But the proof methods of Barwise and van Benthem 1996 provide strong evidence for the following

Conjecture 10.15 *A relational operation $O(R_1, \ldots, R_n)$ definable in $L_{\infty\omega}$ is safe for bisimulation if and only if it can be defined using atomic relations $R_a xy$ and atomic tests (q)? for propositional atoms q in our models, using only the three operations ;, $\bigvee$ and $\sim$, where the unions may now be infinitary.*

10.4 Special Transition Systems

On important special LTSs, axiomatization and definability must be re-analyzed.

10.4.1 Finite Models

Working within the universe of finite models may or may not decrease the complexity of our logical results. (In Finite Model Theory, many standard techniques fall away.)

Proposition 10.16 *Every earlier notion of process equivalence is* EC_Δ.

Proof. All definitions describe model classes closed under isomorphism. Every class of finite models, in a finite vocabulary, with this closure property is EC_Δ-definable, by the negations of all complete first-order descriptions of its finite non-members. $\dashv$

Proposition 10.17 *No process equivalence becomes* EC *which was not* EC *in general.*

Proof. Unlike compactness, Ehrenfeucht Games still work on finite models. E.g., the earlier argument for Finite Trace Equivalence goes through. Another typical case is Complete Finite Trace Equivalence, whose definability by any first-order sentence of quantifier depth N may now be refuted by comparing (i) a chain of immediate successors of length 2^N with $\equiv$ equal to the identity relation, (ii) the same chain plus a disjoint 'large circle'.

(Doets 1996 has alternative model-theoretic arguments, with detours via infinite integer models.) ⊣

Proposition 10.18 *A notion of process equivalence has a finite axiomatization iff it is* EC.

Proof. From right to left, the defining first-order formula is an axiomatization. Conversely, let δ axiomatize $\equiv$. Then, it holds in any model for this equivalence. Conversely, suppose δ fails in some finite LTS $\mathbf{M}$, fully described by the first-order sentence $\alpha_{\mathbf{M}}$. Then δ does not imply $\neg\alpha_{\mathbf{M}}$ on finite models, and so there is a finite model for $\delta \wedge \alpha_{\mathbf{M}}$, which is isomorphic to $\mathbf{M}$ itself. Therefore, δ holds in $\mathbf{M}$. ⊣

Rosen 1995 proves the modal preservation theorem for bisimulation on finite LTSs (despite general failures of such results in Finite Model Theory; cf. Gurevich 1985).

10.4.2 Further Classes of Transition Systems

On *finitely branching models*, here is a typical new outcome (with 'finite trace equivalence' over arbitrary finite traces, whatever the status of their end-points):

Proposition 10.19 *Finite Trace Equivalence and Full Trace Equivalence are equivalent.*

Proof. FullTE implies FinTE. Vice versa, let x, y be finite trace equivalent. Let any infinite trace start at x. Consider the subgraph at y matching all its initial segments. König's Lemma gives an infinite branch at y matching the infinite trace at x. ⊣

Finally, consider LTSs whose transition relations are all well-founded.

Proposition 10.20 *Maximal Bisimulation is* EC *on well-founded models.*

Proof. Consider one atomic action. A good definition here has the first-order definition of bisimulation plus the earlier 'Induction Principle'. Let C be any relation on a well-founded LTS (S, R_a) satisfying these two requirements which does not equal the maximal bisimulation $\equiv$. As C is a bisimulation, it is contained in $\equiv$. Now, by well-foundedness, take some minimal point x where a situation of the type $x \equiv y$, $\neg xCy$ occurs, with some y exemplifying this mismatch. Then, by minimality, it is easy to see that the condition triggering the induction principle is satisfied for all R_a-successors of x and y. (For instance, if $x \xrightarrow{a} z$, then z must have a $\equiv$-matching $\xrightarrow{a}$-successor u of y, because $\equiv$ is a bisimulation. By the minimality of x with respect to $\equiv/C$ mismatches, this implies that zCu. The proof for the converse is analogous.) Thus, the induction principle would imply that xCy after all: which yields a contradiction. ⊣

10.5 Process Algebra

Next, we link up with more elaborate process theories in the computational literature. We shall be concerned with elementary operations, ignoring higher process operations such as recursion, or 'hiding' of vocabulary.

10.5.1 Algebraic Operations and Term Models

To get the flavour of process algebra, here is *elementary process algebra* with finitely many atomic actions (a subset of CCS in Milner 1980), with signature

0	P	zero process
$+$	$P \times P \to P$	alternative composition ('choice')
$a;$	$P \to P$	action prefixing, for each atomic $a \in A$

EPA(A)

$$\begin{aligned} x + y &= y + x \\ (x+y)+z &= x+(y+z) \\ x + x &= x \\ x + 0 &= x. \end{aligned}$$

The *initial algebra*, or *term model* for these axioms is called A^{e}_{ω}. It turns out useful to introduce a first-order axiom system EPT with a relation of inclusion defined as usual:

$$x \subseteq y \Leftrightarrow x + y = y.$$

The theory EPT(A) adds the following principles to EPA(A):

$$\begin{aligned} & 0 \subseteq x \\ (x+y) \subseteq z \quad &\Leftrightarrow \quad x \subseteq z \wedge y \subseteq z \\ a\,;x \subseteq (y+z) \quad &\Leftrightarrow \quad a;x \subseteq y \vee a\,;x \subseteq z \\ a\,;x \subseteq a\,;y \quad &\Leftrightarrow \quad x \subseteq y \\ \neg a\,;x \subseteq b\,;y \quad & \qquad \text{if } a,\, b \text{ are distinct atomic actions} \\ \neg a\,;x \subseteq 0 \quad & \end{aligned}$$

Proposition 10.21 EPT(A) *is sound and complete for validity of both ground equations and inequalities in* A^{e}_{ω}.

10.5.2 The Finite Graph Model

For a more complex process logic, take the full first-order theory of the algebra A^{e}_{ω}. Equivalently, consider the model **G** of all *finite graphs* identified *under bisimulation*, with the following operations (which are well-defined over representatives):

0	single success node
$+$	fusion of graphs at the root
$a;$	prefixing a single incoming a-arrow to the root

Its first-order theory $Th(\mathbf{G})$ may actually be formulated in various ways:

Proposition 10.22 *The following theories are effectively equivalent:*

(i) $Th((\mathbf{G}, 0, +, \{a; \mid a \in A\}))$

(ii) $Th((\mathbf{G}, \subseteq, \{a; \mid a \in A\}))$.

This theory contains a full lattice structure for the above inclusion ordering $\subseteq$. Moreover, it exemplifies special features not found in the general case, such as

(E) $\forall xy\, (\bigwedge_{a \in A} \forall z\, (x \xrightarrow{a} z \leftrightarrow y \xrightarrow{a} z) \rightarrow x = y)$ Extensionality

Indeed, $Th(\mathbf{G})$ encodes a well-known mathematical structure — which we will sketch. (The following technical digression may be skipped without loss of continuity.)

Theorem 10.23 *The first-order theory of the Finite Graph Model under Bisimulation is effectively equivalent to True Arithmetic.*

Proof. Think of $\mathbf{G}$ as a labeled transition system of equivalence classes $\#\mathbf{G}$ of all finite rooted graphs $\mathbf{G}$ under bisimulations connecting their roots. Its relations include identity $=$ as well as transition relations $\xrightarrow{a}$ (for all a in some finite set A), defined via 'root prefixing' (this is well-defined on equivalence classes, being invariant for bisimulation). As will be shown shortly, the first-order theory $Th(\mathbf{G})$ gives us the whole process algebra with the above $\subseteq$, $+$ and all prefix operations $a;$. As usual, True Arithmetic is the complete first-order theory of the model $(\mathbb{N}, +, \cdot)$. Now, there are two directions to the above assertion:

From Graphs to Natural Numbers. Under any effective coding scheme, the following arithmetical predicates are recursive, and hence arithmetically first-order definable:

GRAPH(n)	"n codes some finite graph"
BISIM(m, n)	"m, n code bisimilar finite graphs"
a-TRANS(m, n)	"m encodes a graph having an a-transition from its root to the root of a generated subgraph encoded by n"

Now, $(\cdot)^*$ translates first-order formulas about these graphs into first-order arithmetic:

$x = y$	BISIM(x, y)
$x \xrightarrow{a} y$	$\exists zu\,$BISIM$(x, z) \wedge$ BISIM$(y, u) \wedge a$-TRANS(z, u)
$\neg, \wedge$	Boolean connectives are translated homomorphically
$\exists x$	existential quantifiers become restricted numerical quantifiers over codes of graphs: $\exists x\,(\text{GRAPH}(x) \wedge \ldots)$.

Claim. *For all finite graphs* $\mathbf{G}_1, \ldots, \mathbf{G}_k$ *with numerical codes* $g_1, \ldots, g_k$, *and all first-order formulas* ϕ,

$$\mathbf{G} \models \phi[\#\mathbf{G}_1, \ldots, \#\mathbf{G}_k] \text{ iff } \mathbb{N} \models (\phi)^*[g_1, \ldots, g_k].$$

From Natural Numbers to Graphs. In the opposite direction, we need auxiliary first-order predicates on (equivalence classes of) finite graphs, encoding basic numerical operations. Each natural number n is identified with the (equivalence class of) a finite linear order $\xrightarrow{a}$ over $\{0, \ldots, n\}$. Thus, 'numerical graph objects' are defined by a-NUM(x) saying: "x is a graph with just a-transitions, on which $\xrightarrow{a}$ is transitive, irreflexive and linear". Isomorphism of disjoint intervals on linear graphs amounts to there being a larger graph containing these intervals, with a second relation $\xrightarrow{b}$ serving as a bisimulation between them. More precisely, this involves the following first-order predicates:

BISIM(x, y, s) s has only $\xrightarrow{a}$- and $\xrightarrow{b}$-successors, and $s \xrightarrow{b} x$, the $\xrightarrow{a}$-successors of s form a transitive, irreflexive linear ordering, and $\xrightarrow{b}$ is a bisimulation between $\xrightarrow{a}$-successors of s and points in the closed $\xrightarrow{a}$-interval $[x, y]$ with respect to $\xrightarrow{a}$

EQUI(x, y, z, u) $\exists s\,(\text{BISIM}(x, y, s) \wedge \text{BISIM}(z, u, s))$

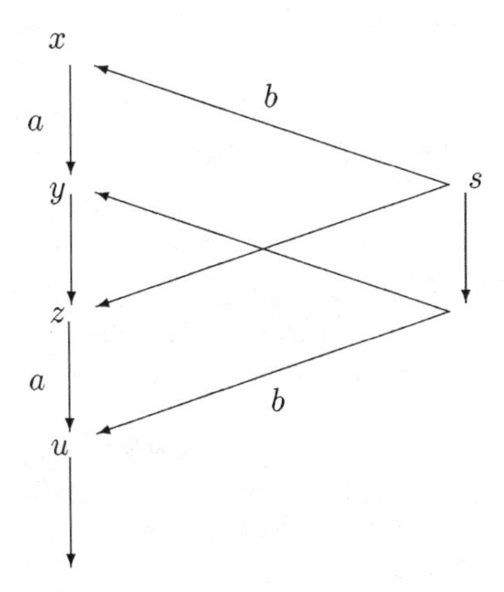

In the Finite Graph Model, the latter predicate holds of intervals on a numerical graph iff they have the same cardinality. Care is needed here, as objects in $\mathbf{G}$ are not graphs, but their equivalence classes under bisimulation. For suitably 'bounded' first-order formulas (cf. Chapter 4), there is no real difference between the two kinds of object.

- One may describe *addition* by a first-order predicate SUM(x, y, z) which first distinguishes some marginal cases (e.g., "$x = 0$" and "$y =$

0" are trivial), and then states that x, y lie in some sequence on the linear order z, say $x \geq y$, where the top interval $[z, x]$ is equinumerous to the bottom interval on z starting from y.

- Multiplication TIMES(x, y, z) is encoded via a division of the linear order z into y intervals, all of equal length (introduce a larger graph with suitable new $\xrightarrow{c}$ transitions marking the boundary points), stating that each interval is equinumerous with x while the subgraph of boundary points (identifiable via the new relation $\xrightarrow{c}$) is equinumerous with y. (Some trivial cases are to be treated separately.)

Finally, we define the relevant first-order translation $(\cdot)^\$$. Without loss of generality, the arithmetical first-order language may be restricted to atoms $x = y + z$ and $x = y \cdot z$. These are translated via the above predicates SUM(x, y, z) and TIMES(x, y, z). Boolean connectives are treated homomorphically, and numerical existential quantifiers are restricted to the above predicate a-NUM defining the numerical graph equivalence classes (cf. (1)). The relevant reduction is then straightforward from the above motivation:

Claim. *For all natural numbers $n_1, \ldots, n_k$, with matching graph (classe)s $N_1, \ldots, N_k$, and all first-order formulas ϕ, the following equivalence holds:*

$$\mathbb{N} \models \phi[n_1, \ldots, n_k] \text{ iff } \mathbf{G} \models (\phi)^\$[N_1, \ldots, N_k]. \quad \dashv$$

10.5.3 From Algebras to Labeled Transition Systems

Any model $\mathbf{M}$ for the above process theories induces a LTS $\mathbf{TS}(\mathbf{M})$:

$$x \xrightarrow{a} y \Leftrightarrow \exists z\, x = a\,;y + z.$$

Several open questions of definability and axiomatizability arise. Are the two classes $\mathbf{TS}$(MOD(EPA(A)), $\mathbf{TS}$(MOD(EPT(A)) EC or EC$_\Delta$? Our conjecture is negative. Both are Σ^1_1-definable in second-order logic, as their defining condition states the existence of predicates or functions verifying the finite list of EPA (EPT) axioms. By general logic, this implies recursive axiomatizability for their first-order theories. One would like explicit axiomatizations. Analyzing the signature $(0, +, \{a; \mid a \in A\})$ via the above stipulation, one finds the following first-order principles for $\mathbf{TS}(A^e_\omega)$:

A_1	$\exists x \bigwedge_{a \in A} \neg \exists y\, x \xrightarrow{a} y$	End-points
A_2	$\forall xy \exists z \bigwedge_{a \in A} \forall u\, (z \xrightarrow{a} u \leftrightarrow (x \xrightarrow{a} u \vee y \xrightarrow{a} u))$	Sums
A_3	$\forall x \bigwedge_{a \in A} \exists y\, (\forall z\, (y \xrightarrow{a} z \leftrightarrow z = x) \wedge \bigwedge_{b \in A, b \text{ distinct from } a} \neg \exists z\, y \xrightarrow{b} z)$.	Predecessors

Proposition 10.24 MOD$(\{A_1, A_2, A_3, E\}) \subseteq \mathbf{TS}$(EPA)

Proof. A1, A2, A3, E follow immediately from EPT under the above stipulation. Conversely, given these principles, one can introduce operations of 'sum' and 'action prefixing' with the right properties. For instance, Associativity follows because both possible ways of computing a 'sum' of three objects lead to points with the same successors, which must be identical by Extensionality. $\dashv$

Often, a process algebra is retrievable from its derived transition system:

Proposition 10.25 *The following first-order theories are effectively equivalent for the model* $\mathbf{M} = (\mathbf{G}, 0, +, \{a; \mid a \in A\})$*:*

(i) $Th(\mathbf{M})$,

(ii) $Th(\mathbf{TS}(\mathbf{M}))$.

10.6 Safety for Process Constructions

In Process Algebra, operations are described by algebraic equations. Examples were *action prefix* $\mathbf{ax}$, *choice* $\mathbf{x}_1 + \mathbf{x}_2$ or *product* $\mathbf{x}_1 \cdot \mathbf{x}_2$. Other typical operators deal with parallel execution of processes, such as the *merge* $\mathbf{x}_1 \parallel \mathbf{x}_2$ and its directed variants. The equations governing these operations are reminiscent of an algebra of relations. But this time, only one distributive law holds:

$$(x + y) \cdot z = (x \cdot z) + (y \cdot z) \text{ but not } x \cdot (y + z) = (x \cdot y) + (x \cdot z).$$

From a modal point of view, this divergence with Chapters 3, 5 may be explained as follows. Process algebra lies 'one level up'. It is not a calculus of binary *relations inside* labeled transition systems, but rather an external calculus of *constructions over* such systems. This shows clearly in the quite different constructions for the two distributive laws in the Finite Graph Model. Even so, the situation is not fully clear. In what follows, we sketch two ways of comparing dynamic logic and process algebra, one more 'internal', and the other more 'external'.

10.6.1 Varieties of Safety

Safety for bisimulation (cf. Chapter 5) makes sense for process-algebraic operations O, too, when viewed as constructions over LTSs. In the literature, the latter are usually taken to satisfy the following minimal requirement of *Respect for Bisimulation*:

whenever $\mathbf{M}_1 \equiv_1 \mathbf{N}_1$, $\mathbf{M}_2 \equiv_2 \mathbf{N}_2$, ..., then $O(\mathbf{M}_1, \mathbf{M}_2, \ldots)$ admits of a bisimulation with respect to $O(\mathbf{N}_1, \mathbf{N}_2, \ldots)$.

This condition is non-trivial: mathematical operations in general only require respect for *isomorphisms* between their values and arguments. Even so, Respect is very liberal. In addition to prefix, sum, product or merge, many other process operations pass this test. Here are two examples — that make good sense as constructions on finite automata:

Process Negation: reverse all success/failure markings in an LTS

Process Iteration: join all success nodes to all successors of the initial node

Respect for Bisimulation suggests a hierarchy of looser or stricter process operations. Usually, the bisimulation between the two labeled transition systems $O(\mathbf{M}_1, \mathbf{M}_2, \ldots)$ and $O(\mathbf{N}_1, \mathbf{N}_2, \ldots)$ is effectively constructed from those for the arguments $\mathbf{M}_i$, $\mathbf{N}_i$. Varying requirements on effectiveness will fine-structure our options. We will proceed by example, in order to motivate more concrete requirements.

10.6.2 Analyzing Specific Constructions

Action Prefix. The operation **ax** takes a rooted LTS x, and gives it a new root y joined to the old only by one a-arrow — as happens in the following picture

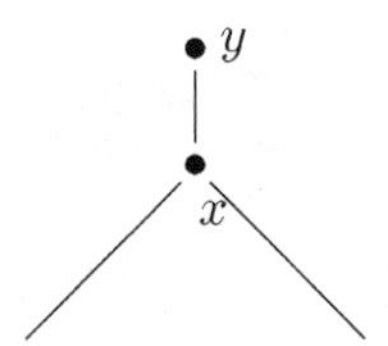

This definition may even be analyzed inside LTSs — via a special relation AP(y, x) which holds between two states just when y heads a submodel of the required form with respect to the submodel headed by x. Here, x serves as a distinguished 'state parameter' (Chapter 5). Thus, the new LTS **ax** may be fully described by an explicit specification of the *new relations* in the submodel generated by y. In the present case, these definitions look as follows:

$$R_a^{\text{new}}uv \quad := \quad R_a^{\text{old}}uv \vee (u = y \wedge v = x)$$
$$R_b^{\text{new}}uv \quad := \quad R_b^{\text{old}}uv \text{ for all other atomic actions } b.$$

Note that y does not occur as a successor of itself. *We make this assumption throughout*. Respect for bisimulation has a straightforward meaning for Action Prefix. Any bisimulation between two models $\mathbf{M}$, $\mathbf{M}'$ connecting x to x' remains a bisimulation when we add one link between y and y'. Let us call this the *Bisimulation Extension Property* (BEP). It may be analyzed in terms of safety after all (Chapter 5). For, what matters is only how the R_a^{new} lie inside the new LTS **ax**. As the generated submodel of x does not change in the construction, we need only specify the unary predicate S_a of being an R_a-successor of y, for each action a. BEP says this predicate is *invariant* for bisimulations with distinguished parameter x. Thus, S_a has a syntactic normal form described in Chapter 5, with just safe operations and resetting to x. The above definition indeed has this format: $S_a(v) = \text{RES}_x(x, v)$. Essentially this same analysis works for our next process operation.

Binary Choice. Consider the following defining picture for $\mathbf{y} = \mathbf{x}_1 + \mathbf{x}_2$

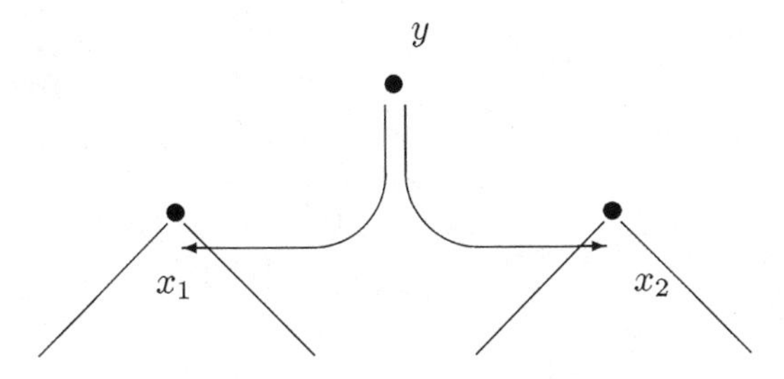

Again, Respect holds in a strict sense. The bisimulation for the sum is the union of those for the arguments, plus one new link $\langle y, y' \rangle$. The preceding analysis gives us a similar defining format, now with both RES_{x_1} and RES_{x_2} allowed in defining R_a^{new} via S_a. As a check, we have indeed that

$$R_a^{\mathrm{new}}uv := R_a^{\mathrm{old}}uv \vee (u = y \wedge (R_a x_1 v \vee R_a x_2 v)).$$

The second conjunct here has the right normal form, viz.

$$(\mathrm{RES}_{x_1} ; R_a) \cup (\mathrm{RES}_{x_2} ; R_a).$$

Negation and Iteration. Similar considerations apply to other process operations. We state some outcomes without proof. Negation leaves a graph 'the same' except for reversing success markings. This yields a different item in the LTS-representation of machine diagrams (or the Finite Graph Model). But the *original* bisimulation serves for the new LTS, as it leaves success marks $\surd$ invariant. This shows in the new relations:

$$\begin{aligned} R_a^{\mathrm{new}}uv &:= R_a^{\mathrm{old}}uv \\ \surd^{\mathrm{new}}v &:= {\sim}((\surd)?) \text{ with the safe strong negation } \sim \\ P^{\mathrm{new}}v &:= P^{\mathrm{old}}v \text{ for all other atomic propositions } P. \end{aligned}$$

For iteration, the same analysis applies (with the slight complication that the root may now also occur as the end-point of a new transition). Here is the resulting definition:

$$R_a^{\mathrm{new}}uv := R_a^{\mathrm{old}}uv \vee ((\surd)? ; \mathrm{RES}_x ; a).$$

Product. Our final example is more complex. In a sequential product $\mathbf{y} = \mathbf{x}_1 \cdot \mathbf{x}_2$, the new relations R_a^{new} consist of R_a^{old} plus all new pairs (u, v) with u a success node in the generated graph for the argument x_1, and v some R_a-successor of x_2:

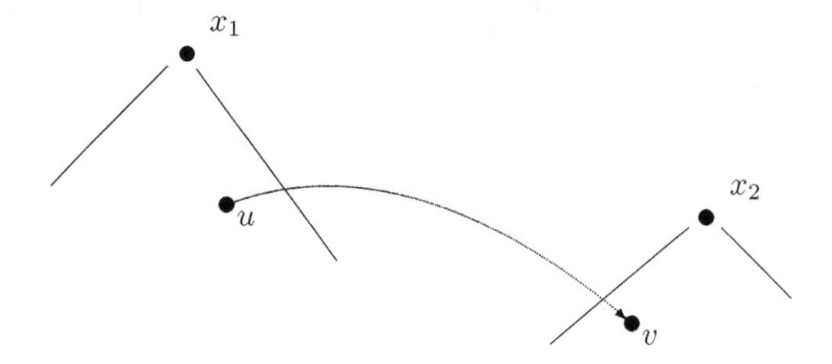

Here, the previous format of definition fails. It cannot enforce that success nodes must lie inside the *submodel generated by* x_1. We lack Respect in the earlier sense. We need predicates D_x defining submodels generated by x. Semantically, we want safety only for bisimulations that leave the latter in place. Still, this can be handled as in Chapter 5, by 'propositional parametrization'. This shows in the actual definitions for this case:

$$R_a^{\text{new}} := R_a^{\text{old}} \cup (D_{x_1})? \,;\, \text{RES}_{x_2} \,;\, R_a^{\text{old}}.$$

10.6.3 Respect for Sequence Extensions

With further process operations, such as *direct products* $\mathbf{x} * \mathbf{y}$ or *merges* $\mathbf{x} \parallel \mathbf{y}$, states themselves change to *ordered pairs* of old states. One cannot just specify new relations. To overcome this, Hollenberg 1995b considerably extends the previous analysis of Bisimulation Extension. Consider a rich first-order language whose vocabulary includes unary predicates S_i for argument domains, binary transition relations $R_{(a,i)}$ (encoding R_a over S_i), unary predicates P_i (encoding P over S_i), constants s_i for roots of the arguments and s for the new root, plus a ternary sequence concatenation predicate C. This language is a powerful format for defining process operations which covers most cases in the algebraic literature. Here are its preferred first-order models. For any finite family of LTSs $(\mathbf{M}_i)_i$ representing n arguments, the *sequence model* $\mathbf{M} = P_C(\mathbf{M}_i)_i$ has a universe of all finite sequences up to length n from the S_i plus a new set C (of distinguished nodes), closed under concatenation. Instead of reciting formal definitions, we show by example how process operations can be defined over sequence models using the above formalism:

Sequential Product

$$\begin{aligned}
\rho_a(x,y) &:= R_{(a,1)}xy \vee R_{(a,2)}xy \vee (\surd_1 x \wedge R_{(a,2)}s_2 y) \\
\delta_{\surd}(x) &:= \surd_2 x \vee (\surd_1 x \wedge \surd_2 s_2) \\
s &:= s_1
\end{aligned}$$

Free Merge

$$\begin{aligned}
\rho_a(x,y) &:= \exists x_1x_2y_1y_2\,(Cx, x_1x_2 \wedge Cy, y_1y_2 \wedge \\
&\quad (R_{(a,1)}x_1y_1 \wedge x_2 = y_2) \vee (x_1 = y_1 \wedge R_{(a,2)}x_2y_2)) \\
\delta_{\surd}(x) &:= \exists x_1x_2\,(Cx, x_1x_2 \wedge \surd_1 x_1 \wedge \surd_2 x_2) \\
s &:= s_1s_2.
\end{aligned}$$

Separate bisimulations for the argument domains S_i may be lifted canonically to their *sequence extension* in the product model. We say that a first-order definition for an operation O (as illustrated in the preceding examples) *respects sequence extensions* if each sequence extension is a bisimulation for its LTS-value $O(S_1, S_2, \ldots)$ in product models. Clearly, this condition will hold if the definition has a suitable modal format. Hollenberg proposes 'safe/invariant' definitions, which are sufficient (though not necessary) to enforce respect — and which covers all ACP-style process operations (Baeten and Weijland 1990). His main result is a modal characterization of safe/invariant process operations as precisely those definable in a suitable extended modal fragment of the above full first-order language. As in Chapter 5, this outcome suggest a natural *Safety Hierarchy* for process operations — though this time, for LTS constructions.

10.7 Appendix I: Modal Logic of Constructions

There are other modal viewpoints on process operations. The above LTS-constructions resemble modal ones. Choice is 'disjoint rooting' (Chapter 4), and sequential product 'glueing' for modal models (cf. Blok 1979, Marx 1995). Also, modal merge products have been studied. (Goldblatt 1988 has a category-theoretic vista.) A main concern has been *preservation*: what truth values for modal formulas in a compound arise from those in components? There is no explicit modal calculus of model constructions as such. We briefly explore the latter, as another promising interface with process algebra.

10.7.1 A Modal Calculus

Take ordinary Kripke models or LTSs $\mathbf{M}$, identifying processes x with nodes via their generated submodels as before. We introduce two new modalities $\{a\}$ and $+$, corresponding to Action Prefix and Choice, on top of the standard modal propositional language with standard modalities $\langle a \rangle$, $[a]$ over these models:

- $\mathbf{M}, x \models \langle a \rangle \phi$ iff there exists some R_a-successor y of x with $\mathbf{M}, y \models \phi$
- $\mathbf{M}, x \models \{a\} \phi$ iff there exists some y which is a unique R_a-successor of x, such that x has no other R_b-links to y, with $\mathbf{M}, y \models \phi$

- $\mathbf{M}, x \models \phi + \psi$ iff there exist states y, z with $\mathbf{M}, y \models \phi$ and $\mathbf{M}, z \models \psi$ such that (for all atomic actions a) the R_a-successors of x are just those of y and z.

10.7.2 Frame Correspondences

The new modal operators satisfy principles, whose semantic content may be gauged by frame correspondences (Chapter 3). We start with three valid modal distribution laws:

$$\begin{aligned} \{a\}(\phi \vee \psi) &\leftrightarrow \{a\}\phi \vee \{a\}\psi \\ (\phi_1 \vee \phi_2) + \psi &\leftrightarrow (\phi_1 + \psi) \vee (\phi_2 + \psi) \\ \phi + (\psi_1 \vee \psi_2) &\leftrightarrow (\phi + \psi_1) \vee (\phi + \psi_2). \end{aligned}$$

The next principle expresses that successors for Action Prefix are unique:

$$\{a\}\phi \wedge \{a\}\psi \rightarrow \{a\}(\phi \wedge \psi).$$

Finally, here are principles regulating the interaction between the new modalities $\{a\}$, $+$ and the standard ones $\langle a\rangle$, $[a]$, which approximate the intended interpretation for their associated accessibility relations Ax, y and Cx, yz:

$$\begin{array}{rl} \{a\}\phi \rightarrow \langle a\rangle\phi & \forall xy\, Axy \rightarrow R_a xy \\ \{a\}\top \rightarrow \neg\langle b\rangle\top \text{ (all other } b) & \forall xy\, Axy \rightarrow \neg\exists z\, R_b xz \\ \{a\}\top \wedge \langle a\rangle\phi \rightarrow [a]\phi & \forall xyzu\, (Axy \wedge R_a xz \wedge R_a xu) \rightarrow z = u) \\ [a]\phi + [a]\phi \rightarrow [a]\phi & \forall xyzu\, (Cx, yz \wedge R_a xu) \rightarrow (R_a yu \vee R_a zu)) \\ \langle a\rangle\phi + \top \rightarrow \langle a\rangle\phi & \forall xyzu\, (Cx, yz \wedge R_a yu) \rightarrow R_a xu \\ \top + \langle a\rangle\phi \rightarrow \langle a\rangle\phi & \forall xyzu\, (Cx, yz \wedge R_a zu) \rightarrow R_a xu \\ \phi + \psi \rightarrow \psi + \phi & \forall xyz\, Cx, yz \rightarrow Cx, zy \\ \phi \rightarrow \phi + \phi & \forall x\, Cx, xx. \end{array}$$

The last lines state algebraic commutativity and idempotence $\mathbf{x}+\mathbf{y} = \mathbf{y}+\mathbf{x}$, $\mathbf{x}+\mathbf{x} = \mathbf{x}$. By contrast, a modal counterpart $\phi + (\psi + \chi) \leftrightarrow (\phi + \psi) + \chi$ to *Associativity* requires the existence of enough 'union' states (Chapter 8) — which need not be valid in all models.

10.7.3 Extended Modal Logic

The preceding principles express only half of the desired behaviour. For Action Prefix, we lack the converse fact that, if there is some unique R_a-successor, R_b-inaccessible for all other b, which satisfies ϕ, then $\{a\}\phi$ holds. Expressing this fact requires a stronger modal language, say, with a 'difference operator' D (cf. Chapter 6):

$$(\langle a\rangle\phi \wedge \neg\langle a\rangle D\phi \wedge \bigwedge_{b \neq a} \neg\langle b\rangle\top) \rightarrow \{a\}\phi.$$

A corresponding converse statement for $+$ will involve backward versions of $\langle a \rangle$, $[a]$.

10.7.4 First-Order Translation

Our modal language admits of translation into a standard first-order one (Chapter 3):

- $ST(\{a\}\phi) :=$
  $\exists y\,(R_a xy \wedge ST(\phi)(y) \wedge \neg\exists z\,(R_a xz \wedge z \neq y) \wedge \bigwedge_{b \neq a} \neg\exists z\, R_b xz)$
- $ST(\phi + \psi) :=$
  $\exists yz\,(\forall u \bigwedge_a (R_a xu \leftrightarrow (R_a yu \vee R_a zu)) \wedge ST(\phi)(y) \wedge ST(\psi)(z))$.

The modal logic inherits RE-ness of first-order logic — but ST does not guarantee *decidability*. (Transcriptions for modal formulas may be outside of the Guarded Fragment of Chapter 4.) The sublanguage with $\{a\}$ is guarded, and hence decidable. Also, the modal language of $\{a\}$ only ends up in a *three-variable fragment* of first-order predicate logic, while $+$ needs a *four*-variable fragment (Chapters 3, 4).

10.8 Appendix II: Concurrent Dynamic Logic

Here is yet a different modal process theory. *Concurrent* PDL (CPDL) has the following syntax.

$$\begin{array}{ll} \text{formulas} & p \mid \neg\phi \mid (\phi \wedge \phi) \mid \langle\pi\rangle\phi \mid [\pi]\phi \\ \text{programs} & a \mid \pi^* \mid (\pi \,;\, \pi) \mid (\pi \cup \pi) \mid (\pi \cap \pi) \mid (\phi)? \end{array}$$

The new operation $\pi_1 \cap \pi_2$ performs π_1 and π_2 concurrently. I.e., several states can be reached by a number of independent processes triggered by a program. Thus, the reachability relation R_π of a program π connects a state s to the *set* of all states reached by these internal processes after successful execution. To model indeterminism, we also allow different output sets here. Here is the truth definition supporting this system (Peleg 1987). Note that the two modalities are no longer duals:

$$\begin{array}{lll} \mathbf{M}, s \models \langle\pi\rangle\phi & \text{iff} & \text{there exists } T \text{ with } sR_\pi T \text{ and for all } t \in T,\, \mathbf{M}, t \models \phi \\ \mathbf{M}, s \models [\pi]\phi & \text{iff} & \text{for all } T, \text{ if } sR_\pi T, \text{ then for all } t \in T,\, \mathbf{M}, t \models \phi. \end{array}$$

Here are the intended operations. Program disjunction is union, as in standard dynamic logic. Program *intersection* corresponds to the following set-theoretic operation $\#$:

$$R \# S := \lambda x, Y.\, \exists Y_1, Y_2\; xRY_1 \wedge xSY_2 \wedge Y = Y_1 \cup Y_2.$$

Sequential product now gets a more complicated description, via this operation $\ominus$:

$$\begin{array}{rcl} R \ominus S & := & \lambda x, Y.\, \exists Z\, xRZ \wedge \exists f : Z \to \mathrm{pow}(S) \\ & & \forall z \in Z\; zSf(z) \wedge Y = \bigcup\{f(z) \mid z \in Z\}. \end{array}$$

For a complete axiomatization, cf. Goldblatt 1992. We merely generalize bisimulation.

Definition 10.26 (Concurrent Bisimulation) A binary relation $\equiv$ between states in models **M**, **N** is a *concurrent bisimulation* if

(i) if $s \equiv r$, then s, r verify the same propositional atoms
(iia) if $s \equiv r$ and sRZ, then there exists U with rRU and for all $u \in U$, there exists a $z \in Z$ with $z \equiv u$
(iib) vice versa in the opposite direction.

This definition is motivated as follows (van Benthem, van Eijck and Stebletsova 1994).

Proposition 10.27 *All CPDL formulas are invariant for concurrent bisimulation. All CPDL operations are safe for concurrent bisimulations.*

10.9 Notes

Process Algebra. Key references are Milner 1980, Bergstra and Klop 1984, Hennessy and Milner 1985. A text book is Baeten and Weyland 1990. Cf. also the survey chapter Stirling 1989. An elegant extension with process iteration is in Fokkink 1994, Bergstra, Bethke and Ponse 1994. Being based on bisimulation equivalence over graphs or trees, rather than flat strings, process calculi often throw interesting new light on regular algebra and other traditional theories. An example is the decidability of strong (bisimulation) equivalence for context-free grammars (Baeten, Bergstra and Klop 1992).

Silence. Many process theories have 'silent steps' $\xrightarrow{\tau}$ for unobservable transitions. These arise by hiding certain transitions in an LTS under one relation. Structures then have the form $(S, A, \rightarrow, \xrightarrow{\tau})$. A binary relation $\equiv$ is a weak bisimulation between transition systems with silent steps if the following conditions hold, whenever $x \equiv y$: (i) atomic markings agree on x and y, (ii) if $x \xrightarrow{a} x'$, then there exists some finite sequence of transitions on the other side of the form

$$y = y_1 \xrightarrow{\tau} \cdots \xrightarrow{\tau} y_n \xrightarrow{a} y_{n+1} \xrightarrow{\tau} \cdots \xrightarrow{\tau} y_{n+k} = y'$$

such that $x' \equiv y'$; and vice versa, (iii) if $x \xrightarrow{\tau} x'$, there exists a finite sequence of transitions on the other side of the form

$$y = y_1 \xrightarrow{\tau} \cdots \xrightarrow{\tau} y_n = y'$$

such that $x' \equiv y'$; and vice versa. De Nicola and Vaandrager 1990, van Benthem, van Eijck and Stebletsova 1994 give variations. It is easy to show by previous methods that weak bisimulation is non-EC_Δ on all LTSs, but EC on so-called 'Δ-saturated' ones. Invariance results make sense here, too. But they are delicate, due to an asymmetry in the above definition. A restricted existential quantifier for a $\xrightarrow{a}$ step on one side need not match with one such quantifier on the other, because of τ-prefixes and suffixes. Quantification must be over successors in compound relations $\tau^* ; a ; \tau^*$. In $L_{\omega_1\omega}$, we conjecture a preservation theorem for invariance under weak bisimulation (Chapter 4).

Process Algebra, Set Theory and Modal Logic. Analogies between Modal Logic and Process Algebra remain elusive. This chapter has by no means settled all. Hollenberg 1996b takes a big leap forward by moving to a modal logic of sequences. Along a quite different route, Janin 1996 links up definability of program constructions in *propositional μ-calculus* with invariance for bisimulation (cf. the relevant passages in Chapters 4, 5), using tree automata techniques. Another route would run via a third party, viz. Set Theory. Process operations look like Δ_0-definable set-theoretic operations, and Milner's process algebra has inspired the non-well-founded set theory of Aczel 1988. Equivalence classes of graphs may be identified with (non-well-founded) sets, which simplifies analysis of the finite graph model (it becomes a non-well-founded analogue of the natural numbers). Thus, some of the theories mentioned here become fragments of set theory. This is the promising research program in Barwise and Moss 1996, d'Agostino 1996.

Alternative Process Theories. One main point of this chapter is that first-order process theories may be attractive alternatives to existing ones. Here are two examples. (1) The first-order theory of *trace models* over branches in *finite trees* is decidable (by an effective reduction to Rabin's Theorem). This should be contrasted with the high complexity found for the finite graph model. (2) The first-order theory of unwound infinite trace models is effectively axiomatizable, using Ehrenfeucht-style 'compression arguments' as in Doets 1987, or Backofen, Rogers and Vijay-Shanker 1995.

Acknowledgment

The work reported here on first-order theories of labeled transition systems and their connections with equational theories in process algebra is mostly due to joint research with *Jan Bergstra* (Programming Group, University of Amsterdam and Department of Applied Logic, University of Utrecht). This is part of a current effort in the Netherlands interfacing the research programs of Dynamic Logic and Process Algebra. Cf. the recent bridge publications van Benthem, van Eijck and Stebletsova 1994, van Benthem and Bergstra 1994 and Ponse 1994 — as well as the state-of-the-art collections van Eijck and Visser, eds. 1994, and Ponse, de Rijke and Venema, eds., 1995.

10.10 Questions

First-Order Theories of LTSs. Some obvious questions occur in the main text. Notably, are the two model classes **TS**(MOD(EPA(A)) and

TS(MOD(EPT(A)) EC or EC_Δ? The referee for this volume has made some further intriguing suggestions. The standard first-order theory for the Finite Graph Model is effectively equivalent with True Arithmetic. But what happens with its *bounded fragments* (defined as in Chapter 4)? More generally, what happens with suitable arrow versions (Chapter 8) of the LTS structures of this chapter? Do the resulting 'core process theories' become decidable?

Complexity. What is the complexity of the various process logics in this chapter?

Special Models. How does the theory of this chapter specialize to finite models?

Safety for Bisimulation. Prove the Safety Theorem for infinitary logic. (Cf. Barwise and van Benthem 1996 for relevant model-theoretic techniques replacing compactness.) Prove such a theorem for LTSs with *silent steps*. Prove a safety theorem for first-order definable operations $\pi(x, A)$ that are safe for *concurrent* bisimulations (CPDL). Find an explicit syntactic characterization of *respect for sequence extensions* in Hollenberg's first-order language. Also, how does the safe/invariant modal first-order language of Hollenberg relate to the Guarded Fragment of Chapter 4? In particular, is it decidable?

Higher Process Operations. Extend the analysis of this chapter to process calculi with axioms for the more complex operations of *recursion* and *hiding*.

Modal Logic of Model Constructions. Is our modal logic of the LTS-constructions $\{a\}$ and $+$ decidable? How can it be completely explicitly axiomatized? Can it be extended to deal also with sequential and parallel *products* of processes? Are there effective embeddings from existing process calculi into such modal construction logics?

11

Imperative Aspects of Logic Programs

In computer science, logic is often associated with declarative matters like representation of data or description of states. Thus, *logic programming* has arisen as a static formalism describing desired outcomes of computations, leaving all procedural details to the machine. But the general dynamic logic of this book focuses on sequential aspects of programs and their induced processes. Is it at cross-purposes with the logical trend in computer science? In this chapter, we are going to show how imperative aspects pervade logic programming. Processes of deductive approximation run its semantics, its computations are run by calculi of dynamic inference, and other imperative aspects abound for those who have eyes to see. Invading a stronghold of declarativism would seem to provide the strongest possible evidence for a dynamic approach.

11.1 Declarative Versus Procedural?

Let us summarize some features of the dynamic paradigm, as introduced in Chapters 1, 2, and developed in Chapters 4–7. It starts from a space of information states, ordered by inclusion or possible extension — over which propositions are (possibly many-valued) procedures F. These may satisfy special *constraints*, such as *monotonicity*, making F preserve informational inclusion, or *idempotence*: which says that cognitive operations on information states need not be repeated. (This is true of teaching machines, not of teaching humans). Other candidates are forms of *continuity*, to which we return later on. Moreover, there are repertoires of *logical operators* forming new transformations out of old. Finally, dynamic propositions suggest several dynamic styles of *inference*:

Test-to-Test $P_1, \ldots, P_n \models_{\mathrm{I}} C$ if $\bigcap_{1 \leq i \leq n} \mathrm{fix}(P_i) \subseteq \mathrm{fix}(C)$
Update-to-Update $P_1, \ldots, P_n \models_{\mathrm{II}} C$ if $\circ_{1 \leq i \leq n} P_i \subseteq C$
Update-to-Test $P_1, \ldots, P_n \models_{\mathrm{III}} C$ if $\mathrm{range}(\circ_{1 \leq i \leq n} P_i) \subseteq \mathrm{fix}(C)$.

In case this criterion is only applied to a state 0 of no information, we get:

Zero-Update-to-Test $P_1, \ldots, P_n \models_{\text{III}\#} C$ if range$(\circ_{1\leq i\leq n} P_i)[0] \subseteq$ fix(C)

Different styles may have different *structural rules*. They live in a two-level architecture for dynamic logic, with declarative and dynamic modules related by modes and projections. Next, we turn to the world of logic programming. The official ideology of this paradigm is as follows. *Programs* P are sets of universal Horn clauses in first-order logic, of the form

$$\forall x_1 \ldots x_k \text{ (conjunction of atoms} \rightarrow \text{single atom).}$$

Special cases are allowed. Thus, quantifier-free cases with an empty antecedent are called atomic 'facts'. Among the models of P, there is a unique *minimal Herbrand model* μ_P, whose universe consists of all closed terms of the language, and whose interpretation for all predicates in the language is inclusion-minimal with respect to verifying P. Intuitively, μ_P is the intended model described by the program. We can think of its true atomic facts as the deductive minimum enforced by P. Now, for conclusions C of the simple form

$$\exists x_1 \ldots x_k \text{ conjunction of atoms}$$

(stating the existence of objects satisfying some goal conditions), the following assertions are easily seen to be equivalent:

- $P \models C$ (i.e., C is true in all models of P)
- C is true in the minimal Herbrand model μ_P
- some quantifier-free substitution instance of C can be derived from P using only the two inference rules of *Modus Ponens* and *Universal Instantiation*.

Thus, for Horn clause programs and suitable conclusions, simple deduction captures the complete truth in a unique intended model. (This is different from general predicate logic.) Analyzing this deductive process still further, we can see it as a stepwise succession of closures under application of all rules, starting from the empty set. (In a first stage, we merely add all atomic facts explicitly available in P: after that, recursion may start). Thus, all deductive consequences become available in finite rounds — a process often described in terms of *closure operators* T_P, where the minimal Herbrand model for P is the smallest fixed point for T_P. But one may also analyze the deductive process in other chunks, being applications of the *Resolution Rule*, which combines Modus Ponens with Instantiation. In what follows, we will not use this rule in its full predicate-logical form, which uses the mechanism of *unification* of terms as a key feature of passing information. Instead, we stay with purely propositional programs, where Resolution amounts to a Cut Rule (Chapter 7):

$$X \rightarrow A \quad Y \wedge A \wedge Z \rightarrow B \text{ imply } Y \wedge X \wedge Z \rightarrow B.$$

In proof search for a goal B from a Horn clause program P containing the clause $X \to A$, this says that an intermediate sequence of goals Y, A, Z may be replaced by a sequence Y, X, Z. This may temporarily increase the stack of goals, but decreases occur when program facts $\to A$ can be adduced. We write $P \Rightarrow C$ for derivable sequents. (Without unification, we lose the *computed answer substitutions* encoding the most valuable practical information. Nevertheless, the propositional case will do for illustrating our main points.)

Example 11.1 (Propositional Horn Clause Derivation) Let $P = \{\to q, p \wedge q \to r, r \wedge p \to s, \to p\}$, and $C = s$. Here is a derivation:

$$\dfrac{\to p \qquad \dfrac{\dfrac{\to p \qquad \dfrac{\to q \qquad p \wedge q \to r}{p \to r}}{\to r} \qquad r \wedge p \to s}{p \to s}}{\to s}$$

This format is still different from the practice of logic programming, where particular proof search strategies are implemented. For instance, PROLOG uses a depth-first refutation strategy involving 'strict linear resolution'. We shall look into its details later on in this chapter. For the moment, we merely note that such strategies guarantee faster search, but at the cost of missing possible solutions. E.g., the simple PROLOG program $\{p \to p, \to p\}$ will not find its trivial logical consequence p. Also, given its fixed proof search mechanism, PROLOG adds further operators, such as the "cut" ! (not to be confused with the above cut rule) allowing the user to skip certain parts of the search space for greater speed. This concludes our brief sketch of logic programming as a declarative paradigm which does its *computation by deduction*. In all this, two footholds appear for dynamic logic (as observed in van Benthem 1992). One lies in the theoretical semantics of Horn clause programs with fixed-point operators — another in the practical proof theory of fixed search algorithms, such as the depth-first backtracker of PROLOG. This shows already with Horn clause programs — and hence we defer predicate logic with variable substitutions to the notes following this chapter.

11.2 Dynamics of Deductive Approximation

11.2.1 Herbrand Transformations

The above semantics of Horn clause programs P makes them operators T_P on Herbrand models over the closed terms of their language. Far from being inherently declarative, this picture exemplifies precisely our dynamic paradigm. Herbrand models provide a natural concrete model for the dynamics of Chapter 1, with sets of atomic facts as constructive *information states* ordered by inclusion, and programs T_P as *update operations* over these. Moreover, the two-level architecture of Chapter 6 applies. Herbrand transformations T_P are really static declarative programs P turned update operations via dynamic *modes* T. Of the latter, there are many possible candidates, of which we mention:

$$T^1_P(X) = X \text{ together with all one-step inferences from it via rules in } P$$
$$T^*_P(X) = X \text{ closed under application of all rules of inference in } P.$$

Both are deductive approximation procedures towards information states satisfying the information in program P — with different properties. Thus, T^* is an idempotent closure operation, whereas T^1 is not. The distinction between the two is one of 'fast' and 'slow' dynamics of a system over time. But they are just two out of a larger repertoire of 'program modes' measuring speed of deductive approximation in the Herbrand Universe. This viewpoint is natural. The fixed points of the transformations T^1_P or T^*_P are just the standard models of the program P over the Herbrand Universe. (A minor deviation arises. The usual operator T_P makes models 'pre-fixed points'; Kowalski and van Emden 1976.) The minimal Herbrand model is the first fixed point reached when iterating from the empty information state. Moreover, various styles of dynamic inference make sense (Chapter 7). Consider one premise P and conclusion C, both Horn clause programs (C need not be an atom). With the slow mode T^1, Test-to-Test consequence is classical consequence, while other dynamic styles put greater demands on the 'processing power' of P to achieve C-effects. E.g., Update-to-Update consequence for T^1 says that each single derivational layer for P must be one for C, too. The 'fast mode' T^* corresponds to 'global updating' of a state to incorporate some piece of information. Here, our dynamic styles work out to:

Proposition 11.2

(i) Test-to-Test is classical consequence over all Herbrand models

(ii) Update-to-Test is also classical consequence over all Herbrand models

(iii) Initialized-Update-to-Test is consequence on the minimal *Herbrand model*

(iv) Update-to-Update expresses classical equivalence *between P and C.*

Proof. (i) Straightforward. (ii) Update-to-Test says that deductive closure of any set of atoms to satisfy P yields a fixed point for C. This is equivalent to classical consequence. (iii) This is obvious. Note that this may be an operation of high complexity. For simple arithmetical programs, the minimal Herbrand model are the standard natural numbers, and hence consequence in this sense measures True Arithmetic. (iv) Inclusion of the transitions for T^*_P in those for T^*_C amounts to *identity* of these two fixed point operations — and this is again easily seen to amount to standard declarative *equivalence* of P and C. ⊣

11.2.2 Horn Clause Consequence Operators

Horn clause programs P have dynamic effects through the associated transformations T_P (of various kinds). Their key semantic property turns out to be the following:

Finite Continuity

$$T_P(X) = \bigcup\{T_P(X_0) \mid X_0 \text{ finite}, X_0 \subseteq X\}.$$

Finite Continuity properly implies *Monotonicity* for set inclusion. This is all there is to Horn clause operators, by a simple converse route. All finitely continuous transformations may be represented in terms of an underlying notion of 'Horn-style inference':

Fact 11.3 *A transformation T satisfies Finite Continuity iff it can be represented by some binary relation XRy between finite sets of objects X and objects y in the form $T(X) = \{y \mid YRy$ for some finite $Y \subseteq X\}$.*

Proof. The relevant relation XRy here is '$y \in T(X)$'. ⊣

The inference relation XRy induced by a program operator T_P derives atoms y from sets of atoms X. It satisfies the standard *structural rules* of Permutation and Contraction, due to the set-like premise argument. Monotonicity is unproblematic, too. But the classical structural rules of Reflexivity and Cut (cf. Chapter 7) are not automatic: they are special features of inference relations underlying special kinds of propositional transformation. This analysis extends to first- order logic, the general paradigm behind logic programming. The first-order effects of Finite Continuity are as follows (van Benthem 1986):

Theorem 11.4 *The predicate-logical formulas that are finitely continuous with respect to some predicate P are precisely those having a 'P-distributive definition' (modulo logical equivalence) using only P-free formulas and P-atoms, with logical operations $\wedge$, $\vee$, $\exists$ over these.*

This is just the format of definition for predicates in so-called 'completed logic programs'. Thus, Horn clauses are the broadest format guaranteeing existence of implicit definitions for the relevant predicates occurring in their heads that can always be computed in at most ω rounds. (These definitions become explicit ones in the *infinitary* language $L_{\omega_1\omega}$.)

11.2.3 Digression: Finitary Modal Logic

There is also a *modal* approach to program transformations and their underlying dynamic inference relations. We briefly digress on this option, out of habit (cf. Chapters 4, 5, 9). A standard existential modality $\Diamond$ induces a transformation on sets X (i.e., extensions of declarative propositions) mapping them to their 'accessible range':

$$\{y \mid \exists x \in X\, xRy\}.$$

This map is *continuous*: it commutes with arbitrary unions of arguments (Chapter 6), without finiteness constraints. Conversely, any continuous transformation T can be represented by a binary alternative relation xRy among worlds, viz. '$y \in T(\{x\})$.' Further properties of a modality may be expressed in axioms like the S4-principles $p \to \Diamond p$ or $\Diamond\Diamond p \to \Diamond p$. By the frame correspondences of Chapter 3, these express mathematical properties of the relation R, such as *reflexivity* and *transitivity*. Now, the preceding analysis of consequence relations R for Horn clause programs generalizes the binary alternative relation to one connecting finite point sets with alternative points. (The latter are rather general 'dependence relations': cf. Chapter 9.) Then, modal T and S4 return as properties of a closure operator T^*_P, and we can derive the following correspondences. Here 'structural rules' canonically extend binary relational conditions to finitary relations.

Proposition 11.5

1. *$T(X) \subseteq X$ corresponds to Reflexivity: $\forall x\, \{x\}Rx$*
2. *$TT(X) \subseteq T(X)$ corresponds to Cut (Generalized Transitivity):*

$$\forall x \forall Y \forall \{Z_y \mid y \in Y\} \left((YRx \wedge Z_yRy\, (\textit{for all } y \in Y)) \to \bigcup\{Z_y \mid y \in Y\}Rx\right).$$

Proof. Ad (1). 'If'. Let $x \in X$. Then $\{x\}$ is a finite subset of X with $\{x\}Rx$, and so $x \in T(X)$ by definition. 'Only if'. Take $X = \{x\}$. Since $x \in T(X)$, either $\emptyset Rx$ or $\{x\}Rx$, where the former implies the latter by monotonicity.

Ad (2). 'If'. Let $x \in T(T(X))$. Then YRx for some finite $Y \subseteq T(X)$, whence each $y \in Y$ has some finite $Z_y \subseteq X$ with Z_yRy. By Cut then, ZRx, where $Z = \bigcup\{Z_y \mid y \in Y\}$. And Z is a finite subset of X, so that $x \in T(X)$ by definition. 'Only if'. Let YRx and also Z_yRy for all $y \in Y$. Now, set $X = \bigcup\{Z_y \mid y \in Y\}$. Then $Y \subseteq T(X)$ by definition, and hence $x \in T(T(X))$. But then also $x \in T(X)$, whence ZRx for some finite $Z \subseteq X$: and hence XRx by Monotonicity. $\dashv$

The preceding correspondences can be computed automatically by a Substitution Algorithm (cf. Chapter 3), this time, producing relational conditions in a *weak second-order logic*. As a further example, Löb Axioms express well-foundedness for dependence (this holds only for special program classes). Also, the dual *universal modality* makes sense:

$$T^{\dagger}(X) \quad = \quad -T(-X)$$
$$x \in T^{\dagger}(X) \quad \text{iff} \quad \forall Y\,(YRx \rightarrow \exists z\,(z \in Y \cap X)).$$

This modality, too, occurs in the theory of logic programming, with inverse 'top-down' approximation procedures for *greatest fixed points*. One rejects successive atoms that have no chance of making it via a program P, starting with $-T_P^1(-\bot)$ and then continuing to apply $T^{\dagger}$ to previous stages (consisting of all those atoms for which each proof attempt through P must contain at least one goal that was already rejected). We conclude with one example of correspondence for a principle relating T and $T^{\dagger}$.

Fact 11.6 *The S5-axiom $T(T^{\dagger}(X) \subseteq X$ corresponds to the relational condition of Generalized Symmetry:*

$$\forall x \forall Y\,(YRx \rightarrow \exists y \in Y\,\{x\}Ry).$$

This natural modal logic over finitary frame relations raises new questions of deductive completeness and expressive power. The minimal base is no longer the usual system K, as Modal Distribution

$$\Diamond(\phi \vee \psi) \leftrightarrow (\Diamond\phi \vee \Diamond\psi)$$

is invalid. Its monotonicity direction holds, but the converse fails. Imposing the latter effects a reduction to the old *binary* case:

> $T(X \cup Y) \rightarrow T(X) \cup T(Y)$ corresponds to the relational condition of $\forall Y \forall x\; YRx \leftrightarrow \exists y \in Y\,\{y\}Rx$.

The minimal finitary base logic for the finitary system is given by two deductive principles: Replacement of Provable Equivalents and Monotonicity.

11.2.4 Imperative Program Constructions

Our discussion so far has related two different program domains:

declarative	imperative
Horn Clause Programs P	Herbrand Transformations T_P.

Both have their own intrinsic operational structure. For instance, Horn clause programs may be combined via set-theoretic unions (modeling informational merge) or intersections. On the other hand, propositional transformations carry our earlier dynamic operations. Deterministic functions support *sequential composition*, *conditional choice* and *guarded iteration*, as in `while` programs (Chapter 3). Static Boolean tests (p)? in the latter two constructions will refer to the presence or absence of certain atoms in information states. Such dynamic constructs may already lie 'embedded' within standard logic programs:

Fact 11.7 *The sequential composition of two operators* $T^1_{P_1}$; $T^1_{P_2}$ *equals* $T^1_{P_1 \# P_2}$*, where* $P_1 \# P_2$ *is the Horn clause program consisting of: 'all clauses from* P_1 *and* P_2 *plus all clauses derived from a clause in* P_2 *by replacing each of its antecedents* A *by a set of antecedents available for* A *via some rule in* P_1*.'*

But not every dynamic transformation has a declarative counterpart. This follows from the characteristic property of Finite Continuity that we discovered earlier on:

Fact 11.8 *The conditional choice function* `if` q `then` $T^1_\emptyset$ `else` T^1_P *is not definable as a transformation* T^1_Q *for any single Horn clause program* Q.

Proof. Finite Continuity is lost under this construction. Let P be the program with one fact p. Our conditional program then has value $\{p\}$ at the empty state (where 'q' does not hold), whereas it has value $\{q\}$ at the state $\{q\}$, so that its corresponding operator is not even *monotone* with respect to set inclusion of its argument. ⊣

More generally, sequential composition preserves Finite Continuity (this explains our first Fact), and so do iterations with a positive atom for their guard. (Note, e.g., that the earlier T^*_P is the iteration `while` T `do` T^1_P.) The other direction has similar questions. Declarative programs have natural operations, too: to which extent is this static structure reflected dynamically? Here are some simple observations.

Proposition 11.9 *The operator* $T^1_{P_1 \cup P_2}$ *has no dynamic definition in terms of the components* $T^1_{P_1}$ *and* $T^1_{P_2}$ *using the regular program operations.*

Proof. Consider the following two programs

$$P_1 = \{p \to q, r \to a\}$$
$$P_2 = \{p \to r, q \to b\}.$$

On an initial state $\{p\}$, their T^1-transformations act as in the following diagram

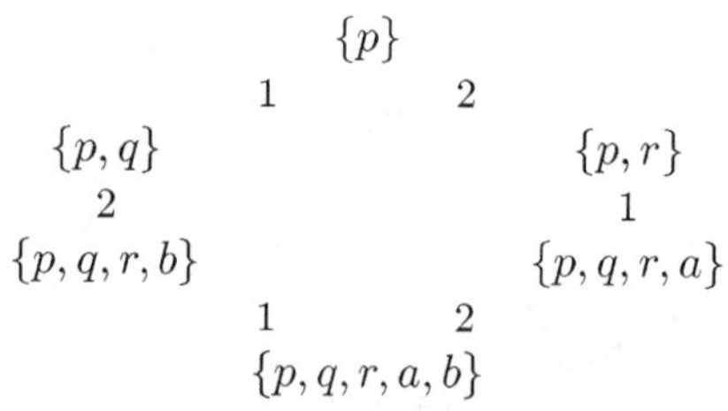

$$\{p\}$$

1 2

$\{p, q\}$ $\{p, r\}$

2 1

$\{p, q, r, b\}$ $\{p, q, r, a\}$

1 2

$$\{p, q, r, a, b\}$$

The value for '1+2' on the initial state is $\{p, q, r\}$, which cannot be reached via successive 1- and 2-transitions: whence it is not regularly definable from the latter. ⊣

Without proof, we mention a positive outcome for the other program transformation (cf. van Benthem 1992):

Proposition 11.10 *The operator $T^*_{P_1 \cup P_2}$ is regularly definable from $T^*_{P_1}$ and $T^*_{P_2}$.*

11.2.5 Modes and Projections

Continuing with our two-level dynamic picture, let us now focus on connections between our two program algebras, declarative and dynamic, i.e., the *modes* and *projections* of Chapter 6. The mode T^1 is fully continuous in its P-argument, in virtue of its shape:

$$\lambda P.\, \lambda X.\, \lambda y.\, Xy \vee \exists \rho \in P \exists_{\text{finite}} X_0 \subseteq X \text{ `} y = \rho[X_0].\text{'}$$

Thus, it completely respects the set-theoretic inclusion structure on logic programs, mapping it onto a corresponding inclusion structure among transformations

$$T \subseteq T' \text{ iff } \forall X\, T(X) \subseteq T'(X).$$

The definition of T^*_P makes it only finitely continuous in its P-argument. Conversely, consider projections taking dynamic transformations to static logic programs. Again, natural structure-preserving maps emerge (cf. Chapter 6). For instance, given any transformation T^*_P and atomic goal A, we can define a *weakest precondition* true in just those Herbrand models from which T^*_P can reach a state where A holds.

Fact 11.11 *All weakest preconditions* $\text{WP}(T^*_P, A)$ *are explicitly definable.*

Proof. $\text{WP}(T^*_P, A)$ has a higher-order definition

$$\lambda X.\, \exists Y \subseteq X\, Y \cup P \models A.$$

For propositional programs, this can be described via all sets of atoms needed to obtain goal A via P, and then enumerating their disjunction. These can be found effectively by repeated application of the following two Hoare correctness rules for logic programs:

$$\vdash \{p_1 \wedge \ldots \wedge p_n\} q := p_1, \ldots, p_n \{q\}$$
$$\{B\} P_1 \{C\}, \{D\} P_2 \{E\} \vdash \{B \vee D\} P_1 \cup P_2 \{C \vee E\}. \quad \dashv$$

Thus, we have a Dynamic Logic over the Herbrand Universe, with basic transformations T^1_P for sets P of Horn clauses. Transformations T^*_P resemble iterations of these, and further dynamic constructions are available too. This framework merges the fixed-point theory of logic programs with standard imperative paradigms. It is flexible, and admits of natural variations, both in its imperative programming repertoire, and in its choice of available Herbrand states. For instance, 'state gaps' (leaving out certain sets of atoms) may be used to encode not just 'single facts', but also more general constraints on information, just as happens in possible worlds models for intuitionistic logic (cf. Chapters 1, 2).

11.3 Proof Theory of Procedures

Practical logic programs employ hard-wired procedures to search large derivation spaces. Fixed process mechanisms pervade the art of programming. To understand the workings of a PROLOG program, one has to be aware of its *topmost search rule* for applicable program rules, and its *left-most selection rule* for dealing with a sequence of intermediate goals — as well as the backtracking associated with this. Only then, one understands why PROLOG finds the trivial logical consequence p from the program $\{p \leftarrow, p \leftarrow p\}$, but not from its relative $\{p \leftarrow p, p \leftarrow\}$. There are even imperative hacks for *explicit control* over derivation, such as a 'cut operator' freezing the current successful proof path. Such concessions to practice, as they are often viewed, may be cognitive assets. Human cognition has just this character. Natural language is a declarative formalism with fixed algorithms processing anaphora or temporal reference, serving as defaults in linguistic communication unless explicitly over-ruled. Also, fixed conventions drive argumentation, witness 'commitment' or 'burden of proof' which are crucial to winning or losing. Even here, the PROLOG experience is to the point. The program rule $p \leftarrow p$, both a semantic triviality and a computational pest, mirrors the argumentative nuisance of 'begging the question'. Even added imperative control and manipulation of a processing algorithm is realistic in dialogue and argumentation, where roles and commitments can be changed by a 'meta-repertoire' of different ways of playing the relevant language games, each with their logical cues. In what follows, we bring out some of these useful features, using our earlier dynamic paradigm.

11.3.1 Proof Search Algorithms and Structural Rules

A fixed algorithmic strategy for a declarative system induces a dynamic style of inference, encoded by more sophisticated structural rules. Consider depth-first back-tracking proof search as in PROLOG, whose 'search rule' looks at topmost eligible clauses first. For the time being, we do not fix any 'computation rule', and demand *simultaneous success* for all antecedents of a procedure call. (In a later section, we shall also investigate the standard 'left-first' selection convention for antecedent goals.) Instead of forward Horn clauses, we use the backwards notation $p \leftarrow \{q_1, \ldots, q_k\}$ for its vivid procedural flavour.

Axiomatizing Depth-First Proof Search: Goal A is sought via the uppermost program clause having A for its head, with 'success' when *all* antecedent goals have been established 'in parallel' (no special processing order is used), 'backtracking' to the next lower eligible program rule when no antecedent goal gets stuck in a loop, but some goal leads to finite failure. In the various sub-searches, again, highest eligible program rules are used first.

Example 11.12 (Getting Stuck in Proof Search) The program $p \leftarrow q$, $p \leftarrow \{r, p\}$ fails on the goal p via the following loop. "Try q, try both r and p: try the latter via the first clause (not the second), and so on ..."

This procedure has program rules $p \leftarrow \{q_1, \ldots, q_k\}$, with a set of antecedents. If desired, 'internal structural rules' state the equivalent derivational power of, say, $p \leftarrow \{q, r, q\}$, $p \leftarrow \{r, q\}$, $p \leftarrow \{q, r\}$. The proof procedure does not satisfy the standard *structural rules*.

Example 11.13 (Failures of Classical Structural Rules) Henceforth, we write $P \Rightarrow C$ if the goal C is derivable from program P. Then we have

non-monotonicity	$p \leftarrow \Rightarrow p$ but not $p \leftarrow p, p \leftarrow \Rightarrow p$
non-contraction	$p \leftarrow$, $p \leftarrow p$, $p \leftarrow \Rightarrow p$ but not $p \leftarrow p$, $p \leftarrow \Rightarrow p$
non-permutation	$p \leftarrow$, $p \leftarrow p \Rightarrow p$ but not $p \leftarrow p$, $p \leftarrow \Rightarrow p$
non-cut	$p \leftarrow p$, $q \leftarrow \Rightarrow q$ and $q \leftarrow$, $p \leftarrow \Rightarrow p$ but not $p \leftarrow p$, $q \leftarrow$, $p \leftarrow \Rightarrow p$.

Inspecting these failures more closely, one finds that our proof procedure does sanction the following *modifications* of earlier structural rules (where X, Y, Z are sequences of Horn clauses, A, B, C single clauses), a phenomenon already observed in Chapter 7:

Rightward Contraction	$X, A, Y, A, Z \Rightarrow C$ implies $X, A, Y, Z \Rightarrow C$
Rightward Expansion	$X, A, Y, Z \Rightarrow C$ implies $X, A, Y, A, Z \Rightarrow C$
Atomic Monotonicity	$X, Y \Rightarrow C$ implies $X, B, Y \Rightarrow C$ for facts B

The first two principles show how structural rules may survive only in special directions appropriate to the processing strategy. The third principle shows how structural movements may be restricted to special kinds of proposition, possibly marked by special operators. (Recall the modal operator strategy from linear logic at the end of Chapter 7.) For instance, we might mark atoms as follows: $\#(A)$, and then allow further principles, such as unlimited Permutation of *adjacent atoms*. As in Chapter 7, 'packages of structural rules' then reflect 'processing strategies for inference'. Unlike before, however, we have not yet been able to find good *representation theorems* to obtain a perfect match between the two.

11.3.2 A Gentzen Style Calculus

A more concrete issue is a *complete proof theory* for propositional inference according to the PROLOG strategy, or its rivals. There is a graph description of successful computations, but we prefer the earlier sequent format. Here is an annotated calculus to this effect. First, nothing changes in program execution if rules come in parallel 'blocks' with the same head. Two occurrences of the same rule may be contracted to the first, highest up in the ordering. Better than this one cannot do, as rules with the same head display 'scoping effects'. This preliminary analysis sanctions the following structural rules of inference:

Rightward Contraction	as above
Rightward Expansion	as above
Permutation	$X, A, B, Y \Rightarrow C$ implies $X, B, A, Y \Rightarrow C$ if A, B are rules with different heads.

These rules reflect the natural ordering of program clauses as a parallel arrangement of sequentially ordered recipes for some goal. Further structural rules elaborate this picture, with Monotonicity describing additions to parallel blocks. Next, we analyze genuine logical inference, i.e., the real computation, starting with instantaneous attainment of a goal. This will take the form of an informal discussion (van Benthem 1992), which provides enough information for an impeccable formal version (Kalsbeek 1995).

Reflexivity $A, X \Rightarrow A$ for atomic goals A

Note that the initial position of the goal A in the program is crucial. The main driving force of derivations is a successful 'procedure call', reflected in a rule of

Modus Ponens $p \leftarrow p, X \Rightarrow q_i$ for all i $(1 \leq i \leq k)$ implies
$p \leftarrow \{q_1, \ldots, q_k\}, X \Rightarrow p$.

The validity of this rule will be clear. The special prefix $p \leftarrow p$ in the premise makes sure that X does not prove some q_i in a way depending on p, which would generate a loop in the search for the conclusion. A counter-example illustrating this point is

$$q \leftarrow p, p \leftarrow \Rightarrow q \text{ versus not } p \leftarrow q, q \leftarrow p, p \leftarrow \Rightarrow q.$$

Modus Ponens would not work behind arbitrary prefixes, witness the following example:

$p \leftarrow p, p \leftarrow p, q \leftarrow \Rightarrow q$ (the search for q will be successful at once)
not $p \leftarrow p, p \leftarrow q, q \leftarrow \Rightarrow p$ (the search for p will get stuck at once).

Example 11.14 (Derivation of $q \leftarrow$, $r \leftarrow \{p, q\}$, $s \leftarrow \{r, p\}$, $p \leftarrow \Rightarrow s$) The premise sequence derives q and p by Reflexivity and Permutation. Then, by Modus Ponens, it derives the goal r (after a Permutation moving the single r-clause up front), and by the same combination, it also derives the goal s.

Our sequent calculus has unorthodox *side conditions* on occurrence of proposition letters. But these are no more complex than variable conditions for predicate logic (Chapter 9). Next, in PROLOG inference some rule may be tried but discarded because of 'finite failure' for some of its antecedents, as in the successful sequent $p \leftarrow q,\ p \leftarrow s,\ s \leftarrow \Rightarrow p$. Here, we need assumptions recording that all paths in the initial search for p ended in a dead end which should be neither q nor p, so that we may proceed to the next available p-clause. We use a small trick to encode this information about 'finite failure' in inferential form:

Prefixing $X \Rightarrow p$ and $p \leftarrow p, X, q_i \Rightarrow q_i$ $(1 \leq i \leq k)$ imply
$p \leftarrow \{q_1, \ldots, q_k\}, X \Rightarrow p$

Validity of the Prefixing Rule. The premise $p \leftarrow p,\ X,\ q_i \Rightarrow q_i$ says that upon evaluation of q_i in $p \leftarrow p,\ X$, either success is encountered or finite failure (after which the last q_i supplies the conclusion in the course of backtracking). In particular, the search for q_i will never lead to a subgoal p, as this starts a loop with the initial clause $p \leftarrow p$. Requiring that this happens for all q_i means that no loops occur, with either success in all cases — and the original program implies goal p by Modus Ponens — or finite failure at least once. In the latter case, the program with a new proof search starting from a second eligible p-clause must derive p, which explains the second premise $X \Rightarrow p$. A small complication occurs. The search for p via its antecedents in the next eligible p-rule (occurring inside X) should take place in the context of the whole original program $p \leftarrow \{q_1,\ \ldots,\ q_k\},\ X$, not just X itself. But we lose no generality by dropping the first clause, since p cannot be called in a successful derivation of the q_i — as this would get us into a loop.

Example 11.15 (Derivation of $p \leftarrow q,\ p \leftarrow s,\ s \leftarrow \Rightarrow p$) What we must show by Prefixing are the following two premises:

$p \leftarrow s,\ s \leftarrow \Rightarrow p$, which works as before via Modus Ponens
$p \leftarrow p,\ p \leftarrow s,\ s \leftarrow,\ q \leftarrow \Rightarrow q$, from Reflexivity and Permutation.

Putting all this together into one calculus, we have the following completeness result.

Theorem 11.16 *Propositional inference via the above parallel goal search mechanism is exhaustively described by a Gentzen calculus with* {*Rightward Contraction, Rightward Expansion, Permutation, and Reflexivity, Modus Ponens, Prefixing*}.

Proof. The completeness direction is by analysis of successful derivations, where the above steps cover all possible cases, maintaining some suitable complexity count. $\dashv$

Judicious bounding of the relevant Gentzen search spaces yields

Fact 11.17 *The above Gentzen Calculus is decidable.*

When formalized more precisely, the above completeness proof gives two-way *effective transformations* between computations and Gentzen proofs, which can be exploited, e.g., to systematically compare program transformations and proof transformations.

11.3.3 Rule Completeness

Another interesting issue is 'rule completeness'. Proof search mechanisms may validate further structural rules, such as

Rightward Rule Monotonicity $X, p \leftarrow \{q_1, \ldots, q_k\}, Y \Rightarrow A$ implies
$X, p \leftarrow \{q_1, \ldots, q_k\}, Y, p \leftarrow \{r_1, \ldots, r_m\} \Rightarrow A$.

This allows us to add rules to 'blocks' in the above description of programs, provided that they come in bottom-most position. Other interesting laws concern internal clause structure:

Inner Contraction	rightward contraction of identical antecedents in a program clause does not affect derivable goals
Subsumption of Rules	$X, p \leftarrow Q, p \leftarrow Q', Y$ and $X, p \leftarrow Q, Y$ have the same consequences if Q is a subset of Q'.

A full description of *admissible* rules for various styles of propositional proof search would be of clear interest — also as an issue in general proof theory. (Rybakov 1985 provides a complete and even decidable description of all admissible rules for the modal logic S4. Kalsbeek 1995 discusses various admissible variants of the Cut Rule in logic programs.)

11.3.4 Axiomatizing an Added Computation Rule

The above proof-theoretic analysis is quite flexible. To see this, consider one particular selection rule in our proof search, in particular, the Left-First convention of PROLOG:

> In the above, try all antecedent goals *from left to right*, continuing when success is encountered, or when a loop is entered, and skipping the whole attempt when finite failure is found for some antecedent goal.

Thus, a call to the current clause will either be successful (all subgoals are satisfied), or enter a loop that vitiates the whole program, or encounter finite failure after success for some initial subgoals, and backtrack to the next eligible program clause for the main goal. This procedure has rules $p \leftarrow \{q_1, \ldots, q_k\}$ whose antecedents form an *ordered sequence*. Again, 'internal structural rules' hold for these. E.g., rightward contraction of antecedents does not affect outcomes. Moreover, earlier 'external' structural rules of Rightward Contraction, Rule Permutation and Atomic Monotonicity remain valid. Valid inferences can only *increase*, as derivations via the first strategy above still succeed under the new regime. Even so, the latter remains incomplete.

Example 11.18 (Deductive Incompleteness) A 'left first' strategy for selecting antecedent goals will not derive goal p from the program $p \leftarrow \{q, r\}$,

$q \leftarrow q$, p, since the loop for q makes the finite failure for r 'inaccessible'. (With a right-first computation rule, though, one does reach the next clause with head p.) The Prefixing Rule of our earlier calculus will not reflect such distinctions, as its premise $p \leftarrow p$, $q \leftarrow q$, $q \Rightarrow q$ is unavailable here.

We can add *new* inference rules capturing the additional power of left-first computation. Again, we encode the relevant 'finite failure' in purely inferential form. The crucial step has

- All subgoals $q_1, \ldots, q_m$ succeed under a search in the remaining program X, with a renewed subgoal p leading to a loop,
- There is finite failure for q_{m+1} which never stops at p.

We give one natural format for this setting. The above computation rule interleaves two basic processes, namely, *proof* and *refutation*. Such a two-sided perspective also has independent proof-theoretic and philosophical interest (cf. Wansing 1993, Jaspars 1994). Thus, we adopt both in our calculus, with a mutual recursion. We introduce *two* sequent arrows $\Rightarrow^+$, $\Rightarrow^-$, letting the former stand for our earlier derivability, obtaining a

A PROLOG-style Proof Calculus

'Proof' all earlier structural rules, plus Reflexivity and Modus Ponens

'Refutation' all earlier structural rules, plus

1. $X \Rightarrow^- p$ if p does not occur in any rule head of X
2. if Y, $X \Rightarrow^- p$ and $Z, X \Rightarrow^- p$, where X has no rule heads p and Y, Z contain only such heads, then Y, Z, $X \Rightarrow^- p$

'Interaction'

3. if $p \leftarrow p$, $X \Rightarrow^+ q_i$ $(1 \leq i \leq k)$ and $p \leftarrow p$, $X \Rightarrow^- q$ and $X \Rightarrow^+ p$, then $p \leftarrow \{q_1, \ldots, q_k, q, q_{k+1}, \ldots, q_m\}$, $X \Rightarrow^+ p$
4. if $p \leftarrow p$, $X \Rightarrow^+ q_i$ $(1 \leq i \leq k)$ and $p \leftarrow p$, $X \Rightarrow^- q$ and p occurs in no rule head in X, then $p \leftarrow \{q_1, \ldots, q_k, q, q_{k+1}, \ldots, q_m\}$, $X \Rightarrow^- p$.

For completeness arguments concerning this calculus, we refer to the techniques of Kalsbeek 1995, who gives a slightly different analysis. Instead, we merely provide one illustration of these rules.

Example 11.19 (Derivation of $p \leftarrow \{q, r\}$, $q \leftarrow a$, $r \leftarrow r$, $p \Rightarrow^+ p$) We need sequents $p \leftarrow p$, $q \leftarrow a$, $r \leftarrow r$, $p \Rightarrow^- q$ and $q \leftarrow a$, $r \leftarrow r$, $p \Rightarrow^+ p$. The latter derives from Reflexivity and Permutation. The former derives from $q \leftarrow q$, $p \leftarrow p$, $r \leftarrow r$, $p \Rightarrow^- a$ (a is no rule head in $p \leftarrow p$, $r \leftarrow r$, p). This is the basic refutation axiom.

11.3.5 Further Extensions

On top of the analysis so far, one may add various features. First, from a logical point of view, a proof calculus invites the addition of *further connectives* than just implication. Indeed, there are three levels where this makes sense. One can have operators between goals inside a Horn clause, manipulating their processing order in selection rules. E.g., a dynamic *conjunction* $A;B$ might require left-to-right selection, and $A \parallel B$ parallel selection. Also, failure naturally invites the introduction of goal *negations*. Then, one may have operators between clauses in a program, manipulating their position in the search rule through the above block structure. And finally, one can have operators over programs, as discussed in our earlier sections on the dynamics of Herbrand models (cf. Miller 1989, Saraswat 1993, Brogi, Mancarella, Pedreschi and Turini 1994). Kalsbeek 1995 presents a number of calculi for the first two levels, including an explicit one for negation as failure. Conversely, one can also take existing constructs from logic programming, and treat them as logical connectives. In particular, Kalsbeek 1995 has a complete calculus for PROLOG with a *cut operator* ! (cf. the notes following this chapter). This is just one instance of recent unconventional proof calculi adding 'procedural' facilities (cf. Hendriks 1993).

11.4 Calculus of Tasks

There is a more general dynamic import to what we have been doing. Logic programs provide a concrete model of something much more general, namely, a *calculus of tasks*. Indeed, one persuasive initial motivation was the 'procedural interpretation' of Horn clause formalisms (Kowalski 1979). A sequence of Horn clauses is a bunch of procedures that can be invoked to solve a task "from input data via the rules to outputs". There are several strategies here. The earlier *modes* T^1_P and T^*_P were just two, viz. 'try all rules once, and add their results, if any, to the input' and 'repeat all rules on the input until no new output appears'. But other natural conventions exist for procedure calls and data handling. Are resources re-usable or consumed by application of a procedure? Different styles of resource management produce different logics here. E.g., one-rule programs of the form $P = \{\rho\}$ have at least the following three natural options for making their atomic data transitions:

$$A\,P_1\,B \quad \text{iff} \quad B = \{\rho[A_0]\} \text{ for some suitable } A_0 \subseteq A$$
$$A\,P_2\,B \quad \text{iff} \quad B = A \cup \{\rho[A_0]\} \text{ for some suitable } A_0 \subseteq A$$
$$A\,P_3\,B \quad \text{iff} \quad B = (A - A_0) \cup \{\rho[A_0]\} \text{ for some suitable } A_0 \subseteq A.$$

Various strategies aggregate larger programs $P = \{P_1, \ldots, P_n\}$. Some use single steps, like the dynamic procedures of Part II, others repeat until nothing new appears, like the mode T^*_P, or go through rules in several passes according to some search mechanism. 'Problem solving' then becomes computation of a weakest precondition WP(P_1, ..., P_n, T). To conclude, we briefly point out some general aspects of the task perspective, which all call for further refinement of our dynamic paradigm so far.

Inference as Ternary Transduction. The task view changes the format of inference from a binary premises/conclusion schema $P - C$, to ternary data/rules/results: $D - R - C$. This recalls the five-component schema for argumentation in Toulmin 1958:

$$\text{Data} \quad \underset{\text{Warrant}}{\rightarrow} \quad \text{Qualifier}_{\text{Rebuttal}} \quad \text{Claim.}$$

where the 'qualifier-rebuttal' element signals a style of inference (cf. Chapters 1, 7). Here, our account of structural rules has to be refined. Components may show independent behaviour, say, with classical rules for the R-component, but others for input and output. Moreover, new kinds of structural rule emerge, connecting various components.

Resource-Sensitive Inference. Resource-conscious logics (categorial, linear or relevant) keep track of occurrences of premises and linguistic items generally (Chapter 12 below). This fits in well with the task analysis. Computation rules for proof strategies depend on an occurrence order of assertions. Thus, finitely continuous transformations may be refined to deal with *sequences* or multi-sets.

Example 11.20 (Consuming Input with Arbitrary Procedure Calls) Consider the above task strategy P_3, with arbitrary calls to rules in P. Take *multi-sets* for D, C and ordinary sets for P. (This is resource-sensitive for input data, but not for processing time.) Contraction and Monotonicity hold for P, but not for D, C. Similar observations apply to other structural rules, such as Cut. Here is how this works, with data $D = \{p, p\}$, and program $P = \{q \leftarrow \{p, p, r\},\ p,\ r \leftarrow p\}$:

$$\{p, p\} \text{ goes to } \{p, p, p\} \text{ goes to } \{p, p, r\} \text{ goes to } \{q\}.$$

Complex Data Structures. One can do inference over any structured object: trees, program lists, databases, smiles, clothes. Our analysis supports this point. Specifically, *structural rules* reflect intended formats of representation. Moreover, these formats need not be homogeneous. For instance, the PROLOG experience told us that inferential structures may mix 'parallel' and 'sequential' aspects, manipulating unordered blocks for identical goals, inside which there was a sequential order of program clauses.

Dynamic Task Operations. New styles of inference may induce new logical operations. This generalizes the point already made about further operations in logic programs. Our task strategies high-light the role of the *comma* that signals putting together data, program rules or conclusions. This might be union, composition or any general 'merge'. E.g., sequential combination of logic programs suggests a directed non-commutative conjunction $P \bullet Q$ concatenating two programs P, Q to be processed in this sequential order. Such a conjunction naturally engenders two further non-equivalent 'directed implications'

$$P \text{ implies } A\backslash B \quad \text{iff} \quad A, P \text{ 'derives' } B$$
$$P \text{ implies } B/A \quad \text{iff} \quad P, A \text{ 'derives' } B$$

The resulting repertoire of direction-sensitive operations $\{\bullet, \backslash, /\}$ is similar to that found in categorial and relation-algebraic calculi for natural language (cf. Chapters 3, 12).

At this point, we hope to have amply demonstrated that Logic Programming, far from being a stronghold of declarative orthodoxy, is a hotbed of dynamic agitation.

11.5 Notes

Logic Programming. Lloyd 1985 is a theoretical classic exposition, Bratko 1986 a practical one, Apt 1991 is a sophisticated survey (including fine-structure of programs, such as their possible 'stratification'), Doets 1994 is an up-to-date logic-oriented textbook. As for variations and extensions, joint calculi of proof and refutation are in Wansing 1993 (referring to a PROLOG-style format by David Pearce), Jaspars 1994 (treating Nelson's logic of constructive falsity as a form of epistemic logic). Kanovich 1993 compares Horn clauses with other task calculi, such as Petri Nets. Van Benthem 1996b elaborates on connections with Argumentation Theory, Aliseda-Llera 1996 on those with abduction.

Changing Model Classes. Chapter 2 viewed new information as elimination of alternatives, decreasing ranges of worlds. Thus, propositions were transformations on model classes X: $T_P(X) = X \cap \text{MOD}(P)$. This is reminiscent of Herbrand transformations. But the latter *increase* sets of atomic facts, which encode classes of models over the fixed Herbrand universe. It is of interest to compare the two perspectives. The *minimal* Herbrand model suggests the following new operation, taking only minimal models for P (Shoham 1988):

$$T_P^{\mu}(X) = \mu(X \cap \text{MOD}(P)),$$

where the operator μ selects appropriate minimal models. (Chapter 2 has its properties.) What do the earlier varieties of dynamic consequence mean in this new setting?

Fact 11.21 *For minimal transformations, style I expresses standard consequence, style II again expresses equivalence between premises and conclusion, and both styles III, III$^{\#}$ express:*

$$\mu(\text{MOD}(P_1)) \cap \text{MOD}(P_2) \cap \cdots \cap \text{MOD}(P_n) \subseteq \text{MOD}(C).$$

Ordinary preferential inference has

$$\mu(\mathrm{MOD}(P_1) \cap \mathrm{MOD}(P_2) \cap \cdots \cap \mathrm{MOD}(P_n)) \subseteq \mathrm{MOD}(C).$$

Dynamics of Vocabulary. A noticeable phenomenon in information flow is changing vocabulary. New data often come in different terms from old, reflecting the cognitive process of concept formation in addition to information growth. There is dynamics of 'form' as well as of 'content', and the two are hard to separate. This richer dynamics also occurs in natural language (cf. Section 12.3.2 on 'context change'). Shifting vocabulary occurs in logic programming when new programs change the term base for the Herbrand Universe, so that (minimal) models change, and old models for subprograms in a smaller similarity type can only be recovered through 'projection'. Thus, intended models of a program may change depending on its total environment. Little attention has been paid to this kind of dynamics, which has its own logical constants and varieties of inference. Logical operations include projecting a program P in language L to its restriction $P \upharpoonright L'$ in a smaller similarity type L', or merging programs in different similarity types to some common expanded type. The formal philosophy of science studies scientific theory change emphasizing shifts in 'observational' and 'theoretical' vocabulary (van Benthem 1989c). Also, Bernard Bolzano's pioneering work in the nineteenth century (cf. Chapter 13) mixes inference with changing vocabulary. The computational literature on abstract data types involves different similarity types for algebras, and their connections. 'Module algebra' (Bergstra, Heering and Klint 1990) provides various algebraic laws for such operations.

Axiomatizing Prolog Cut. Prolog computation via Gentzen-type proof systems was discovered independently by Mints 1990. (Cf. also Stärk 1994, 1995 for new connections with classical proof systems.) Kalsbeek 1995 compares several approaches, including recent uses of linear logic. Complete Gentzen calculi exist for PROLOG with the cut !, which freezes success branches so far, and forbids further backtracking beyond them. Here are some typical rules from Kalsbeek 1995, with a Gentzen arrow $\Rightarrow$ encoding ordinary derivability and $\Rightarrow^*$ derivability with a *finite search tree*. The negation operator $\neg$ indicates finite failure. Each rule describes an intuitive case in applying the cut construct.

$$\frac{A \leftarrow \{A\}, P \Rightarrow B_i \ (i \in I)}{A \leftarrow \{B\text{'s-with-interpolated-cuts}\},\ P \Rightarrow A}$$

$$\frac{A \leftarrow \{A\},\ P \Rightarrow B_i \ (i \in I) \qquad A \leftarrow \{A\},\ P \Rightarrow^* C_j \ (j \in J)}{A \leftarrow \{B\text{'s-with-interpolated-cuts, !, } C\text{'s}\},\ P \Rightarrow^* A}$$

$$\frac{A \leftarrow \{A\},\ P \Rightarrow B_i \ (i \in I) \quad A \leftarrow \{A\},\ P \Rightarrow^* C_j \ (j \in J) \quad A \leftarrow \{A\},\ P \Rightarrow \neg D}{A \leftarrow \{B\text{'s-with-cuts, !, } C\text{'s, } D\text{, } E\text{'s-with-cuts}\},\ P \Rightarrow^* \neg A}$$

Kalsbeek 1995 also provides complete Gentzen calculi for logic programming styles with partial connectives of negation, conjunction and disjunction, in clause-internal position. Her arguments are effective, and hence transfer detailed information between proofs and computations. Other topics in this connection include intermediate substructural calculi with deductive power between PROLOG and classical logic, and 'frugal' variants of PROLOG computation.

Variables, Unification and Resolution. *Predicate logic* is the proper habitat of logic programming. Here, analogies with imperative programming are only reinforced. Substitutions are the driving force in computation for logic programs. We start with their *procedural semantics.* $[t/x]$ is a procedure shifting assignments in models $\mathbf{M}$ (Chapter 9):

$$[\![[t/x]]\!]^{\mathbf{M}} = \{(a, a^x_{\mathrm{val}(\mathbf{M},t,a)}) \mid a \text{ any assignment in } \mathbf{M}\}.$$

This is a transition for an imperative assignment $x := t$, as in the Substitution Lemma:

if term t is free for x in ϕ, then $\mathbf{M} \models [t/x]\phi[a]$ iff $\mathbf{M} \models \phi[a^x_{\mathrm{val}(\mathbf{M},t,a)}]$.

The latter states the correctness axiom for assignment statements in a Hoare calculus, being $\{[t/x]\phi\}x := t\{\phi\}$ (where the equivalence arises by considering also the case with $\neg\phi$). Full predicate logic used random assignments for existential quantification (Chapter 2):

$$[\![\eta x]\!]^{\mathbf{M}} = \{(a, a^x_d) \mid \text{ for some arbitrary } d \in D\}.$$

This dynamic semantics for substitutions suggests an imperative analysis of inferential power for rules in logic programs, via Hoare-style correctness statements, such as

$\{a \wedge A\}A \to B\{a \wedge B\}$, where A is any conjunction of atoms and B any atom, while α is an arbitrary predicate-logical statement.

To get the force of universal quantification over variables x in this implication, one can prefix it with random assignments ηx, leaving the others as 'free parameters'.

Example 11.22 (Deriving an Instance of a Rule) The rule $\forall x\,(Ax \to Bx)$ implies its instance $ASy \to BSy$ roughly as follows

$$\{ASy \wedge Sy = Sy\}x := Sy\{Ax \wedge x = Sy\}Ax \to Bx\{Bx \wedge x = Sy\},$$

where ASy implies the first precondition, and the last postcondition implies BSy. Passing from explicit to random assignment plus the Hoare rule for sequential composition derive the desired correctness statement $\{ASy\}\eta x.\, Ax \to Bx\{BSy\}$.

The correctness format suggests computation of *weakest preconditions* for goals with respect to certain rules. Then 'most general unifiers' make their appearance.

Example 11.23 (Weakest Preconditions and Unification) Calculating the weakest precondition $\mathrm{WP}(\eta x\,\eta y.\ Axy \rightarrow BSxy, Buf(vu))$ results in the statement $\exists x_1 \exists y_1\, Ax_1y_1 \wedge u = Sx_1 \wedge f(vu) = y_1$, whose solution for the two identities is the most general unifier for the two atoms $\{BSxy,\ Buf(vu)\}$.

Our analysis of propositional logic programs needs revision in predicate logic. The account of rules via Finite Continuity must be extended with quantifier rules and term substitutions. The underlying domain of terms under 'substitutional specification' must enter essentially. Also, rules of inference will come to include successive computation of *answers*

$P \vdash \sigma : A$ $\quad$ P derives goal A with computed answer substitution σ.

This is a type assignment calculus modifying current substitutions by procedure calls. Such moves may change complexity. E.g., we found a decidable proof calculus for propositional PROLOG. But its predicate-logical version might involve undecidable 'loop avoidance'.

11.6 Questions

Imperative Constructions over Logic Programs. Which logical operations over functions preserve finite continuity? Do indeterministic constructs also make sense?

Finitary Modal Logic. Develop the correspondence theory of finitary modal logic.

Dynamic Logic of Herbrand Models. Axiomatize the propositional dynamic logic of the Herbrand model with all `while` programs over the family of update operators T_P for all Horn clause programs P. Same question for models allowing just certain sets of atoms as states, but not necessarily all of them.

General Completeness Theorems. Develop a general transformation between proof search procedures and Gentzen calculi, based upon the samples provided in this chapter.

Representation Theorems for Proof Procedures. Find a combination of modified structural rules which precisely determines the inference relations that can be represented via the proof search strategy implemented in PROLOG. This is related to the next question.

Admissible Proof Rules. Find a complete description of all *admissible rules* (cf. Došen 1992) for the proof search procedures in this chapter. In its effective version, this extends the completeness theorem for derivable sequents, representing successful computations, to a description of admissible transformations between computations.

Imperative Constructs. Give a general proof theory of constructs pruning search spaces. (Hendriks 1993 uses Lambek Calculus with such a construct to enforce a unique correspondence between proofs and different semantic readings, thus solving the 'spurious ambiguity problem'. Kanazawa 1993a does something similar for dynamic inference.)

Classical Proof Theory. Develop more general connections between classical proof theory and computations through our completeness theorems. For instance, are there computational versions of 'cut elimination' corresponding to the weaker cut rules of Kalsbeek 1995? Also, compare the substructural landscape of structural rules for PROLOG-style computation procedures with that known for categorial and linear logics.

Task Calculus. Develop the more general task calculi sketched here, including an account of structural rules that may operate differently in different arguments.

Predicate Logic. Find predicate-logical versions for all the results obtained in this chapter.

12

Understanding Natural Language

This chapter provides only some brief forays into a vast area. Its aim is to show how logical dynamics occurs across various parts of Linguistics. This insight yields no ready-made solutions for specific queries, but rather a fresh perspective on many existing descriptive topics, ranging from syntax to semantics and pragmatics. Thus, we discuss connections between categorial grammar and relational algebra or arrow logic, or between dynamic semantics and modal or dynamic logic, but we also propose new formats for theories of context change. At the end, we touch upon other logical paradigms in the field, such as proof theory and feature-value logics, and indicate analogies with the main dynamic approach of this book. Given the complexity of natural language, no unified story is told in this chapter, and the reader may wish to consume only bits and pieces at a time.

12.1 Syntax

Our starting point is one long-standing logical approach to the syntax of natural languages.

12.1.1 Categorial Grammar in a Nutshell

Categorial Grammar equates key grammatical categories with mathematical functions of types $A \to B$, taking A-type arguments to B-type values. (Cf. van Benthem 1991b, Moortgat 1996.) Language structure starts with some 'base categories' denoting ground level objects. For instance, a Verb Phrase is a syntactic functor taking Noun Phrases to form Sentences, or an Adverbial takes Verb Phrases to (complex) Verb Phrases. Thus, categorially, expressions in most linguistic categories denote procedures that change states of some sort. Interpretation of compound expressions then corresponds to computation of the associated complex procedures. This view may be described as a process of derivation in some suitable logic L, treating base categories as propositional atoms and the above function arrows as logical implications. Derivability of a sequent

$$A_1, \ldots, A_k \Rightarrow B$$

in a categorial logic says that L provides the resources for combining any sequence of objects of consecutive categories or types $A_1, \ldots, A_k$ into an object that has category or type B. The implicational calculi that drive grammatical combination are usually fragments of the full classical implicational logic. The best-known ones lie in the so-called *Categorial Hierarchy*, consisting of implicational calculi such as

Ajdukiewicz Calculus	*Modus Ponens* only
Lambek Calculus	add *Conditionalization*
Linear Logic of implication	add *Permutation*
Relevant Logic of implication	add *Contraction*
Intuitionistic Logic of implication	add *Monotonicity*
Classical Logic of implication	add *Peirce's Law*.

Example 12.1 (Ajdukiewicz Derivation of a Sentence)

$$\dfrac{\begin{matrix}\textit{Mary}\\ \text{NP}\end{matrix} \qquad \dfrac{\begin{matrix}\textit{travelled}\\ (\text{NP} \to \text{S})\end{matrix} \qquad \begin{matrix}\textit{fast}\\ ((\text{NP} \to \text{S}) \to (\text{NP} \to \text{S}))\end{matrix}}{(\text{NP} \to \text{S})}\,\textbf{MP}}{\text{S}}\,\textbf{MP}$$

Example 12.2 (Lambek Derivation of a Negated Verb Phrase)

$$\dfrac{\dfrac{\dfrac{\begin{matrix}*\\ \text{NP}\end{matrix} \qquad \begin{matrix}\textit{travels}\\ (\text{NP} \to \text{S})\end{matrix}}{\text{S}}\,\textbf{MP} \qquad \begin{matrix}\textit{not}\\ (\text{S} \to \text{S})\end{matrix}}{\text{S}}\,\textbf{MP}}{(\text{NP} \to \text{S})}\,\textbf{CON}, \text{withdrawing } *$$

Intuitively, in these derivations, Modus Ponens steps encode function applications, while Conditionalization encodes abstraction over certain argument positions. In actual syntactic analysis, there are both *backward* and *forward* arrows, indicating different positions where a functor wants to consume its argument. Thus, one also encounters operators $B \leftarrow A$ with a rightward-looking Modus Ponens pattern $B \leftarrow A, A \Rightarrow B$. The preceding calculi are 'substructural'. I.e., they do not postulate any of the usual *structural rules* of classical logic (Chapter 7) — except Reflexivity. (*Cut* is an admissible rule in most cases.) Structural poverty is a key feature of natural language. Categorial logics manipulate *occurrences* of formulas, as befits syntax- or instruction-oriented calculi. E.g., when conditionalizing, only one occurrence of an assumption may be withdrawn at a time. In the current jargon, such logics are 'resource-sensitive'. Here is the Lambek

Calculus in a sequent format, starting from axioms $A \Rightarrow A$, and working with both directed implications $\rightarrow$, $\leftarrow$ as well as a categorial *product* $\bullet$ storing two successive occurrences of a category:

Lambek Calculus **L** *in Sequent Formulation*

$$\rightarrow \qquad \frac{X \Rightarrow A \quad Y, B, Z \Rightarrow C}{Y, X, A \rightarrow B, Z \Rightarrow C} \qquad \frac{A, X \Rightarrow C}{X \Rightarrow A \rightarrow C}$$

$$\leftarrow \qquad \frac{X \Rightarrow A \quad Y, B, Z \Rightarrow C}{Y, B \leftarrow A, X, Z \Rightarrow C} \qquad \frac{X, A \Rightarrow C}{X \Rightarrow C \leftarrow A}$$

$$\bullet \qquad \frac{X, A, B, Y \Rightarrow C}{X, A \bullet B, Y \Rightarrow C} \qquad \frac{X \Rightarrow A \quad Y \Rightarrow B}{X, Y \Rightarrow A \bullet B}$$

The key result in the celebrated paper Lambek 1958 was that this calculus enjoys Cut Elimination — where Cut is the following admissible structural rule (cf. Chapter 7):

$$\frac{X \Rightarrow A \quad Y, A, Z \Rightarrow B}{Y, X, Z \Rightarrow B}$$

Remark 12.3 (Occurrence Counts) The occurrence character of this calculus shows in certain proof invariants that fail for stronger implicational logics. E.g., for atoms p, define the *p-occurrence count* of a formula as its number of positive occurrences of p minus that of the negative ones. For finite sequences of formulas, one adds all separate counts. All **L**-derivable sequents have the same p-occurrence count for their premise sequence and conclusion. Cf. van Benthem 1991b, and Roorda 1991 for finer directed versions of this invariant. By contrast, classical derivabilities like $p, p \Rightarrow p$ lack this property.

Deductive Power. The *Categorial Hierarchy* of grammar logics arises by reintroducing classical structural rules, in case not all information in the presentation of premises is needed in our deductions. The calculus **LP** is Lambek Calculus with an added permutation rule (this is linear logic of premise 'bags', with implication only), **LPC** is linear logic plus contraction (which is relevant implicational logic), **LPCM** is relevant logic plus monotonicity (this is intuitionistic implicational logic of premise 'sets'), and finally, **LPCM** + P is intuitionistic logic plus Peirce's Law: which is classical implicational logic. We emphasize this multiplicity, as it resembles what we found in the dynamic landscapes of Part II. From **L**, one can also go downwards, and curb deductive power. Notably, the above sequent notation presupposes that categorial combination is *associative*. There is no information in binary grouping of premises (the commas in sequents form a 'flat product'). But in recent syntactic theories, more discrimination is found in the *Non-Associative Lambek Calculus* **NL** with obligatory binary products $A \bullet B$ that group premises (Kandulski 1988, Kurtonina 1995). This move is similar to that creating Arrow Logic from relational algebra (cf. Chapter 8).

Expressive Power. A further dimension of change has been the categorial *vocabulary*. Lately, operations have been added to the classical repertoire $\{\rightarrow, \leftarrow, \bullet\}$, such as Boolean *disjunction* leaving a choice between categories, or *conjunction* assigning two categorial roles at the same time (as in current type theories). Another expressive extension are *modal operators* inspired by analogies with linear logic (Morrill 1994, Moortgat 1996). Thus, a fixed repertoire of structural rules may support varying vocabularies. The two degrees of freedom are not independent. Kurtonina and Moortgat 1996 show how calculi in the Categorial Hierarchy can be mutually embedded through the addition of suitable modal operators — a technique illustrated in Chapter 7.

12.1.2 Categorial Semantics

'Semantics' in Categorial Grammar has two different senses. One is the semantics of proofs themselves, i.e., grammatical derivations. The latter carry all information needed to interpret linguistic utterances. Starting with Montague, truth definitions for natural language have followed this format. Its technical tool is the *Curry-Howard-deBruyn isomorphism* between categorial proofs and terms in a *typed lambda calculus* describing denotations. Different categorial calculi correspond to different fragments of full lambda calculi (van Benthem 1991b has the full theory). This mathematical model has a dynamic flavour, as lambda terms denote functions.

Example 12.4 (Semantics of Derivations) Consider the Geach Composition Rule, a principle from the Lambek calculus driving the earlier example of verb phrase negation. It is expressed in the derivable sequent $A \rightarrow B, B \rightarrow C \Rightarrow A \rightarrow C$. We compute the meaning of its obvious derivation. The crux is this. The Curry-Howard-deBruyn algorithm lets function applications encode instances of Modus Ponens, and lambda abstractions instances of Conditionalization. Then, we arrive automatically at *function composition* as the intended meaning:

$$\begin{aligned} x : A, y : A \rightarrow B &\Rightarrow y(x) : B \\ x : A, y : A \rightarrow B, z : B \rightarrow C &\Rightarrow z(y(x)) : C \\ y : A \rightarrow B, z : B \rightarrow C &\Rightarrow \lambda x.\, z(y(x)) : A \rightarrow C. \end{aligned}$$

But, standard logical semantics abstracts away from proof structure, and merely fits derivable sequents of categorial calculi with some model-theoretic notion of validity. This requires *models* interpreting type-forming operators. Many of these are reviewed in van Benthem 1991b: including syntactic 'language models' as well as numerical ones. Of most interest to us, however, are dynamic models consisting of *binary relations*.

Definition 12.5 (Binary Relational Interpretation) The Lambek Calculus can be interpreted in relational set algebras. Atomic categories stand for binary relations over some set. Then, the type-forming operations become:

$A \bullet B$	$\lambda xy.\, \exists z\, (Axz \wedge Bzy)$	relational composition
$A \rightarrow B$	$\lambda xy.\, \forall z\, (Azx \rightarrow Bzy)$	left inverse
$B \leftarrow A$	$\lambda xy.\, \forall z\, (Ayz \rightarrow Bxz)$	right inverse.

Thus, each categorial term A denotes a binary relation $\underline{rel}(A)$. Of course, one can interpret further type-forming operations in the same fashion, such as Booleans. Moreover, we define *validity for sequents* as follows:

$A_1, \ldots, A_k \models B$ if the corresponding inclusion of binary relations $\underline{rel}(A_1) \circ \cdots \circ \underline{rel}(A_k) \subseteq \underline{rel}(B)$ is true in all families of relations.

This interpretation fits the peculiarities of categorial deduction very well. For instance, both forms of Modus Ponens are valid, as is the Geach Rule. On the other hand, none of the classical structural rules are valid. This is not surprising. The *Update-to-Update* style of Section 7.1 is just this notion of validity, and its representation theorem said that Reflexivity and Cut are the only structural principles that hold for it. We even have

Fact 12.6 *The Lambek Calculus is sound for the binary relational interpretation.*

A major recent advance has been the converse result.

Theorem 12.7 *The Lambek Calculus is complete for the binary relation interpretation.*

Proof. Cf. Andréka and Mikulás 1993, Kurtonina 1995. ⊣

This shows that the basic categorial calculus may be viewed as a fragment of relational algebra, and hence as a simple theory of combination of procedures. Nevertheless, in the long run, there is a simpler, yet more general and illuminating dynamic perspective. Here is the interpretation for the *Non-Associative Lambek Calculus*.

Definition 12.8 (Ternary Relational Interpretation) One can interpret categorial terms as conditions on objects in ternary frames carrying a relation Rx, yz of 'composition'. This encodes combination of syntactic structures, or information pieces. This time, the interpretation of the basic type-forming operations is

$A \bullet B$	$\lambda x.\, \exists yz\, (Rx, yz \wedge Ay \wedge Bz)$
$A \rightarrow B$	$\lambda x.\, \forall yz\, ((Ry, zx \wedge Az) \rightarrow By)$
$B \leftarrow A$	$\lambda x.\, \forall yz\, ((Ry, xz \wedge Az) \rightarrow By)$.

There is an obvious analogy with models for Relevant Logic (Dunn 1985), and the *arrow frames* of Chapter 8 — where Rx, yz says arrow x is a composition of y, z, in that order. Ternary semantics is generalized dynamic interpretation for categorial deduction. This gives more general models than binary relation set algebras — with the same benefits as the move to Arrow Logic. Validity may be defined between formulas A, B as inclusion of A-arrows in B-arrows, in every model of this kind.

Theorem 12.9 *The Non-Associative Lambek Calculus is sound and complete for validity in the ternary relational semantics.*

Proof. Modal completeness arguments for categorial calculi are like representation arguments for styles of inference on binary transition relations (Chapter 7). Here is the heart of the proof (Kurtonina 1995). Consider sequents with at most two premises. Let $\Rightarrow$ be a style of inference satisfying Reflexivity, and Cut in the following special form:

$$\frac{A \Rightarrow B \quad C \Rightarrow D \quad B, D \Rightarrow E}{A, C \Rightarrow E.}$$

Here is a concrete ternary representation. For A, B, C, define

$$RA, BC \quad \text{iff} \quad \text{for all } D,\ E, \text{ if } B \Rightarrow D \text{ and } C \Rightarrow E \text{ and } D, E \Rightarrow F, \text{ then } A \Rightarrow F.$$

Let $A^{\#} := \{B \mid B \Rightarrow A\}$. A simple calculation shows that

$$A, B \Rightarrow C \text{ iff } \forall D \in A^{\#}, E \in B^{\#} : \forall F \text{ with } RF, DE\text{: } F \in C^{\#}.$$

From left to right, this is immediate from the definition of R. Conversely, one uses that $A \Rightarrow A$ and $B \Rightarrow B$, applying Cut. $\dashv$

Completeness for implicative sequents even holds over *finite* ternary models. (For binary relational models, this is still open.) On top of this, one can compute semantic properties for further categorial principles, such as additional structural rules, by *frame correspondences* (Chapter 3).

Example 12.10 (Ternary Frame Correspondences)

Permutation	$\forall xyz\,(Rx, yz \leftrightarrow Rx, zy)$
Associativity	$\forall xyzu\,((Rx, yz \wedge Ry, uv) \rightarrow \exists w\,(Rw, vz \wedge Rx, yw))$
	$\forall xyzu\,((Rx, yz \wedge Rz, uv) \rightarrow \exists w\,(Rw, yu \wedge Rx, wv))$
Contraction	$\forall x\, Rx, xx$

The latter condition says that information pieces admit of 'miraculous multiplication': which is just what destroys the cues in a resource logic.

Further representation results for inferential styles allowing additional structural rules may be proved along the same lines. For instance, Permutation in the form $A, B \Rightarrow C$ iff $B, A \Rightarrow C$ makes the above ternary relation R symmetric in its two arguments. For the case of Associativity (as in the original Lambek Calculus), see Section 13.3.1.

12.1.3 Categorial Fragments of Modal Logics

Categorial languages may be viewed as fragments of modal ones. This is achieved by defining the slashes using the following binary existential modalities (cf. Chapter 7):

$$\begin{array}{lll} \mathbf{M}, x \models A \bullet_1 B & \text{iff} & \exists yz\ Rx, yz \wedge \mathbf{M}, y \models A \wedge \mathbf{M}, z \models B \\ \mathbf{M}, x \models A \bullet_2 B & \text{iff} & \exists yz\ Rz, yx \wedge \mathbf{M}, y \models A \wedge \mathbf{M}, z \models B \\ \mathbf{M}, x \models A \bullet_3 B & \text{iff} & \exists yz\ Rz, xy \wedge \mathbf{M}, y \models A \wedge \mathbf{M}, z \models B \end{array}$$

Now, introducing Boolean negation, we can define

$$\begin{array}{lll} A \to B & := & \neg(A \bullet_2 \neg B) \\ B \leftarrow A & := & \neg(A \bullet_3 \neg B). \end{array}$$

Conversely, one can define these two products using categorial arrows plus negation. Moreover, product $\bullet_1$ is ordinary categorial product. Thus, with Booleans added, categorial logics are almost modal logics over three dual existential modalities. Complete axiomatizations for the latter are well-known. The minimal system has only Modal Distribution (Chapter 3) plus conversion principles relating the three products (Venema 1991). E.g., categorial Modus Ponens becomes the conversion principle:

$$A \bullet_1 \neg(A \bullet_2 \neg B) \to B.$$

The non-associative Lambek Calculus **NL** corresponds to the minimal modal logic here. By contrast, Geach's Composition Rule depends on the associativity of the Lambek calculus **L**, which needs an extra modal axiom of Associativity. Such distinctions are crucial. The minimal modal logic of three dual products is decidable, but its associative version is *undecidable* (Andréka et al. 1994). This fact highlights another difference. The Lambek Calculus is decidable again — having a finite search space for its Gentzen proofs. The point is that categorial logics are *non-Boolean fragments* of modal logics. This accounts for their special properties (such as invariance for occurrence counts). Kurtonina 1995 shows how this simplifies completeness arguments, explaining the above ternary representation. But, fragments may also engender complications. Boolean structure is pervasive in standard logic. Standard notions or arguments need no longer work for fragments without Booleans. A case in point is our analysis of modal languages in terms of *bisimulation invariance* (Chapters 3, 4). A similar account of categorial languages cannot use bisimulations $\equiv$: their symmetry will automatically respect Booleans. The remedy is an interesting new notion (Kurtonina 1995).

Definition 12.11 (Directed Bisimulation for Categorial Logic) A *directed bisimulation* between ternary models **M**, **N** consist of directed links $\Rrightarrow$, $\Lleftarrow$

between states pointing either way, obeying the following four conditions — and their converses:

1. if $\mathbf{M}, x \models p$ and $x \Rrightarrow s$, then $\mathbf{N}, s \models p$
2. if $x \Rrightarrow s$ and Rt, us, then there are y, z with Ry, zx, $z \Lleftarrow u$ and $y \Rrightarrow t$
3. if $x \Rrightarrow s$ and Rt, su, then there are y, z with Ry, xz, $z \Lleftarrow u$ and $y \Rrightarrow t$
4. if $x \Rrightarrow s$ and Rx, yz, then there are t, u with Rs, tu, $y \Rrightarrow t$ and $z \Rrightarrow u$.

Invariance is then replaced by preservation of categorial formulas in the direction of the simulation link. The following may be proved by a straightforward induction:

Fact 12.12 *If* $\mathbf{M}, x \models \phi$ *and* $x \Rrightarrow s$, *then* $\mathbf{N}, s \models \phi$ *— and vice versa.*

For axiomatic *completeness*, similar points apply. Completeness proofs for categorial fragments tend to be simpler — but they do need special care because Boolean maximally consistent sets are not available as worlds. Sometimes, simple formula-based models do the job. These involve representations like the above. This is like axiomatizing dynamic styles of inference (Chapter 7). Finally, as for *complexity*, categorial logics may be much simpler than their modal counterparts (Buszkowski, to appear). As we saw above, **L** drops from undecidable to decidable. And even with decidable modal logics, non-Boolean modal fragments may drop an order of magnitude. The minimal modal logic is known to be PSPACE-complete, but its implicative categorial fragment **NL** is decidable in polynomial time.

12.1.4 Connections with Dynamic Logic

The preceding excursion into modal logic shows that categorial logics are like dynamic logics in many ways. More generally, we conclude by summarizing how themes from earlier chapters have emerged so far. First, an emphasis on *occurrences* is typical for both syntactic resources and procedural instructions. Typically, the classical structural rule of Contraction is invalid. Thus, the *landscape of substructural logics* for categorial combination is like the landscape of dynamic styles of inference. Chapter 7 provided examples of this resemblance, including modal translation strategies relating various dynamic styles of inference. In particular, there is the intriguing completeness of the Lambek Calculus for the natural $\{\rightarrow, \leftarrow, \bullet\}$-fragment of relational algebra, i.e., a dynamic core logic of procedural composition. Also, both fields had a move from binary to ternary models, decreasing complexity. We do not yet know the full extent of these analogies. For instance, there are various *translations* between arrow logics (Chapter 8) and categorial logics — both over ternary models.

Example 12.13 (Categorial Logic and Arrow Logic) Here are arrow transcriptions of the two directed functional slashes, using Boolean negation and converse of arrow-style relations:

$$\begin{aligned} A \to B &:= \neg(A^{\smile} \bullet \neg B) \\ B \leftarrow A &:= \neg(\neg B \bullet A^{\smile}). \end{aligned}$$

Categorial product goes to arrow-style ternary relation composition $\bullet$. Using frame correspondences, the two basic categorial laws then express earlier interaction principles for composition-plus-reversal triangles on arrow frames (Chapter 8):

$$\begin{array}{ll} A \bullet (A \to B) \Rightarrow B & \forall xyz\ Cx, yz \to Cz, r(y)x \\ (B \leftarrow A) \bullet A \Rightarrow B & \forall xyz\ Cx, yz \to Cy, xr(z). \end{array}$$

This gives us the two implications

$$\text{if } X \Rightarrow A \to B, \text{ then } A \bullet X \Rightarrow B \qquad \text{if } X \Rightarrow B \leftarrow A, \text{ then } X \bullet A \Rightarrow B.$$

Adding their converses (generating all of **L**) yields no further semantic constraints. Thus, Basic Arrow Logic contains the Lambek Calculus, and that even faithfully, thanks to the earlier completeness theorem. One can also investigate this connection proof-theoretically, locating weak arrow logics co-interpretable with categorial ones. (This also involves converse translations, mapping converse to categorial surrogates.) E.g., Kurtonina 1995 presents an arrow-style counterpart of the Lambek Calculus extending its proof-theoretic count invariants to deal with converse and negation.

Another feature in common to both fields is proliferation of *new operations*. Thus, are there notions of *safety* for directed bisimulations that might do for categorial operations what we did for dynamic ones in Chapter 5? Here is another analogy. Dynamic arrow logic (Chapter 8) may be compared to a categorial logic extended with Kleene iteration.

Example 12.14 (Derivation of $(A \to A)^* = (A \to A)$) Here is a principle from Ng 1984, derived in dynamic arrow logic:

$$\begin{array}{ll} (A \to A) \Rightarrow (A \to A)^* & \text{axiom of DAL} \\ (A \to A) \bullet (A \to A) \Rightarrow (A \to A) & \text{by Lambek rules in arrow logic} \\ (A \to A)^* \Rightarrow (A \to A) & \text{by the star rule of DAL.} \end{array}$$

Finally, here is another modal perspective. Like modal logics, categorial logics may be effectively *translated* into a first-order language over binary or ternary models (cf. Chapter 3). A crucial point is then in which *fragments* they land. E.g., binary relational semantics leads to bounded formulas, but not necessarily *guarded* ones (cf. Chapter 4). But ternary semantics works as follows. Only 3 variables are needed for translating categorial formulas, and ternary bounding atoms then create guarded formulas. This reduces decidability for categorial logic over ternary models to the known decidability of the first-order Guarded Fragment. Again, this connection is as yet largely unexplored.

12.2 Semantics

12.2.1 Sources and Frameworks

Linguistic semantics shows a broad spectrum of dynamics, amply demonstrated in Chapter 2. Anaphora changes denotations for pronouns, temporal narrative induces stepwise construction of representations (Kamp 1979, Naumann 1995), quantifiers shift local domains of discourse. New illustrations emerge all the time. These include the object dependencies of Meyer Viol 1995, who uses dynamic stack mechanisms with epsilon terms that are related to generalized assignment models (cf. Chapter 9), the dynamic plural dependencies of van der Berg 1996 (cf. also Chapter 9), or the account of indexicality in Perry 1993 via context change (cf. Section 12.3.2). A growing number of computational architectures implements these ideas. Even before the semantics community caught on, computational linguistics was replete with dynamic paradigms, witness Grosz and Sidner 1986. In the same tradition, Kameyama 1993, 1994 has a process model changing attention, focus, and epistemic preferences, interleaved with static grammatical knowledge. Other models are proof-theoretic (Gabbay and Kempson 1991, Meyer Viol 1995), or use unification methods, suitably generalized, from logic programming (Prüst et al. 1994). The survey Muskens et al. 1996 has a non-partisan panorama. These trends have produced elegant formal systems, such as *Dynamic Predicate Logic* (Groenendijk and Stokhof 1991; Chapter 2). Also influential is *Discourse Representation Theory* (Kamp 1984, Kamp and Reyle 1993, van Eijck and Kamp 1996). The two approaches are converging, especially in the Utrecht School (Vermeulen 1995, Visser 1994a, 1994b). Historically, the 'file change semantics' of Heim 1982 is a common ancestor to both. We continue with a short discussion of DPL, from the standpoint of this book.

12.2.2 DPL and Dynamic Logic

Let us summarize the logical picture of DPL that came out of the preceding chapters.

Proposition 12.15 *Over standard first-order models, dynamic predicate logic is the dynamic propositional logic of atomic tests* $(At)?$ *and variable-value reassignments* $\exists x$. *Briefly,* $\mathbf{DPL} = \mathbf{PDL}((At)?, \exists x)$.

Proof. Here are two-way embeddings that show this (cf. Chapters 2, 3). In one direction, DPL consequence amounts to a modal statement

$$[\phi_1 \circ \cdots \circ \phi_k]\langle\psi\rangle\top :$$

every successful processing chain for the premises achieves a state where the conclusion can be processed. Viewed as programs, the formulas inside here involve atomic tests, random reassignments, as well as the two relation-algebraic operators $\sim$, $\wedge$. These could be unpacked in PDL by the following equivalences (cf. Chapter 3):

$$\begin{array}{rcl} \langle{\sim}\phi\rangle A & \leftrightarrow & A \wedge \neg\langle\phi\rangle\top \\ \langle\phi\wedge\psi\rangle A & \leftrightarrow & \langle\phi\rangle\langle\psi\rangle A. \end{array}$$

Conversely, for each formula α in this propositional dynamic logic, with only atoms $\top$ ("true") and program operations $\circ$, $\cup$ (we omit Kleene iteration), there is a poly-modal formula without program operations which has the same declarative meaning. Thus, it is enough to show how DPL provides tests for each atomic modality:

$$\begin{array}{rcl} \underline{test}(\langle\exists x\rangle A) & := & {\sim}{\sim}\exists x\,\underline{test}(A) \\ \underline{test}(\langle P\mathbf{t}\rangle A) & := & P\mathbf{t} \wedge \underline{test}(A). \quad \dashv \end{array}$$

Remark 12.16 As it happens, DPL is closed under Kleene iterations (although $*$ is not explicitly definable in it) — but we will not elaborate this technical point here.

These connections remain relevant when switching from standard semantics for relational algebra and propositional dynamic logic to the *generalized semantics* of Chapters 8, 9. The same reduction formulas will produce *decidable* versions of DPL, without impairing its effectiveness for the description of linguistic anaphora. (Indeed, it would be very strange if the latter phenomenon were to require undecidability.) Nevertheless, there are some subtleties. For instance, an existential quantifier now means there exists some available state where the following formula can be processed. Moreover, we lose general relation-algebraic laws like Associativity of composition: $((\exists x \circ \exists y) \circ \exists z)$ need no longer be equivalent to $(\exists x \circ (\exists y \circ \exists z))$. We do retain the **CRS** laws, of course, such as $\exists x \circ \exists x = \exists x$. Over generalized assignment models, many observations in Chapter 2 are to be rechecked. E.g., the alternative translation

$$\mathit{Trans}(\phi)(x_1^{\mathrm{in}}, \ldots, x_n^{\mathrm{in}}, x_1^{\mathrm{out}}, \ldots, x_n^{\mathrm{out}})$$

no longer works without further ado, as it heavily depended on existential assumptions about availability of variable assignments. Finally, the discussion of **CRS** also suggested the introduction of dynamic semantic operations for syntactic *substitutions* $[t/x]$. From a logical viewpoint, this rounds off DPL rather nicely, without changing its behaviour in any essential way. But also from a linguistic viewpoint, one needs these further operations to deal with anaphoric cases like "<u>Paul</u> entered the audience hall. <u>He</u> noticed the Padisha Emperor at once..." .

We conclude this discussion with a recent logical result. Van Benthem and Cepparello 1994 observed that DPL-substitution instances seem sufficient to refute any algebraic identity in $\{\sim, \circ\}$ which is not valid over all binary relational set models. Therefore, they conjectured that schematic validity in DPL is *complete for relational set algebra*. This conjecture was proved in Visser 1995, whose core is an elegant short argument.

Theorem 12.17 *Schematic validity in DPL is complete for relational set algebra.*

Proof. Let the relation-algebraic identity E in the vocabulary $\{\sim, \circ, \cup\}$ be falsified on a family of binary relations over some carrier set S. This gives a transition model $(S, \{R_a \mid a \in A\})$, where A is the set of atomic relation symbols in the equation E. Now, we consider another transition model consisting of 'ordered pairs' over S (encoded by functions from two variables x, y):

$$(S^{\{x,y\}}, \{\underline{R}_a \mid a \in A\}) \text{ where we set } \underline{R}_a(s_1, s_2)(s_3, s_4) \text{ iff } R_a s_1 s_3.$$

It is easy to show that the projection map sending pairs (s_i, s_j) to their first element s_i is a *bisimulation* (indeed, a p-morphism). Therefore, the refuted equation E is also refuted on the second transition model (by the results on bisimulation in Chapter 3). But then, it suffices to show that the relations $\underline{R}_a$ are definable in a DPL-model over the first structure (with the states in S as the individual objects, and relation symbols referring to the original relations R_a). Let the new relation symbol I denote identity in this model. Then indeed, the following formulas have have precisely these transition relations between assignments for their DPL denotations:

$$\exists y\, R_a xy \wedge \exists x\, Ixy \wedge \exists y\, \top. \quad \dashv$$

12.2.3 DPL and Fragments of Predicate Logic

There are at least two strategies for decreasing complexity of logical systems. One is to *redesign* their semantics, as we did with the generalized semantics of Chapters 8, 9. The other is to restrict the use of a system to suitable *fragments* of its language, where all intended applications take place. This was the theme of Chapter 4, where powerful decidable 'bounded fragments' of predicate logic were found. The second strategy of fragments also make sense in dynamic semantics. One obvious issue in representing meanings of linguistic expressions is our choice of corresponding logical formulas. Intuitively, natural restrictions apply here. One does not encounter full predicate logic. E.g., vacuous quantification $\exists x\, \phi$ (with x not free in ϕ) seems to make little sense. Indeed, restrictions on occurrences of variables and quantifiers are vital to the way in which DPL is supposed to match natural language. For, the syntactic distribution of variables will determine the dynamic behaviour of a text. Since we know that DPL is mutually interpretable with standard predicate logic, finer points of its efficiency must lie precisely in details such as these. The precise merits of a logical calculus can only be established in conjunction with an account of representation. The total package matters. Clever formalizations might avoid the pitfalls of a potentially very complex calculus. Here is a very reasonable restriction when using DPL in transcribing texts.

Each new quantifier uses a *new* variable, different from all earlier ones.

This is a designer's decision. Arguably, natural language itself wants to be a finite-variable fragment, given its scarcity of pronouns (Joshi and Weinstein 1981). The formal language resulting from the above restriction is no longer *context-free*, unlike full predicate logic. But despite this syntactic twist, it does change DPL's behaviour for the better. Consider its non-monotonic Update-to-Domain consequence $\models_{\text{dyn}}$ (Chapter 2).

Fact 12.18 *Without repetitions of quantifier variables, Monotonicity holds unrestrictedly.*

Proof. Assume that A, $B \models_{\text{dyn}} C$. We show that A, D, $B \models_{\text{dyn}} C$, when the sequence A, D, B, C obeys our quantifier restriction. In particular, no quantifier $\exists u$ in D occurs in A, B, C. Now, consider any transition sequence

```
      >> s1 >>      >> s2 >>      >> s3
  A             D             B
```

The assignments s_1, s_2 agree on all variables in A, B, C (D at most involved tests for these). Rearranging assignments then, there is another successful transition sequence

```
      >> s1 >>      >> s4
  A             B
```

where s_4 agrees with s_3 on all A, B, C variables. By the above assumption, then, there must be some C-continuation in s_4. But then, there is also one starting in s_3, since s_4, s_3 differ only in 'inessential variables' u. ⊣

The above argument presupposes 'semantic irrelevance of non-occurring variables'. This no longer holds in the generalized assignment models of Chapter 9. (Lowering complexity of logics may come at a semantic price elsewhere.) Interactions between standard structural rules and variable restrictions on the first-order language may be investigated more systematically. Possible restrictions include:

I *no repeated quantifier variables*

II *no repeated variables at all from left to right* (this excludes backwards anaphora as in $\neg P\underline{x} \wedge \exists \underline{x}\, Px$).

One may impose restrictions on single formulas, but also on sequences of premises, or on whole 'texts' plus conclusions. Structural rules include all those of Chapter 7.

Example 12.19 (Transitivity) Transitivity fails even with the strongest restriction II (at least, in standard semantics):

$$\sim\sim\exists y\, Ay \models_{\text{dyn}} \exists x\, Ax \models_{\text{dyn}} Ax \text{ but not } \sim\sim\exists y\, Ay \models_{\text{dyn}} Ax.$$

Example 12.20 (Permutation) Permutation fails with the weaker variable restriction I:

$$\exists x\, Px \wedge \sim Px \models_{\text{dyn}} \bot \text{ but not } \sim Px \wedge \exists x\, Px \models_{\text{dyn}} \bot.$$

It does hold with the stronger version II, though:

Fact 12.21 *With no repeated variables left-to-right, Permutation holds.*

Proof. If A, B satisfies the left-to-right restriction II, then we can interchange as in the following picture

```
      >> s1 >>       >> s2      to          >> s2 >>        >> s1
    A                 B                    B                 A
```

as the assignments were to different variables. Tests are unproblematic, too. A test in the A-sequence is on variables **x** (from the previous context) plus **a** (from A). Performing it later has the same result (as it does not involve **b** variables). Tests in the B-sequence may involve **x**, **a**, **b**: but performing them earlier only creates problems if some quantifier $\exists a$ in A is relevant. But then, sequence B, A violates our variable restriction. (This 'cautious monotonicity' keeps additions inside our fragment.) ⊣

Restrictions I, II rule out Reflexivity. We retain alphabetic variants:

$$\text{not } \exists x\, Ax \models_{\text{dyn}} \exists x\, Ax, \text{ but } \exists x\, Ax \models_{\text{dyn}} \exists y\, Ay.$$

There are two general strategies for dealing with deviant aspects of dynamic inference. When a classical structural rule fails, one may look for a 'modified version' which should hold for *all formulas* (e.g., Left Monotonicity for $\models_{\text{dyn}}$), or one can retain the full principle for *special formulas*. Thus, DPL has unrestricted monotonicity for all *test* formulas (Chapter 7 has this observation and its converse). Many examples of the latter strategy exist for allowing unrestricted monotonicity: modalized shriek-formulas in linear or categorial logics, or purely universal formulas in circumscription. As for the *complexity* of dynamic inference, it would be of interest to analyze all these initial observations on DPL fragments in the first-order Guarded Fragment of Chapter 4.

12.2.4 Digression: Discourse Representation Theory

Another major dynamic paradigm is Discourse Representation Theory. We sketch its features very briefly. From the viewpoint of this Book, DRT is a rich dynamic logic over constructive states (structured databases called *discourse representation structures*) performing both anaphoric shifts and informational updates via a *discourse algorithm* taking previous discourse representation structures to new ones incorporating an update for the next sentence. DRSs are much like files with information written on them, with an appealing graphics (called "BOXese").

Example 12.22 ("A woman catches a cat. It scratches her.") Here are DRSs for this two-sentence discourse. Variables at the top of these boxes are called *discourse referents*, the open sentences underneath *conditions*.

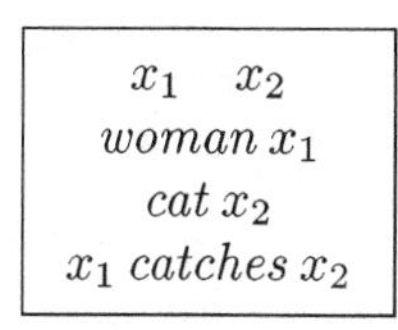

$x_1 \quad x_2$
woman x_1
cat x_2
x_1 *catches* x_2
x_2 *scratches* x_1

Boxes can also directly represent universal or conditional information.

Example 12.23 ("Every woman catches a cat") Quantifier conditions require relations between subboxes:

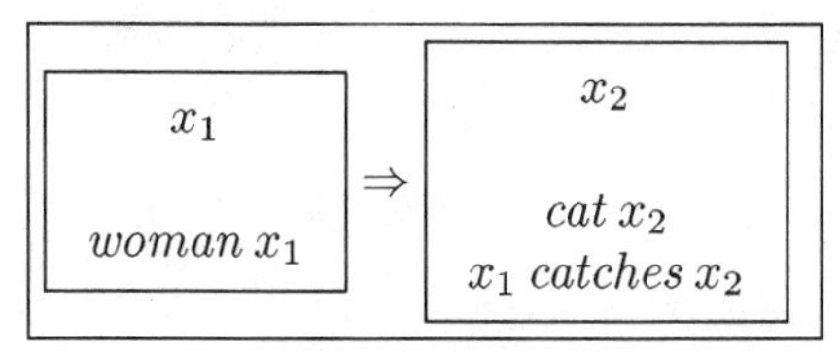

Further information is provided in the earlier references, including various *accessibility conditions* on DRS construction that give the theory its descriptive bite. DRSs may be psychologically realistic mental representations, but we can also view them as a formal language. The following rules define the basic constructs, *conditions* (γ) and *boxes* (K):

$$\begin{aligned} \gamma &:= Px \mid x_1 R x_2 \mid x_1 = x_2 \mid \neg K \mid K_1 \vee K_2 \mid K_1 \Rightarrow K_2 \\ K &:= [x_1 \ldots x_n \mid \gamma_1, \ldots, \gamma_m]. \end{aligned}$$

For DRSs K, one can give dynamic update conditions much as for DPL in Chapter 2. These usually involve transitions of *partial assignments*, while conditions γ denote sets of finite assignments in a model. Here is a truth/update definition of the intended meanings (with $a[x_1 \ldots x_n]b$ for: $a \subseteq b$ and $\text{dom}(b) = \text{dom}(a) \cup \{x_1, \ldots, x_n\}$).

$$\begin{aligned} \|Px\| &= \{a \mid x \in \text{dom}(a) \wedge a(x) \in I(P)\} \\ \|x_1 R x_2\| &= \{a \mid x_1, x_2 \in \text{dom}(a) \wedge \langle a(x_1), a(x_2)\rangle \in I(R)\} \\ \|x_1 = x_2\| &= \{a \mid x_1, x_2 \in \text{dom}(a) \wedge a(x_1) = a(x_2)\} \\ \|\neg K\| &= \{a \mid \neg \exists b \, \langle a, b\rangle \in \|K\|\} \end{aligned}$$

$$\begin{aligned} \|K_1 \vee K_2\| &= \{a \mid \exists b\,(\langle a,b\rangle \in \|K_1\| \vee \langle a,b\rangle \in \|K_2\|)\} \\ \|K_1 \Rightarrow K_2\| &= \{a \mid \forall b\,(\langle a,b\rangle \in \|K_1\| \rightarrow \exists c\,\langle b,c\rangle \in \|K_2\|)\} \\ \|[x_1 \ldots x_n | \gamma_1, \ldots, \gamma_m]\| &= \{\langle a,b\rangle \mid a[x_1 \ldots x_n]b \wedge b \in \|\gamma_1\| \cap \ldots \cap \|\gamma_m\|\}. \end{aligned}$$

We have not investigated DRT from the viewpoint of our general logical dynamics. But there are many links. Van Eijck and Kamp 1996 show how referent systems and other dynamic features may be incorporated into the formalization. Muskens 1994 proposes a translational methodology for DRT into dynamic logics, thereby providing it with a compositional dynamic semantics. Muskens et al. 1996 discuss natural *merge* operations on DRSs, with a category-theoretic flavour. In line with earlier chapters, one wants an independent account of natural logical operations over DRSs, and the dynamic styles of inference supported by these.

12.3 Pragmatics

Linguistic pragmatics is the study of language *use*, beyond the structural sentence level. Pragmatic phenomena are evidently dynamic. A typical example are speech acts like *questions*. These call for answers that update one's current information state. Discourse models have long been computational: witness DRSs, or stack-based models from the computational tradition. A typical case where pragmatics has deeply influenced dynamic semantics is the study of *presupposition* as a process of changing utterance contexts. Our discussion follows Beaver 1996.

12.3.1 Dynamics of Presupposition

Presupposition is not easy to define. Factive verbs are one example. "Anna regrets that Bill is sad" presupposes that "Bill is sad". Definite descriptions carry presuppositions. "The king is bald" presupposes that "There is a king". These are not ordinary logical inferences. "Anna does *not* regret that Bill is sad", too, presupposes that "Bill is sad". But $A \models B$, $\neg A \models B$ imply logical validity of the conclusion B: quod non. Sources of presupposition are diverse — though both examples involved 'definiteness' (the first regrets *the fact that*). One can delimit presupposition from ordinary inference, and also from Gricean 'implicature' — the discourse phenomenon which makes us infer from "Mary saw some birds" that "Mary did not see all birds". There is a whole spectrum of inferential mechanisms in natural language, which cooperate in human performance. This fits with our plea for logical plurality in Chapters 7, 13.

For a dynamic definition of presupposition, we turn to the classical *Projection Problem*. Presuppositions are inherited from atomic expressions to complex linguistic ones in a systematic fashion. But, they may also be *canceled*. The conjunction "Bill is sad and Anna regrets that Bill is sad" no longer presupposes that "Bill is sad" — and the same holds for the implication "If Bill is sad, then Anna regrets that Bill is sad". But in antecedent positions, presuppositions go through. "If Anna regrets that Bill is sad, then Claudia is desolated" presupposes that "Bill is sad". The Projection Problem asks for a systematic algorithm computing complex presuppositions systematically. Kartunen and Peters 1979 propose this mutual recursion on 'assertions' and 'presuppositions':

$$\begin{aligned} \text{pres}\,\neg A &= \text{pres}\,A \\ \text{pres}\,A \wedge B &= \text{pres}\,A \wedge (\text{ass}\,A \to \text{pres}\,B) \\ \text{pres}\,A \to B &= \text{pres}\,A \wedge (\text{ass}\,A \to \text{pres}\,B) \end{aligned}$$

plus recursive clauses for assertions.

The same ground is covered by the dynamic semantics of Beaver 1995, extending the eliminative update semantics of Chapter 2. Implications $A \to B$ are read as $\neg(A \wedge \neg B)$. A novelty are *partial functions*, via a test operator as in dynamic logic (cf. Chapter 3):

$$\delta A(X) = \begin{cases} X & \text{if } A(X) = X \\ \uparrow \text{ (i.e., undefined)} & \text{otherwise.} \end{cases}$$

Beaver's proposal reads presupposition $A \gg B$ ('A presupposes B') as follows (Chapter 7 had a corresponding notion of inference for arbitrary binary relations):

Definition 12.24 (Presupposition) $A \gg B$ if, in all eliminative update models, whenever $A(X) = Y$, then $B(X) = X$:

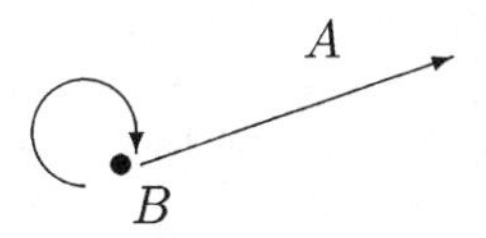

Applications to natural language must first encode raw data by suitable insertions of δ. E.g., in the above "Anna regrets" sentences, one inserts a conjunct "δ Bill is sad". The calculus itself then projects presuppositions compositionally to complex formulas. We can test this in a formal way. Positive predictions rest on valid principles like the following, which are quite easy to check:

Fact 12.25 *The following principles are valid:*

$$\begin{array}{lll} \delta A \gg A & & \\ A \gg B & \textit{if} & \neg A \gg B \\ A \gg B & \textit{implies} & A \wedge C \gg B \\ A \gg B & \textit{implies} & A \to C \gg B \\ A \gg B & \textit{implies} & C \to A \gg C \to B. \end{array}$$

The implication clauses follow from those for conjunction by the above definition of $\rightarrow$. By contrast, we do not get what we did not want:

Fact 12.26 *The following principles are invalid:*

$$A \gg A$$

$$A \gg B \quad \textit{implies} \quad C \wedge A \gg B$$

$$(\textit{e.g.}, \ \delta A \gg A, \textit{ but not } A \wedge \delta A \gg A)$$

$$A \gg B \quad \textit{implies} \quad C \rightarrow A \gg B.$$

This logical calculus derives all Kartunen-Peters rules. It also makes new predictions, when combined with dynamic epistemic modalities. E. g.,

$$A \gg B \text{ iff } \textit{may } A \gg B.$$

Like every empirical theory about language, the above is not wholly correct. Stronger presuppositions remain beyond it. "If I go to Amsterdam, my sister will meet me at Schiphol" presupposes that "I have a sister", not that "If I go to Amsterdam, I have a sister" (intriguing though it may be). To deal with these, Beaver introduces complex 'information orderings', whose dynamics recall the changing preferences of Chapter 2. The complete logic of the above system is in Chapter 7. Dynamic presupposition is an inverse of Update-Test Consequence, and so its complete logic may be found by a simple transformation. Here are its characteristic structural rules:

Right Monotonicity	$X \Rightarrow B$ implies $X, A \Rightarrow B$
Right Cut	$X \Rightarrow A$ and $A, X \Rightarrow B$ imply $X \Rightarrow B$.

This duality suggests a broader conjecture, concerning directions in ordinary reasoning. In natural language, presupposition and inference might be two sides of the same coin, one computing 'preconditions' and the other 'postconditions' (cf. Chapters 2, 6).

12.3.2 Dynamics of Context

We conclude with a quite different example, of changing contexts and shifting *forms of linguistic expression* accessing these (McCarthy 1993, Perry 1993). Perry is concerned with the flexibility and efficiency of mechanisms of reference in natural language, McCarthy rather with the design of efficient modular systems in artificial intelligence. The main principle at work is one of *Minimal Representation.* In communication, we use minimal linguistic code, taking maximal advantage of regularities available in the relevant context. But when changing contexts (to take someone else's point of view, to change the subject, or to access background information), this linguistic form may be modified — lifting relevant information from its original setting. Perry has Pacific islanders discuss their weather. For their aims, it suffices to refer to events and times — while place indicators are redundant ("rain now", "sunshine yesterday"). But in a larger world, place indicators become necessary to convey the local facts ("it is raining here, but not over there" or "it is raining at latitude X and longitude Y"). Simply put, one may switch between local evaluation of a simple statement p in context $\mathbf{c}$ and global evaluation of a more complex explicit statement $P(c)$ in a larger environment $\mathbf{M}$:

$$\mathbf{c} \models p \text{ iff } \mathbf{M} \models P(c).$$

Conversely, retreat into a smaller context skips indicators whose values are understood. McCarthy has switching formulas relating implicit and explicit local positions of objects in a spatial context. This is reminiscent of Correspondence Theory for modal languages (Chapter 3), which translates modal assertions into first-order ones over modal models, making reference to states or contexts explicit. For instance, let α be an assignment setting the linguistic context variable c to the real context $\mathbf{c}$:

$$\begin{aligned} \mathbf{M}, \mathbf{c} \models p \quad &\text{iff} \quad \mathbf{M}, \alpha \models Pc \\ \mathbf{M}, \mathbf{c} \models \Diamond p \quad &\text{iff} \quad \mathbf{M}, \alpha \models \exists c' \, (Rcc' \wedge Pc'). \end{aligned}$$

The right-hand formulas may be computed algorithmically from the left-hand ones. This may be realistic, as linguistic context switches also seem virtually automatic. Modal correspondences are often seen as a tool without linguistic or philosophical import. But in the above light, they lie within one language, where *both* modes are available to us in communication. We do not have to choose: natural language is, in Tarski's phrase, *semantically universal* — and as such, it transcends all separate formal languages.

Here is a connection with our main concerns. Recall the dynamic variations (Chapter 2) on Tarski's static interpretation for first-order logic:

$\mathbf{D}, \mathbf{I}, \mathbf{s} \models \phi$ formula ϕ is true in domain $\mathbf{D}$, given interpretation function $\mathbf{I}$ and variable assignment $\mathbf{s}$.

Evaluation of a linguistic assertion ϕ may involve changes in all semantic parameters. What stayed the same, however, was the linguistic assertion ϕ *itself*. But the above context changes make even the form of the assertion under evaluation a modifiable dynamic degree of freedom. There are many technical mechanisms for context switches with changing assertions. These may be quite diverse. Context change is not a unique, nor a simple phenomenon! For instance, making the universe of discourse explicit is the standard operation of *relativization*. This equates evaluation of a formula ϕ in the restriction $\mathbf{M} \upharpoonright D$ of a first-order model $\mathbf{M}$ to some subdomain D with evaluation of the corresponding relativized formula $(\phi)^D$ in some suitable expansion $\mathbf{M}^+$ of the whole model $\mathbf{M}$:

$$\mathbf{M} \upharpoonright D \models \phi \quad \text{iff} \quad \mathbf{M}^+ \models (\phi)^{\underline{D}},$$

where $(\phi)^D$ has all its quantifiers $\exists x\,\psi$ from ϕ relativized to $\exists x\,(\underline{D}x \wedge \psi)$ and $\underline{D}$ is a new unary predicate set to denote D in the expanded model $\mathbf{M}^+$. This somewhat pedantic formulation emphasizes what is really going on in this shift. A similar move drives the earlier-mentioned dynamics of domains for generalized quantifiers, whose standard ternary bounded forms $Q_E\,AB$ indicate obligatory shifts in the arguments E, A. Quantifier relativization is just one instance of a more general operation. This comes out in a well-known formula schema in set theory, which states the adequacy of Tarski's truth definition for models viewed as set-theoretic structures:

$$\text{for all } \mathbf{M}, \phi\text{: } \mathbf{M} \models \phi \leftrightarrow (\phi)^{\underline{M}}.$$

The operation on the right-hand side 'relativizes' all syntactic items interpreted by the model $\mathbf{M}$, thus displaying also the role of its interpretation function I. For instance, one specific equivalence reads as follows:

$$\mathbf{M} \models \exists x\, Qjx \quad \text{iff} \quad \exists x\,(D_{\underline{M}}x \wedge I_{\underline{M}}(Q)(I_{\underline{M}}(j)x),$$

adding explicit parameters for the predicate Q and the individual constant j. Further forms of context dependence emerge when we move from extensional to intensional logic. Consider again first-order translations of possible worlds semantics for *modal propositional logic*. Slightly restated, the Perry example looks like this:

$$\mathbf{M}, \mathbf{c} \models p \text{ iff } \mathbf{M}, [c := \mathbf{c}] \models Pc.$$

The modal index $\mathbf{c}$ stands for a whole package of things: time, place, possible world, 'reference group', or whatever else is relevant to determining what is being asserted. Indeed, modal operators themselves require evaluation dependent on the current context, via accessibility among worlds, witness the following clause:

$$\mathbf{M}, \mathbf{c} \models \Diamond\phi \text{ iff } \mathbf{M}, [c := \mathbf{c}] \models \exists c'\,(Rcc' \wedge [c'/c]\phi).$$

More subtle forms of context-dependence arise in *modal predicate logics*, which involve interactions between modalities, predication and quantification. This shows in the various relativizations needed for its translations. Consider the following example

$$\mathbf{M}, \mathbf{c}, a \models \exists x\, Qjxy \text{ iff } \mathbf{M}, [c := \mathbf{c}], \alpha \models \exists x\,(\underline{D}_c x \wedge Qcjxy).$$

Here, the domain of quantification gets restricted by the local context c, like the predicate Q and the individual constant j — because we relativized the predicate Q to a new argument c. But this translation is not forced upon us. It reflects certain decisions that we could have made differently. Consider the following alternative:

$$\mathbf{M}, [c := \mathbf{c}], \alpha \models \exists x\,(\underline{D}_c x \wedge I_c(Q)jxy).$$

Here the individual constant j is not relativized to its value in the local context $\mathbf{c}$: it serves as an absolute *rigid designator* across contexts.

This multiplicity is no problem. Both (more) rigid and (more) context-dependent names occur in natural language. Finally, there is a less obvious choice point. Standard semantics for modal predicate logic keeps the current world and the variable assignment separate. But a 'context' may package these into one object containing all that is needed for evaluation.

12.3.3 Digression: Context Logics

In this digression, we discuss a two-level context formalism highlighting the empirical phenomena at issue, following van Benthem 1996h. (Other options are in Buvac and Mason 1995, Buvac 1995.) The limit is a *two-sorted first-order logic* with variables over contexts and over objects. This can be interpreted in a standard manner, over models with two separate domains. In particular, it contains mixed atoms $Q\mathbf{cx}$ which relate tuples of contexts and worlds. Usually, one has cases $Qc\mathbf{x}$ where object predicate Q holds of objects $\mathbf{x}$ in a single context c — but cross-contextual comparisons are also conceivable ("I am happier *here* than you are over *there*"). This rich two-sorted language swamps the intuitive interplay between local and global contexts. Hence, we take a first-order language with object and context variables, while restricting occurrences of the latter. It has predicate-object atoms $Q\mathbf{x}$, individual quantifiers $\exists x$ and constants j, but these also have indexed versions $Q_c\mathbf{x}$, $\exists_c x$ and j_c. The latter are marked for their context of evaluation, while the former are not: by default, they are governed by the local context. We can make the same distinction for modalities $\Diamond$ and $\Diamond_c$. This may be implemented in a fixed two-sorted model $\mathbf{M}$. Let A be an assignment to object and context variables, $\mathbf{c}$ a local context which is like a standard model, with a domain and interpretation for predicate letters and individual constants. First, we define semantic values for terms:

$$V_{\mathbf{M}}(A, \mathbf{c}, j) = I_{\mathbf{c}}(j) \qquad V_{\mathbf{M}}(A, \mathbf{c}, j_{c'}) = I_{A(c')}(j) \qquad V_{\mathbf{M}}(A, \mathbf{c}, x) = A(x).$$

Definition 12.27 (Truth definition for Context Logic) Here are some self-explanatory truth conditions (we suppress the fixed model $\mathbf{M}$):

$$A, \mathbf{c} \models Qjk_{c'}x \quad \text{iff} \quad I_{\mathbf{c}}(Q)(V(A, \mathbf{c}, j), V(A, \mathbf{c}, k_{c'}), V(A, \mathbf{c}, x))$$

$$A, \mathbf{c} \models Q_{c''}jk_{c'}x \quad \text{iff} \quad I_{A(c'')}(Q)(V(A, \mathbf{c}, j), V(A, \mathbf{c}, k_{c'}), V(A, \mathbf{c}, x))$$

Boolean operations $\neg, \wedge, \vee$ are interpreted as usual

$$A, \mathbf{c} \models \exists x\, \psi \quad \text{iff} \quad \text{for some } u \text{ in } D_{\mathbf{c}}, A[x := u], \mathbf{c} \models \psi$$

$$A, \mathbf{c} \models \exists_{c'} x\, \psi \quad \text{iff} \quad \text{for some } u \text{ in } D_{A(c')}, A[x := u], \mathbf{c} \models \psi$$

$$A, \mathbf{c} \models \Diamond\psi \quad \text{iff} \quad \text{for some } \mathbf{c}' \text{ with } R\mathbf{cc}', A, \mathbf{c}' \models \psi$$

$$A, \mathbf{c} \models \Diamond_{c'}\psi \quad \text{iff} \quad \text{for some } \mathbf{c}'' \text{ with } RA(c')\mathbf{c}'', A, \mathbf{c}'' \models \psi.$$

This language switches back and forth. We can make a local context explicit by tagging on a context variable. Conversely, we can fix a context, remove all variables referring to it, and interpret the resulting slimmer formula.

Thus, the two mechanisms from the preceding section find their precise expression. Proof theory, too, switches between contexts, just as needed. Evidently, this language is a *fragment* of the full two-sorted first-order logic over contexts and objects (its context component has a modal flavour). Such modal fragments have a chance of decidability, even when the full first-order language does not (cf. Chapter 4). Choosing between options may be a subtle matter, witness similar issues in *temporal semantics* for natural language (cf. Kamp 1971, van Benthem 1983a). Also, contexts have more structure than that provided above. When identified with models, a natural relation is 'domain extension' with new objects. Logics for their modalities will reflect structural properties of such special context relations. We also need an algebra of natural operations on contexts, such as *merges*.

We conclude with a program for a more radical context logic.

(1) First, as in the generalized semantics of Chapter 9, we may drop the assumption that all model triples (D, I, a) are *available* as contexts. Dropping this tacit assumption will give us *decidable* context logics. But we can push the dynamic stance even further.

(2) Compared with actual discourse, there is one more unrealistic idealization in standard Tarski semantics. Contexts do not interpret predicate letters or quantifier symbols in a text uniformly, but each *occurrence* one by one (van Deemter 1991). Two occurrences of an existential quantifier may easily get a different range of individuals, depending on where we are in processing the sentence. Likewise, two occurrences of a predicate symbol may be ambiguous, either blatantly, or via a changed reference set (think of degree adjectives like "small"). Interpretation by occurrences changes the standard format of semantics. Our conjecture is that the minimal predicate logic of such a scheme is *decidable*. (For a special case, cf. Alechina 1995b.)

(3) Finally, consider again the semantics of modal predicate logic. This is fraught with conceptual difficulties, whose full extent is only gradually becoming clear (Ono 1987, Ghilardi 1991). These difficulties all assume the usual possible worlds modeling. But context logic suggests redesign. Instead of the usual semantic format $\mathbf{M}, \mathbf{w}, a \models \phi$, one may interpret modal predicate logic as follows: $\mathbf{M}, \mathbf{c} \models \phi$, where the context $\mathbf{c}$ has absorbed the individual variable assignment. This leads to atomic clauses like this: $\mathbf{M}, \mathbf{c} \models Qx$ iff $\mathbf{c}(Q)(\mathbf{c}(x))$. The clause for an existential modality then reads simply as in the propositional case: $\mathbf{M}, \mathbf{c} \models \Diamond\phi$ iff *for some* $\mathbf{c}'$ *with* $R\mathbf{cc}'$, $\mathbf{M}, \mathbf{c}' \models \phi$. These truth conditions no longer enforce cross-world identity of objects: $\mathbf{c}(x)$, $\mathbf{c}'(x)$ need not be the same. But, they make validity of all predicate-logical laws unproblematic, unlike in standard semantics.

We forego such radical avenues here. Even with this brief digression, though, we hope to have shown that context change, implicit in much of standard logic, is a suggestive new source of dynamic ideas.

12.4 Alternative Frameworks

Other mathematical paradigms for natural language have their own brands of dynamics. We link up with two conspicuous instances, namely, proof theory and feature logic.

12.4.1 Proof Theory for Natural Language

In recent years, Proof Theory is coming to the fore as a paradigm for natural language. This was clear with the Categorial Grammar of Section 12.1. But we found a problem, of *too much* dynamics. Categorial calculi had relation-algebraic models making them fragments of dynamic logic (Chapters 2, 3). But they also had natural proof dynamics, via the Curry-Howard-DeBruyn isomorphism. How are these two perspectives related, if at all? This issue is raised more systematically in van Benthem 1996f, who finds other instances, e.g., in the comparison between proof dynamics for discourse and dynamic logics dealing with the same phenomena. Speaking generally, this is a real question, which emerges in many settings. For instance, how are the proof accounts and the Kripke model accounts of constructive reasoning really related (cf. Chapter 2)? Without achieving a Grand Unification, we can at least make a few useful remarks.

We will use *labeled deductive systems* (Gabbay 1992), a format motivated as follows. Binary 'labeled statements' $\mathbf{l} : \mathbf{A}$ encode richer information than what is usually manipulated in inference, combining logical syntax ($\mathbf{A}$) with explicit semantic indices of evaluation or other useful items ($\mathbf{l}$) that do not show up in surface syntax. Labels remove conventional constraints on the expressive power of logical formalisms, allowing us to construct and reason about explicit proofs or semantic verifiers for statements. After all, most information passed in natural language is heterogeneous, including both linguistic code and conversational or physical context. This point seems convincing, and it can be motivated from many different angles. But it does not commit us to any particular choice of labels, or any particular calculus. Concrete examples are model-theoretic statements $\mathbf{w} : \mathbf{A}$ from modal logic (world w verifies statement A), or proof-theoretic ones $\boldsymbol{\tau} : \mathbf{A}$ (τ has type A, τ proves proposition A). Thus, the whole enterprise may be summed up in the meta-equation

$$\mathbf{LDS} = MGU[\mathbf{ML}, \mathbf{TT}].$$

That is, labeled deductive systems form a *most general unifier* over two systems: modal logic and type theory. What becomes of the above issue in this framework? Consider categorial calculi and their dynamic models.

As with modal logic, one can *translate* such systems into first-order logic, transcribing their semantics (Chapter 3). Namely,

primitive types	$T(a) = R_a xy$
product types	$T(A \bullet B) = \exists z\,(T(A)(x,z) \wedge T(B)(z,y))$
left implications	$T(A \to B) = \forall z\,(T(A)(z,x) \to T(B)(z,y))$
right implications	likewise.

Validity of a categorial sequent $X_1, \ldots, X_k \Rightarrow B$ is equivalent to first-order validity of the corresponding implication

$$T(X_1 \bullet \cdots \bullet X_k)(x,y) \to T(B)(x,y).$$

In this way, we could analyze categorial validity in a first-order metalanguage over transition models. There is some slack in this translation. The Lambek Calculus is decidable — unlike first-order logic. As observed before, we are dealing with first-order *fragments* (translated categorial formulas needed only 3 state variables) and in the proof calculus needed to drive the above equivalence: a decidable sublogic suffices. Here LDS comes in. We can also analyze categorial reasoning via labeled statements $\mathbf{xy} : \mathbf{A}$ and decidable calculi in between the Lambek Calculus and full first-order logic.

Now, back to our main issue. Can we meaningfully merge *type-theoretic* statements $\tau : A$ and model-theoretic ones $w : A$? Consider the following labeled versions of Modus Ponens — coming, respectively, from standard logic, relational categorial semantics and lambda calculus:

$$x : A \quad x : A \to B \ \vdash \ x : B$$
$$xy : A \quad yz : A \to B \ \vdash \ xz : B$$
$$x : A \quad y : A \to B \ \vdash \ y(x) : B.$$

The most natural labeled generalization covering all these runs as follows:

$$x : A \quad y : A \to B \quad Rz, xy \vdash z : B,$$

where Rz, xy is some *ternary* condition relating z, x, y. The condition Rz, xy can be '$z = x + y$' (z is the supremum of x and y in some Kripke model; cf. Chapter 2) or 'z is the composition of the arrows x and y' (again, if one exists) or 'z is the result of applying y to x' (if defined). Keeping this analysis in mind, we now analyze the matching introduction rules of Conditionalization. The outcome is that they all exhibit the following format:

$$\tau : X \quad x : A \quad Rz, xy \vdash z : B \text{ implies } \tau : X \vdash y : A \to B.$$

For instance, consider the specific case of lambda abstraction:

$$\tau : X \quad x : A \vdash \sigma : B \text{ where } x \text{ does not occur free in the term } \tau,$$
$$\text{implies } \tau : X \vdash \lambda x.\sigma : A \to B.$$

This becomes an instance of the above by reading Rz, xy as the true ternary application condition $z = \lambda x.\sigma(x)$ $(= \sigma)$, with $y = \lambda x.\sigma$. In full detail:

$$\tau : X\ x : A \vdash \sigma : B \text{ is equivalent to } \tau : X\ \ x : A\ \ z = \lambda x.\sigma(x) \vdash z : B.$$

Thus, the 'most general unifier' that we were looking for turns out to be a *ternary* transcription of implicational logic, which reads, e.g.,

$$A \to B \text{ as } \forall xz\,((Rz, xy \wedge A(x)) \to B(z)).$$

This is *relevant implication* (Chapter 8), which was complete for the Non-Associative Lambek Calculus (Section 12.1.2) — which is then arguably the basic labeled calculus unifying modal logic and type theory.

Remark 12.28 (Binary Formats) Ternary relevant semantics is a decidable common ground for lambda calculus and dynamic logic. But it has hardly any computational specifics left. Can one do better? Modal logic has a *binary* semantic format. Its rules for implication have this shape:

$$x : A\quad y : A \to B\quad Ry, x \vdash x : B,$$

where Ry, x is some binary condition relating x, y, and

$$\tau : X\quad x : A\quad Ry, x \vdash x : B \text{ implies } \tau : X \vdash y : A \to B.$$

The typed lambda calculus should match up, as it matches intuitionistic implication. Can this be explained? The answer is a common perspective for proof dynamics and for Kripke models (cf. Chapter 2). Lambda rules specialize the above schema as follows (using upward heredity of intuitionistic formulas along the information ordering $\subseteq$):

- Ry, x becomes $y(x) \subseteq x$ in the partial order of information extension
- the rule of lambda abstraction involves the premise
  $\tau : X\quad x : A \vdash \sigma : B \Rightarrow \tau : X\quad x : A\quad \sigma \subseteq x \vdash x : B$
  $\Leftrightarrow \tau : X\quad x : A\quad \lambda x.\sigma(x) \subseteq x \vdash x : B \Rightarrow \tau : X \vdash \lambda x.\sigma : A \to B.$

12.4.2 Feature Logics

There are many other logical paradigms for linguistic analysis. *Unification grammars* and their associated *feature logics* take linguistic structures as labeled graphs carrying syntactic, semantic, morphological and phonetic information. An excellent survey is Rounds 1996. On this view, the purpose of a linguistic theory is to describe admissible graphs underlying texts or utterances. E.g., a context-free grammar is a device for accepting derivation trees (satisfying 'node admissibility conditions' corresponding to its production rules). These conditions may be formulated in many formalisms. Standard first-order logic is one candidate, but in practice, much can be done with modal fragments describing the possible successors for the nodes. Interestingly, this logical approach works both for Chomskyan grammars and for its latter-day rivals.

Example 12.29 (Context-Free Grammars) Consider the rewrite rules:

N ⇒ N PP PP ⇒ Prep Det N
N ⇒ mother, child Prep ⇒ with, without Det ⇒ a, no.

This produces complex noun phrases such as

"mother without a child without a mother".

These may be ambiguous, unless we supply derivations. Here are two:

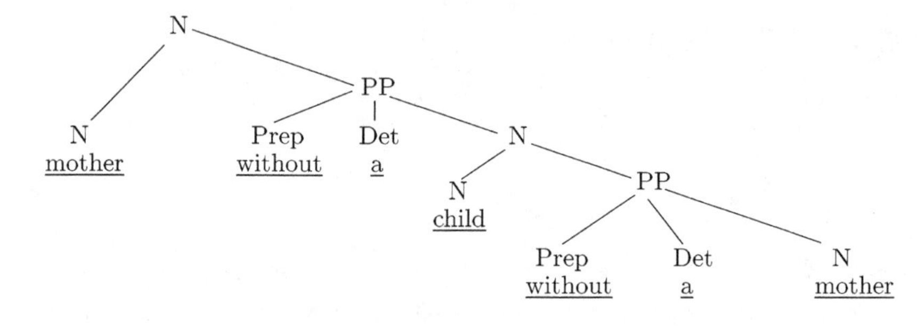

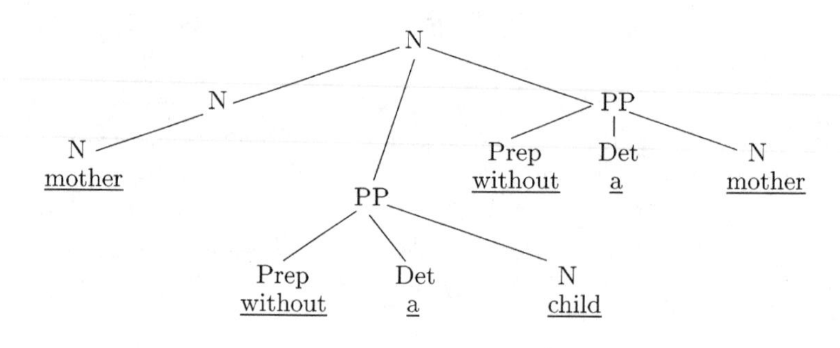

Modal Phrase Structure Grammar. Evidently, such analysis trees may be viewed as finite modal models, with proposition letters for grammatical categories or lexical items. Thus, each grammar has its strong recognizing capacity encoded by a class of trees, and we can ask for a suitable modal definition, with polyadic modalities referring to finite branchings. Thus, grammars correspond effectively to modal formulas. This analogy leads to further questions. Modal languages come in a *hierarchy* of expressive power (Chapter 4). How do its layers relate to the usual Chomsky Hierarchy of grammars, running from regular through context-free to all languages recognized by Turing Machines? Does the jump from context-free to context-sensitive correspond to a natural leap in modal expressivity? Also, what identifications are needed between trees to get at the appropriate level of 'linguistic form'? For instance, we may want to identify trees that are 'synonymous' in some intuitive sense. (The above trees are not.) This is similar to earlier concerns with process equivalences, providing the right abstract processes underlying graphical representations (Chapter 4). Finally, what deductive power is needed? There are *decidable* modal logics over finite trees, or even decidable monadic second-order languages. From the latter perspective, modal formalisms are *fragments*, just as in Chapters 3, 4, 5 above.

Feature-Value Structures. This is not the end of the matter. Nodes in derivation trees may have internal structure, provided by *features* for gender, number, 'head', or other morphological and grammatical information. This yields complex linguistic 'signs'. E.g., in rewrite rules, grammars may state that parts of an expression should show gender agreement. Feature information is given by labeled arrows indicating its sort, and pointing at a relevant value. This picture is modal again, being a deterministic dynamic logic over labeled transition systems (Chapters 3, 6). Thus, we have a stacked modal semantics, with different accessibility relations at different levels — which may allow zooming in at any level of detail. Modal and dynamic logics of feature structures have been studied extensively: cf. again Rounds 1996. Questions like the above return. In addition, we have issues like the following. Feature logics are often *non-Boolean fragments* of modal logics, as we found above with categorial languages. Moreover, we need a combined modal architecture, which should not mix the two levels in undesired ways.

From the viewpoint of this book, all this raises issues of intellectual kinship. For, the linguistic tradition of rewrite rules and automata is dynamic, oriented towards processes that produce linguistic structure. In an interesting inversion, while logic is becoming dynamic, the move towards tree logics in linguistics is static. But matters are more delicate. Apart from the vagueness of the terms 'static' and 'dynamic' (Chapter 13), motivations remain mixed in various historical phases. Technically, the modal hierarchy up to the first-order language is very similar to what we found for process formalisms. Also, finite automata and other language recognizing devices are paradigmatic examples for process theories. It would be of interest to compare our process equivalences (Chapters 4, 5) with structure levels in linguistics. E.g., over finite automata (whose machine diagrams are finite LTSs) finite trace equivalence measures weak recognizing power, while bisimulation and more sensitive process equivalences measure various forms of strong recognizing power. To be sure, there is a difference. A branching point in a graph is a *choice* from a process point of view, and a sequence of *parallel* tasks to be performed from a linguistic one. Things are not exactly the same — but they are alike.

12.5 Notes

Categorial Grammar. The paradigm goes back to Ajdukiewicz and Husserl. A classic paper from our viewpoint is Lambek 1958. Montague 1974 reinforced the categorial trend. An author responsible for much of the revival is Buszkowski 1996. The anthologies Buszkowski et al. 1988, Oehrle et al. 1988 contain much old and newer material. For the modern phase, see van Benthem 1991b, Morrill 1994, Carpenter 1996, Moortgat 1996. Other substructural logics are in Girard 1987, Došen and Schröder-Heister 1994. Background in type theory is found in Barendregt 1992, Turner 1996. As for completeness proofs for conventional semantics, simple formula-based representations do the job (cf. Došen 1985 for the non-associative Lambek Calculus, or van Benthem 1991b on associative Lambek Calculus, using finite sequences of formulas as worlds). The approach has been extended recently by the use of labeled calculi: cf. Kempson 1995. There is no single system doing the job of linguistic description. Kurtonina 1995 shows how different labeled calculi may be complete for the same categorial logic. (Orlowska 1991 has labeled calculi for modal logic and relational algebra.)

Semantics of Proofs. For Categorial Grammar, the first proof analysis via restricted lambda terms with a single occurrence property was given in van Benthem 1983b. This may be generalized to connections with Category Theory (another major paradigm for dynamics). Cf. Lambek and Scott 1986, Girard et al. 1990, and Lambek's contribution to Došen and Schroeder-Heister 1994. Abrusci 1995 gives a novel term semantics for objects in directed functional types $\{\rightarrow, \leftarrow\}$.

Logical Syntax. Rule-based or automata-based descriptions are the traditional mathematical model for the syntax of for natural languages (Hopcroft and Ullman 1979). Such recognition procedures have also been given for the syntax of formal logical languages. These are usually context-free, but some syntactic modifications lead to varying complexity. Cf. van Benthem 1988b.

Partial Dynamic Semantics. For descriptive purposes, DPL can be improved. It has not yet provided a definitive treatment of *negation* or *disjunction* in natural language, and its *conjunction* misses non-sequential uses (Dekker 1993, van der Berg 1996). Also, *partiality* must be accommodated at different levels (van der Berg 1996). Dynamic operators may be undefined, inducing *three* values in procedures. (Tests may result in "true", "false" or "inappropriate", while variable-value updates may be "successful", "failed" or "irrelevant".) This requires a three-valued generalization of Relational Algebra, which may be used to define further negations in natural language. Van den Berg has proposals mixing the counter-domain test negation of DPL with a partial polarity reversal $\sim R$ accepting those transitions which R rejects and vice versa (while agreeing with R on the irrelevant transitions). Three-valued atomic actions arise as follows. Tests only have a definite truth value if their presupposition of no-state-change is satisfied (otherwise, they are undefined):

$$s_1[\![Pt]\!]^+s_2 \quad \text{iff} \quad s_1 = s_2 \text{ and } \mathbf{M}, s_1 \models Pt$$
$$s_1[\![Pt]\!]^-s_2 \quad \text{iff} \quad s_1 = s_2 \text{ and not } \mathbf{M}, s_1 \models P_t$$
$$s_1[\![Pt]\!]^*s_2 \quad \text{iff} \quad s_1 \neq s_2.$$

Next, variable updates cannot be 'false', intuitively:

$$s_1[\![\exists x]\!]^+s_2 \quad \text{iff} \quad s_1 =_x s_2$$
$$s_1[\![\exists x]\!]^*s_2 \quad \text{otherwise.}$$

This requires a three-valued procedural repertoire, too. From the literature on partial logic (Langholm 1988, Thijsse 1992, Jaspars 1994), one would expect something like:

$$s_1[\![\phi \wedge \psi]\!]^+s_2 \quad \text{iff} \quad \text{for some } s_3,\ s_1[\![\phi]\!]^+s_3 \text{ and } s_3[\![\psi]\!]^+s_2$$
$$s_1[\![\phi \wedge \psi]\!]^-s_2 \quad \text{iff} \quad \text{for all } s_3,\ s_1[\![\phi]\!]^-s_3 \text{ or } s_3[\![\psi]\!]^-s_2$$
$$s_1[\![\phi \wedge \psi]\!]^*s_2 \quad \text{otherwise}$$
$$s_1[\![\neg\phi]\!]^+s_2 \quad \text{iff} \quad s_1[\![\phi]\!]^-s_2$$
$$s_1[\![\neg\phi]\!]^-s_2 \quad \text{iff} \quad s_1[\![\phi]\!]^+s_2$$
$$s_1[\![\neg\phi]\!]^*s_2 \quad \text{iff} \quad s_1[\![\phi]\!]^*s_2$$

Van den Berg 1996 has complex alternatives, mixed with strong dynamic negation.

States themselves can be partial, too, assigning values only to variables that matter, or are known. This requires models with extensions or contractions of states. These have two distinct atomic variable updates. One changes a value for a variable already in use, the other introduces a value for this variable for the first time (cf. Chapter 9).

Modal Dynamic Predicate Logic. Natural language has many dynamic mechanisms. DPL changes anaphoric bindings, while Update Semantics (Chapter 2) deals with incoming propositional information. Combination of the two is desirable. Now, DPL is an algebra of relations over assignments, but US a family of functions on sets of valuations. Various proposals have been made for their mathematical unification (cf. Dekker 1993, van Eijck and Cepparello 1994), but the most sophisticated follow-up is Groenendijk et al. 1994, 1996b. These papers start from the usual linguistic evidence behind *modal predicate logic*. Consider the pair of sentences

(1) A man is walking in the park *who* might be wearing a blue sweater.

(2) A man is walking in the park. He might be wearing a blue sweater.

The relative clause in (1) states a property of the man in the main clause: he might be wearing a blue sweater. The authors argue this is not the role of the second sentence in (2). The latter states the possibility that some individual introduced in the antecedent is wearing a blue sweater. A combined dynamic semantics must account for these facts. Since discourses (1), (2) are equivalent in standard DPL, a departure is needed from it, in which antecedent existentials need no longer scope over free variables in succedents. The combined semantics (inspired by Vermeulen 1994) has information states with three components: (1) an assignment of variables to 'pegs' (discourse individuals; cf. Landman 1986), (2) an assignment of pegs to individuals in a standard domain, (3) a set of possible worlds over the latter (the current range of descriptive uncertainty). Updating combines elimination of possibilities with enrichment of assignments. A new feature is the update for an *existential quantifier*. In DPL, $\exists x$ was a single instruction for random assignment. In the present models, this would enrich the current state with every possible assignment of objects to (the peg for) the variable x. Compound formulas $\exists x\, \phi$ then compose this move with an ordinary update for ϕ. This yields unintuitive results on a modal statement $\exists x\, \Diamond Px$: since the resulting state might retain assignments to x denoting objects which lack property P. The new update condition makes $\exists x\, \phi$ syncategorematic after all: "*take the union of all actions* $x := d\, ;\, \phi$ *for all objects* d *in the domain*". Then, an update for $\exists x\, \Diamond Px$ ends up with x assigned only to those objects that have P in some possible world. This also provides a dynamic cut on the traditional puzzles of modality and identity in the philosophical literature. (The above-mentioned papers also discuss consistency and discourse coherence.)

Discourse Representation Theory. There is a vast literature on this paradigm. Cf. Kamp and Reyle 1993, Van Eijck and Kamp 1996, Visser 1994a, Zeevat 1991. Despite the grand name, there is little on intrinsic discourse phenomena in DRT. For the latter, one must rather consult the computational tradition (cf. Polanyi and Scha 1984).

Computational Linguistics. DRT tells us little about which referent must be chosen if more than one is accessible. Linguistic clues restrict antecedents for anaphoric elements, such as gender and number agreement, but the structure of discourse itself puts important further constraints. Thus, theories of discourse as in Polanyi and Scha 1984, Grosz and Sidner 1986 are a natural complement to DRT or DPL. Their greater wealth also serves as a healthy reminder how rarified an air we have been breathing. Grosz and Sidner distinguish three parts of discourse structure. Linguistic structure cuts up any discourse into *discourse segments*. Experimental speech data suggest that segmentation of this kind is uniform. Certain cue phrases signal discourse boundaries: "in the first place", "in the second place", "anyway", and so do changes in tense and aspect. Discourse segments have *purposes*, which can stand in relations of 'dominance' and 'satisfaction-precedence'. These constitute the *intentional state*. The third part, *attentional state*, is most truly dynamic. It consists of a stack of *focus spaces* with the discourse referents, properties, relations and discourse purposes that are salient at any given moment. The closer a focus space is to the top of the stack, the more salient its objects. Anaphoric expressions pick up the referent on the stack that is most salient. Change is brought about by pushing and popping the stack. Entering a discourse segment causes its focus space to be pushed onto the stack and leaving a segment causes its space to be popped. This model of discourse resembles the structure of programs in an imperative language such as PASCAL. Nested segments are like the loops and subloops that we find in a typical program. We can also compare the nested structure with procedures calling subroutines, which may in their turn call subroutines. Then the stack of focus spaces for attentional state resembles the usual computer stack.

Dynamics of Presupposition. Some basic references on context change in presupposition are Stalnaker 1972, van der Sandt 1988, Beaver 1996. Van Eijck 1991 treats presuppositions $A \gg B$ in a partial dynamic logic with an 'error state', as follows: $\langle A \rangle \top \to \text{fix}(B)$. Aliseda-Llera 1996 has an account in terms of *abduction*. In this area, as in all empirical inquiries, one may dispute the evidence. The initial argument against identification with straight inference might rest on unclarity in the scope of the negation. Also, concerning the Projection Problem, a warning is in place. If presupposition is at least *like* logical inference, this might be a very unrealistic query: there just is no recursive definition of all valid consequences from a formula — though there is one for the *underlying* notion of truth. But in natural language, there might be a fast compositional subroutine pre-computing simple presuppositions. (Algorithmic approximations are probably also available for ordinary inference.)

Context Change. Similar motivations have been put forward in robotics. Thus, minimality principles might govern robots that must reason, and yet perform on-line (cf. Rosenschein and Pack Kaelbling 1986, Barwise 1989).

Feature Logics. There is a flourishing computational literature on feature logics, connecting up with domain theory on the one hand, and logic-programming-style unification theory on the other: cf. Carpenter 1992, Rounds 1996. In particular, algorithmic complexity for non-Boolean feature systems has been studied extensively.

Tree Logics. For connections with Government Binding Theory, cf. Rogers 1996, Kracht 1995. Backofen et al. 1995 give connections with standard model theory, including one with Ehrenfeucht game techniques used in Doets 1987 to obtain complete monadic Π^1_1-theories for linear and well-founded. The latter involve 'compression' arguments for models (w.r.t.

formulas up to some fixed quantifier depth), allowing one to prove also *finite model properties*, and hence decidability.

Modal Grammars. This viewpoint has been elaborated by many authors, including Blackburn and Meyer Viol 1994, Kracht 1995, Blackburn et al. 1996. Several approaches are compared in Volume 4 of the *Journal of Logic, Language and Information* (1995, P. Blackburn, ed.). Conservative combinations of modal phrase structure logics and modal feature logics may be obtained by the 'fibering method' of Finger and Gabbay 1992. Cf. also the sensitive historical discussion in Blackburn 1995.

Situation Theory. Yet another take on graphs and bisimulation is provided in Situation Theory (Moss and Seligman 1996). Here, bisimulation equivalence serves to indicate the right 'parts' for information structures. As for further dynamics over these, cf. the Channel Theory of Barwise and Seligman 1996.

Linguistics. We stress some general points from the above. Complexity of linguistic description is not only measured by automata or rewrite rules, but also by thresholds of expressive power in logical languages over linguistic structures. (Immerman 1996 proposes a similar move in Complexity Theory.) E.g., in the 'modal phrase structure grammar' of Blackburn and Meyer Viol 1994, adding modalities is the unit of expressive power. Blackburn and Spaan 1993 show how apparently innocuous steps may lead to undecidability. Similar proposals occur in the tree logics mentioned above. E.g., Rogers 1996 shows how many first-order logics encoding Chomskyan syntactic constraints over finite trees remain decidable — while certain acyclicity conditions lead to undecidability. Thus, a hierarchy of logical fragments may be the yardstick of linguistic description. In addition, the proliferation of frameworks demonstrates the importance of optimal *representation*, letting one get away with maximal simplicity of code and shallowness of inference over it. The amazing thing about natural language is not fire-power, but the fact that we owe so much to so little.

Semantic Automata. The traditional dynamics of automata has been used in the 80s to define procedural semantics where denotations *themselves* are dynamic objects. A typical example are the *semantic automata* of van Benthem 1986, which analyze generalized quantifiers as model checking procedures of various complexities. Cf. van Benthem and Westerståhl 1995 for the current state of research along these lines. There is a strong connection with the first-order and modal tree logics mentioned above. E.g., there is a *fixed-point theorem* extending the well-known De Jongh-Sambin Theorem of modal provability logic, showing that many recursive first-order definitions for finite classes of trees (of the sort produced by rewrite grammars) can be made explicit.

12.6 Questions

Dynamics Logic of Categorial Grammar. Settle the Finite Model Property for the Lambek Calculus with respect to binary relational semantics. Prove a Safety Theorem with respect to directed bisimulations. (Added in Print. De Rijke and Kurtonina 1996 present a solution.) Develop the full duality between arrow logics and categorial logics.

Dynamic Semantics. Specialize our general theory to DPL. E.g., what is a complete safe repertoire of operations over its (generalized) assignment models? (DPL might be closed under operations like $\cap$ or $-$, which are not safe in general ...) Develop the model theory of partial Relational Algebra and Propositional Dynamic Logic. In particular, determine complete safe repertoires, in an appropriate partial sense of safety (cf. the proposals in van der Berg 1996). Develop a general proof theory for fragments with variable restrictions encountered in practice. Find a complete axiomatization for Dynamic Modal Predicate Logic.

Dynamics of Questions. Give a dynamic update semantics for linguistic questions (various hints occur in Groenendijk and Stokhof 1996).

Dynamic Analysis of DRT. Give a dynamic analysis of DRT addressing analogies with semantic tableaus, natural operations for combining DRSs, and styles of dynamic inference. (A presupposition of this question is that standard accounts of DRT are too conservative in this respect, taking traditional logic as the role model.)

Context Logics. Axiomatize our context logic, and determine the complexity of the mentioned modal fragments. Also, axiomatize predicate logic over occurrences. Likewise, for modal predicate logic over contexts combining 'world' and 'assignment'.

Lambda Dynamics and Modal Dynamics. Extend and exploit the analogies found in our text. What do they tell us about lambda calculus, or dynamic logic?

Finite Model Theory. Given the interest in *finite grammar* and feature models, do a finite-model-theoretic version of modal logic. Our conjecture is that, unlike the situation with first-order logic in general, all major metaproperties of modal logic (including Interpolation and Łoś-Tarski) will go through. (Cf. Chapters 4, 10.)

13

Philosophical Repercussions

In this final chapter, we discuss broader backgrounds and repercussions of the dynamic stance developed in this book. Admittedly, twelve chapters so far have not produced a clear definition of the opposition "dynamic" vs "static". Like the opposition "procedural" — "declarative" in computer science, it has immediate appeal, but very elusive content. Is a labeled transition system, or a process algebra, static or dynamic? Is an arrow more dynamic than a dot? Mathematical representations seem static, even for dynamic systems. The distinction seems to apply rather to the reality denoted, or to our intentions when using language. We have no further definition to offer, and the remainder of this chapter takes a round-about route. We are dealing with an intuitive distinction, on a par with that between "algebraic" vs "geometric" in mathematics. It describes a duality between two useful and complementary perspectives, sometimes on the same phenomenon. And we may define it further by example and application, rather than continued formal analysis. Given this point of departure, we do not pretend to any deep philosophical contribution — but rather to highlighting some broader themes from the preceding, and ways in which they might affect philosophical discussions. Even this little seems timely. Discussions in the philosophy of language or logic are still centered around the realities of the thirties, and far removed from the much greater sophistication of logical research to-day.

13.1 Meaning and Cognition as Activities

13.1.1 Activities and Products

Let us recall our general argumentation for getting the distinction across (Chapter 1). There is a pervasive duality between human activities and their static forms or products, which is reflected in a benign ambiguity of natural language. Thus, the word "judgment" is ambiguous between an act and its content, and so are "reasoning", "dance" or "move". Mainstream classical philosophy shows a bias towards finished products. Judgment revolves around static propositions, and reasoning around static proofs. This is clear in textbooks on Theory of Knowledge, which emphasize structural features of knowledge (a hard currency in the bank, as opposed to the more doubtful IOUs issued by belief). Now, the contemporary literature has a shift in emphasis "from knowledge to cognition". The latter involves interaction between more static 'representation' and more dynamic 'procedure'. Illustrations abound (Chapter 2). In the history of mathematics, one finds a distinction between 'theorems' (flat truths) and 'constructions' (algorithms for producing geometrical objects) right in Euclid's "Elements". And in the foundations of mathematics, intuitionism has emphasized creative mathematical activity, even though its formalization focused on products (definitions, proofs), and not yet on the actions that produce these. The latter are at centre stage in computational theories of cognitive 'updating', 'retraction' and 'revision' (Chapters 2, 6), which we see, therefore, as carrying on the constructive enterprise. A similar movement is under way in the semantics of natural language, with meanings described in terms of dynamic update conditions rather than static truth conditions (amply demonstrated in the Chapter 12). Our book is meant to provide some 'strategic depth' for these developments, from a logical point of view.

13.1.2 Historical Predecessors

The movement described is not unique to this period in time. There is a long history of philosophers attempting to put cognitive activities at centre stage. We mention just a few. Many classical philosophers have been action- or process-oriented. Whitehead 1929 is an example in metaphysics. Another instance are the American pragmatists. The procedural aspect of human reasoning was emphasized eloquently in Toulmin 1958, a highly critical study of modern logic, insisting that the essence of good reasoning lies in adherence to *juridical formalities* rather than to mathematical form. *Learning* by trial and error and the induced updates of belief states have been prominent in the philosophy of science since Popper, especially in its probabilistic tradition. Also, the time is not long past when the philosophy of language was not dominated by logical or grammatical form, but by Wittgenstein's *language games*. Games are an even older philosophical paradigm, witness Huizinga 1938 and the references given there. They reached logic in the form of various game theories, inspired by a variety of epistemic concerns (Kamlah and Lorenzen 1967, Hintikka 1973). The main difference between modern dynamic paradigms and these predecessors may be the inspiration provided in the meantime by *computer science*. The latter is especially clear in philosophical logic, witness the action logics of Belnap 1977, 1989, or the general influence of AI (cf. Dunn and Thomason 1988). And of course, since the thirties, mathematical logic has developed proof theory and recursion theory, which study proofs, rules and algorithms in both static and dynamic senses.

13.1.3 Rethinking Classical Issues

Thus, at least, it might be of interest to re-examine philosophical history in this light. Many issues change shape when viewed dynamically. For instance, in the Theory of Knowledge, there has been a long search for a "missing ingredient X" which would turn justified true belief into real knowledge. Could it be that one needs, not another static feature, but an account of the cognitive actions producing justified true belief? Likewise, traditional vexed issues like the status of "analytic" versus "synthetic" judgments may be relocated to a distinction between cognitive activities, and only in a secondary sense to one between propositions. (This move also seems congenial to Kant's writing, whose key term of 'judgment' breathes the above-mentioned benign linguistic ambiguity.) Further examples can be chosen almost at random. Consider Frege's crucial distinction in the theory of meaning between "reference" and "sense". The latter received a surprising dynamic analysis in an intensional theory of 'algorithms' (Moschovakis 1991). As for modern theories of meaning, key contributions like Kripke 1971 are being re-examined from a dynamic view on the role of individual objects in modal discourse (Groenendijk et al. 1996a, 1994, 1996b, Cepparello 1995). These are all scattered examples. More ambitiously, one would want to *explain* traditional static representations by their role in human cognition. Some glimpses of this drove our discussion of context change (Chapter 12). Presumably, representations are just those flexible encodings that enable us to act in a changing environments, operating from different perspectives.

13.2 Three Models for Cognitive Dynamics

13.2.1 Proofs, Programs and Games

When thinking about cognitive dynamics, several possible paradigms spring to mind, each with its own range of intuitions. The first model is that of *proofs*. Although modern logic is mainly concerned with meaning and truth, there has always been an interest in the precise arguments that show how the truth lies. Proofs are constellations of inferential steps reflecting basic moves in human competence and performance. The abundance of different proof formats in logic shows a rich source of intuitions. The discipline of proof theory studies, amongst others, when two proof systems may be counted 'equivalent', or how proofs within one format can be effectively transformed into another. For present purposes, what matters here is not extensional equality of input-to-output encoded in premise-conclusion transitions $P_1, \ldots, P_k \Rightarrow C$, but *intensional* equality of proof formats for styles of reasoning. The cognitive import often lies in the 'little things'.

Proof theory has been proposed as a philosophical paradigm for meaning in Dummett 1976. The second model is that of *programs*. Many people think of cognitive processes as mental computation triggered by pieces of text. On this view, natural language is a programming language for cognition, and the art is to analyze cognitive activities (discourses, arguments, plans) in terms of navigation across information states. This leads to a study of available basic moves as well as the global programming structure of cognitive activities, in particular, their expressive power and computational complexity. The program view has mathematical backgrounds in recursion theory, but even more in various branches of computer science. It has been the inspiration for the present book. A third cognitive model is that of *games*. These, too, come with vivid intuitions. There are logical games for evaluating statements, comparing model structures, or carrying on debates, with suitable roles for players, and winning conventions. Winning strategies in evaluation or debating games have provided novel notions of truth and consequence, and indeed, games have been defended for their general philosophical import in Kamlah and Lorenzen 1967, Hintikka 1973. Formal background is found in mathematical game theory, and also in logic itself (witness the Ehrenfeucht-Fraïssé games of Chapter 3).

13.2.2 Comparisons

Common to all three models is movement through a space of relevant states: partial stages of argumentation, information states, or game configurations. This is achieved via a repertoire of atomic moves, that can be combined into higher programs through the use of dynamic 'logical constructions'. What are natural constructions depend on the model. Proofs combine tree-like patterns: 'combination' via conjunction, 'case distinctions' via disjunction and 'hypothesizing' via implication, creating a dynamic block structure of nested reasoning. Programs employ the usual constructions of 'sequential composition', 'indeterministic choice' or 'iteration', possibly guided by 'control assertions'. Game operations reflect involvement of players, such as 'choices of continuation' (signalled by conjunctions or disjunctions indicating their rights of choice and duties of response) or 'role change' (signalled by negation). There is a certain unity to this logical repertoire, that comes out at a suitably abstract level of analysis. E.g., the general dynamic logic of this book describes back-and-forth movement through the 'epistemic time' of a universe of information states ordered by inclusion. Cognitive actions then denote sets of state transitions, i.e., binary transition relations over states. But there are alternatives. Actions might be sets of state transition sequences ('traces'), with a *branching time logic* evaluating at both states and trajectories representing 'epistemic histories'. A noticeable feature of this framework is its explicit interplay between two levels: activities changing states, and standard declarative statements about the states traversed (cf. Chapter 6). This design makes sense in all three models. Programming involves interaction of static tests and dynamic actions, with results described by static preconditions and postconditions. Likewise, playing a game involves the interaction of moves with tests concerning their intermediate results. (Most games depend on periodic checks "whether the ball was out".) Thus, at this abstract level, our three models for cognition are indeed similar. They will begin to differ, though, when we look more closely — as we shall see in a moment.

13.2.3 Virtues of Coexistence

There are several reasons for emphasizing this threefold perspective. Not the least important one is historical justice. Dynamic intuitions have often been expressed in other vehicles than current programming paradigms, and this should be recognized. Another is that their juxtaposition may provide a new light on the traditional division of Logic into its mathematical subdisciplines of model theory, proof theory and recursion theory. These still have a common cause. Finally, juxtaposition raises interesting new questions. Cognitive models derive their philosophical interest mostly from serving as vehicles for expressing famous *cognitive claims*. Of the above paradigms, the program model comes with the most intriguing philosophical programme. *Church's Thesis* claims that

"Any effective computation can be programmed on a Turing Machine."

or some equivalent recursion-theoretic device. Indeed, this Thesis covers all cognitive activities, encoded in Turing machines and their ilk. The term "machines" rather than "programs" is unfortunate here. The hardware of computing devices is irrelevant, of course, the crucial feature is the software — but with that correction, Church's Thesis gives the program model its bite and appeal. But then, similar claims arise for proofs or games. Hintikka's work on logical games suggests a parallel claim, appealing to our intuitions concerning roles and winning strategies:

"Any rational human activity can be played via Logical Games."

And David Israel (p.c.) has suggested a 'Gentzen Thesis' in a proof-theoretic framework:

"All rational inference admits of a Natural Deduction formalization."

perhaps even one imposing proof-theoretic normal forms. But in a dynamic perspective, further clarification is needed with all three assertions. Church's Thesis is the *extensional* statement that the input-output behaviour of every effective function can be programmed on a Turing Machine. It also has a stronger *intensional* version, stating that any algorithm can be rendered faithfully in the programming repertoire of the latter. As for the evidence, there are indeed similarities between computation on Turing and register machines, or logic programs, but there are also obvious differences in Turing programs with "goto" loops and higher programming languages. This intensional viewpoint affects proofs and games, too. What are natural repertoires of logical constructions for faithful modelling of rational inference or cognitive play? The intensional Church Thesis then relates to philosophical analyses of *functional completeness* for operator repertoires in proof theory (Sundholm 1986) — and to our study of complete logical repertoires for procedural constructions on programs (Chapter 5). From the latter, we conclude that there are natural levels down below the recursive functions, where various viewpoints converge — so that discussions of cognitive power can be stated in much more sensitive terms. Somewhat provocatively, the question is not whether or not humans are Turing machines, but where we are located in a hierarchy of automata, and, wherever that is, which process equivalences govern our actions.

13.2.4 Differences in Fine-Structure

Differences between our three paradigms come out when we consider actual activities. What are the *natural moves* in proofs, programs or games? What are the appropriate logical constants, or notions of consequence? We discuss some differences. Not all *logical constants* are equally at home in our three models. The proof-theoretic perspective is unhospitable towards *negation*, unless one treats 'refutation' on a par with 'proof'. (Cf. our discussion of logic programs in Chapter 11.) Negation as program complement is a marginal operation ("avoidance") — but negation as a 'role switch' is crucial to games. The proper framework for 'logical structure' of proofs may be nested data structures, for programs, it may be relational or process algebra, while for games, no general analysis of operations is available yet. Analyzing conventional notions in these models produces unconventional outcomes. E.g., quantifiers are no longer on a par with propositional connectives, even though the two are often lumped together in standard expositions. Quantifiers signal atomic moves of establishing a binding or drawing some object, plus perhaps some further propositional test. The syntax of formal languages is misleading. One now sees a formula prefix $\exists x$ as an instruction by itself: "pick an object ($\exists$), and assign it a temporary name x", different from y, even when we compare such pairs as $\exists x\, Px$ and $\exists y\, Py$. Likewise, 'ill-formed formulas' acquire a clear dynamic sense. The string $(p \wedge q$ tells us to push p and q, and we are now waiting (perhaps) for a pop instruction. Cf. Chapters 2, 12 for further examples of this enhanced sensitivity to 'little things' of syntax. The different flavours of our models may also be sampled by looking at inference. Proofs justify an inference by composing it out of basic moves.

Example 13.1 (Disjunctive Syllogism I) $A \vee B, \neg A \Rightarrow B$ requires combination of argument by cases and one basic negation step

$$\dfrac{A \vee B \qquad \dfrac{A \quad \neg A}{B} \qquad B}{B}$$

Example 13.2 (Disjunctive Syllogism II) With programs, this inference will be viewed as an updating instruction (Chapter 7):

> *updating any information state by $A \vee B$ and then by $\neg A$ (by suitable decomposition of these compound instructions) leads to a new information state which may be tested to validate B.*

Example 13.3 (Disjunctive Syllogism III) With games in the 'agonistic' Lorenzen style, one would rather say that:

> *defending the claim B in a dialogue against an opponent who has granted the 'concessions' $A \vee B$, $\neg A$ admits of a winning strategy.*

Differences will also show in the *structural rules* governing inference (cf. Chapters 1, 7). Permutation is reasonable in proofs and games, but not for programs, as sequential order of instructions is crucial to their intended effect. Monotonicity may be defended in games (the more concessions from one's opponent the better), but not on the other accounts. But even these outcomes will depend on further specification. If one takes the ordering of premises in a game to convey the priority of commitments incurred, then Permutation will lose its appeal in this model too. For a concrete example where the three models may be put to work, compare the discussion of *semantic tableaus* at the end of Chapter 2, viewed as a meeting ground for proofs and games which can still generate new developments.

13.3 Cognitive Fine-Structure

13.3.1 Lessons from Computer Science

This Book shows not just a technical influence from computer science in modelling cognition. Broader 'cultural trends' from this field are equally important. In one sense, the basic theory of computation was finished before computer science started. Church's Thesis had been formulated, based on an impressive array of equivalences, the central undecidability results had been obtained (e.g., for the Halting Problem), and recursion theory was well under way. And yet, all this was of only marginal relevance to the actual developments, which were about hierarchies of languages (upward from the machine level to programming languages for human users), distinctions between potential and feasible computation, and various architectures. In short, computation turned out to have a rich *fine-structure*, which had remained completely invisible from the logical Olympus. Moreover, both the art and the science of computation are about the delicate balance between *representation* of data and performing *computation* over them. The latter issues seem central to understanding cognition, too (rather than theological debates about the import of Gödel's Theorem). Much of this has shown here. We have tried to 'sensitize' the reader to fine-structure, noting dynamic aspects of even first-order syntax, or moves in tableaus or proofs. In the process, we found life underneath classical systems, such as fragments of standard formalisms, or reduced forms of semantic and algorithmic complexity. Another way of 'loosening up' was historical. We showed how 'classical' logical systems have implicit *choice points*, which may be set differently to 'parametrize' them into families that can be fine-tuned for applications. For instance, Chapter 2 showed that many different notions are lumped together as classical conjunction: viz., sequential composition of actions, but also parallel execution or mere unordered listing of tasks. Likewise, implication covered many dynamic variants ("if condition P holds, then perform action A", "if action A has been performed, then condition P will hold", etc). One can then recognize such options elsewhere. (Categorial logic supports different implications, and different conjunctions occur in relevant or linear logic — Chapter 12.)

13.3.2 Digression: Artificial Intelligence

One objection is the following. We are advocating the use of computational models for the study of cognition. But is not that precisely what *Artificial Intelligence* is all about? The actual history is complicated. Dynamics of cognition in our sense has played little role in the development of AI. By and large, this field has payed much more attention to issues of knowledge representation than to the algorithmics of computation. Moreover, many theories in AI are wed to declarative programming languages, while we set out mainly in an imperative direction. Action structure *as such* is absent from the agenda, even in influential theories of epistemic logic (Fagin et al. 1995). Likewise, context theories (McCarthy 1993) do not deal with context-changing actions (updates of information, perspectival shifts). Of course, like in all other fields that we surveyed, in retrospect, one sees many cues for dynamics in the AI literature after all. Examples are Allen 1983 on updates for temporal databases, or default reasoning with semantic networks. But Artificial Intelligence still has to undergo the same dynamification that we have proposed for logic and linguistics — no doubt, with similar benefits.

13.3.3 Resources

A prominent aspect of logical fine-structure are inferential *resources*. Cognitive activity has no unlimited supplies of information or deductive energy, and logical analysis must bring out just which mechanisms are adopted, with which cost in resources. This comes out in management of *occurrences* of assertions or instructions in proofs, programs and games. Repeating the same formula twice in a proof means two calls to its evidence, repeating the same instruction in a program calls for two executions and repeating it in the course of a game signals a new obligation for its defense or attack. Unlimited energy or unfailing commitment or 'standing obligation' must be encoded as a matter of logic. (Recall the tableau operator ! signalling commitment in Chapter 2). Many recent systems of logic employ occurrences, operating with finer detail than classical calculi. Another broad aspect of fine-structure is *dependence*. Standard inference freely calls the individuals under discussion. But in practice, objects may depend on others — notably, in the course of dynamic interpretation. This observation was a major motivation for the generalized assignment models in Chapter 9. Proof-theoretically, dependence requires conventions as to what objects can be introduced when. As a bonus, one will see a much richer garden of quantifiers — with classical $\exists$, $\forall$ as extreme cases of 'random access'.

13.3.4 Taking Fine-Structure Seriously

There are many further aspects to fine-structure. For instance, all three paradigms had *universal* computational strength. But underneath lies a computational hierarchy with natural levels of inhibition. Thus, Kanovich 1993 provides a match between computation in restricted automata, derivation in Horn clause calculi, linear logic based games and Petri Nets, which mimicks analogies predicted by Church's Thesis much further down. Does all this have any consequences for genuine philosophical issues? Or are the above observations just epi-cycles to the insights of our founding fathers, reflecting (undue) attention to computational implementation? The answer of this book will be clear. Imperative aspects of cognition are on a par with declarative ones. This is no break with earlier theories of meaning, but a refinement. 'Fine-structure' involves re-analysis, separating general insights in existing accounts of meaning from accidental features of the logics that inspired them. Thus, new questions move up on the philosophical agenda. E.g., design of a repertoire of logical operations is not done once and for all: it needs reappraisal with changing tasks. Traditional philosophical theories of meaning focussing on standard connectives and quantifiers are light-years removed from natural reasoning. More positively, fine-structure shows that there is more to 'standard' logical systems than meets the eye. One reason why this is sometimes hard to see is the mind-set induced by the success of logical *completeness theorems*. These have become the hallmark of logical theorizing. The thrill of seeing independently motivated syntax and semantics converge is irresistible. But a completeness theorem only finds different approaches to consequence equivalent *at some level of detail*. At other levels, there might be no convergence at all. E.g., Gödel's completeness theorem for predicate logic connects semantic validity and proofs in a purely extensional sense. It does not provide any significant link between natural proof structure and natural semantic structure. Indeed, natural proof structures of dependence lack semantic counterparts, and proof-theoretic accounts of logical constants match up only incidentally with semantic ones. The lust for completeness theorems has even harmed some paradigms. Lorenzen's dialogue games naturally support linear logic, but this discovery was missed. Natural game intuitions were twisted to fit the mold of classical or intuitionistic logic — trading immortality for respectability.

13.4 Contents and Wrappings

13.4.1 In Praise of Lightness

Given the dynamic stance, a crucial aspect of fine-structure is complexity, expressive and computational. This theme has been developed in Chapters 8, 9, showing how stepping throug the looking-glass of classical logics takes us to a whole landscape of calculi with varying computational strength. In this section, we return to the underlying broad issues. Any description of a subject carries its price in terms of complexity. To understand what is being described, one has to understand the mechanism of the language or logic employed, adding the complexity of the encoder to the subject matter being encoded. More succinctly, "complexity is a package of subject matter plus analytic tools". This price is inevitable, and scientific or common sense insight does result all the same. But, one should never pay more than is necessary. Aristotle already formulated the necessary intellectual 'lightness' (a virtue emphasized in Calvino 1988): "it is the hall-mark of an educated mind to give a subject no more formal structure than it can support". Critics of formal logic would agree, pointing at cases in philosophy, linguistics, computer science and even the foundations of mathematics where general mathematical sophistication, or essayistic common sense, is the more appropriate road towards insight than elaborate logical formal systems. But also inside formal logic, this seems a legitimate concern. Are standard modelings really appropriate for certain phenomena of reasoning, and are received conclusions about complexity of phenomena in our field (using qualifications 'undecidable' or 'higher-order') warranted, or rather an artefact of those modelings? More disturbingly, could it be that much of the respectable literature in our journals, proving difficult theorems as a sign of academic worthiness of a topic, mostly derives its continuity from the fact that one encounters the same issues over and over again, precisely because they reflect the formalisms employed, not the subject matter at hand? These are serious questions. But, how to distinguish the two sources of complexity? There may not be any systematic method — but one answer is logical practice itself, which generates many alternative formal models. Thus, wrappings becomes visible by comparison with alternatives (e.g., set-theoretic versus algebraic or category-theoretical formulations of the same problem). Although logical models may be a cause of the complexity problem, logicians mitigate this drawback by producing so many of them!

Received views on semantic complexity spot 'thresholds' where complexity is increased. A well-known danger zone is the transition from the finite to the infinite, or from first-order to higher-order objects. The latter produce undecidability or non-axiomatizability in describing computational or linguistic phenomena. In program semantics, undecidability may arise through infinitary operations of recursion. Temporal semantics for concurrency has higher-order branches or histories, with quantification over sets. Natural language has similar thresholds. Many quantificational phenomena seem second-order. 'Branching quantifiers' trigger parallel rather than serial processing, beyond first-order logic. Other sources of higher-orderness are plural and polyadic quantification (Chapter 9). At least undecidability has been claimed for the linguistic process of anaphora (Chapters 2, 12). New thresholds of complexity have emerged in Artificial Intelligence, with corresponding received views. Non-monotonic reasoning is higher-order and complex — and the same has been claimed for any human cognitive task in Kugel 1986. Fool-proof strategies are scarce. E.g., a blanket restriction to *finite models* has been advocated in natural language. But this restriction may also block useful results, as in Finite Model Theory (Chapter 10). Infinitary logics increase complexity in some ways, but decrease it in others. Whatever their merits, these technical outcomes need not square with plausible initial estimates that we may have of the intrinsic complexity of a subject described, prior to formalization. We discuss some specific cases, drawing upon earlier parts of this book.

13.4.2 From Higher-Order to First-Order

Many phenomena call for higher-order modelling, even at the cost of employing a formal system whose logical validities are non-axiomatizable (even non-arithmetically definable). Upon inspection, such cases suggest a distinction. It is one decision to employ a higher-order *language*, referring to non-first-order objects like sets, choice functions or branches — but quite another to insist that this language should have set-theoretic *standard models*. The latter is a voluntary commitment to one particular mathematical implementation of our language, whose complexities will 'pollute' the laws of a logic which was supposed to describe the core phenomenon. An insidious term often confuses this issue, namely the "concreteness" of set-theoretic models. Is a set-theoretic model really more concrete than an algebraic structure or a geometrical picture? Intuitively, the opposite seems true. Another insidious groove of thought is the 'separation of concerns' favoured by many proponents of semantic innovations. They prefer 'standard set-theoretic modelling' as their vehicle. (This is the dominant style of presentation in 'dynamic semantics', cf. Chapters 2, 12.) But this standard background may not be an orthogonal concern. It can be detrimental, repeating hereditary sins of old paradigms — just when dynamic semantics should provide cognitive relief from these.

For higher-order logic, a neutral approach uses *general models* with restricted ranges for set quantification (Henkin 1950). Standard models remain, as the limit case where all mathematically possible sets are present (whether needed or not for the object of study). This move makes higher-order logic a *many-sorted first-order logic* treating individuals, sets and predicates on a par. This is no mere tactical opportunism. Independent justification for restricting predicates ranges comes from philosophical property theories. Thus, one combines philosophical virtue with practical advantage, trading set-theoretic complexity for new sorts of individuals. A common objection here is lack of canonicity. No unique class of admissible predicates is specified. But thus, general models do the right thing, replacing Platonic complacency by honest work. If one wants new objects ('sets', 'branches'), then the task is to find all principles germane to our subject. Just what about 'admissible sets' is relevant to programs or texts? Answers may compromise. Standard set-theoretic models are too rich — but admitting all general models is too bleak. We need something in between. Second-order logic over branches may be replaced by many-sorted theories of states and 'branches' or 'paths' — but one wants to see the key axioms concerning paths and points on them. Such principles are *geometrical conditions*, on a par with standard geometrical axioms about points and lines. Geometry is as good mathematics as set theory, and perhaps more basic (witness Frege's later philosophy). Another example is intuitionistic logic. Kripke models are first-order, with truth conditions over possible worlds and accessibility, whereas second-order 'Beth models' involve branches. Beth models can be redefined as a geometry of information states and search paths, with worlds and branches on a par. One can then analyze what (little) theory about branches explains validity in constructive reasoning.

There is again fine-structure. Different geometrical assumptions reflect different styles of reasoning. This was the point of Arthur Prior's work on temporal reasoning (van Benthem 1988c), which may be brought out by *correspondence* techniques (Chapter 3). Some correspondences for putative temporal principles turn up general desiderata on state-branch models, such as linearity for branches. Others turn up negotiable options. E.g., a quantifier shift from $F\Diamond$ to $\Diamond F$ expresses this condition: time-travel along my current history followed by a switch to another future history may also be done by first switching to another history and then traveling into its future. Other modal-temporal quantifier shifts reflect set-theoretic axioms of choice. If genuine logical laws should be free from existential import (cf. Etchemendy 1990), one might admit just purely *universal* first-order requirements as semantic core conditions, shifting conditions with existential commitment to negotiable mathematics (as in Chapters 8, 9). Even so, our approach does not require that any reasonable semantics should be simply first-order. If phenomena are really complex, that should show. These points also arise in natural language. *Branching quantification* looks like a second-order mechanism, quantifying over Skolem functions beyond individual objects. But, this move only shows the need for a family of *relevant* choice functions — not a full function space. 'Functional answers' to questions motivate functional objects as first-class citizens in our semantics. ("Whom does every man love? His mother.") These issues also arise with complexity in Artificial Intelligence. E.g., non-monotonic reasoning styles like circumscription might just employ minimization in general models, over restricted families of 'computationally relevant predicates'.

13.4.3 From First-Order to Decidable

The move from higher-order to first-order brings clear gains in complexity. But, though effectively axiomatizable, predicate logic is *undecidable*. This may still import external complexity into subjects whose 'natural complexity' would be decidable. Again one feels the lure of 'concrete set-theoretic models': standard semantics seems the essence of simplicity. On the other hand, working logicians in linguistics or computer science have gut feelings that the phenomena at hand are largely decidable. But it is hard to give any mathematical underpinning to these intuitions. Can one find well-motivated decidable versions of predicate logic, by first locating and then giving up semantic prejudices? As it happens, there are many suitable strategies. One may use *fragments* of predicate logic (Chapter 4), or linear *occurrence logics*. A powerful strategy is *algebraization*, which works modulo modest requirements on the base logic. But as with general models, the interesting possibilities lie in between. This was achieved in Chapters 8, 9, which explore arrow logics and predicate logics, identifying 'semantic parameters' to be varied. Modal frame correspondences provided fine-structure, computing frame conditions for first-order axioms. As we saw before, this move has radical philosophical repercussions. What remains the point of the Completeness Theorem, stating a 'natural fit' between Tarskian semantics and Fregean axiomatization? From our current stance, Gödel's result ties just one choice of predicate-logical validity to 'standard set-theoretic modelling'. Kreisel 1967, Etchemendy 1990 have stressed the beauty of ensnaring 'intuitive validity' by means of exact mathematical notions. We must disagree. From equally natural semantic points of view, other logical equilibria arise, with different sets of validities.

13.4.4 Digression: Alternative Strategies

The preceding discussion by no means exhausts the topic of logical complexity. Other technical approaches exist for high-lighting core content. *Oracles* may supply extraneous information to be bracketed out — as in completeness theorems in program semantics. Another way of coping is that of the moderate drinker: "ingest only small amounts". Second-order complexity may be avoided by sticking to small *fragments* of a language, and the same holds for first-order logic. (By contrast, the above strategies retain the full language, lowering the over-all complexity of its logic.) Finally, *statistical* approaches to average-case complexity of reasoning take the sharpest edges off worst- case problems. In political terms, these strategies do not change the world of logic, but offer ways of adjustment to its complexity. Finally, there is a next desideratum down the road, where our book has hardly anything to offer. How to perform a reduction from *decidable* to *tractable*? Given that human cognition *works*, there may be some philosophical interest, even here.

13.5 Managing Diversity

13.5.1 Pluralism

A final motive throughout this book has been logical *pluralism*. We have not advocated any specific system of dynamic logic as the rallying point of the new paradigm. Indeed, our view is that the heart of the matter is the landscape of choices, and their connections. This diversity has occurred with our hierarchies of process equivalences (Chapter 4), logical operators (Chapter 5), dynamic styles of inference (Chapter 7) or decidable predicate logics (Chapter 9). No single 'correct calculus' exists for the laws of thought, nor one 'universal calculus' providing worst-case proof power for all cognitive tasks. Rather, we want delicate tools for describing mechanisms in reasoning at the appropriate level of exactness. This is why linguistics provides hierarchies of grammars, or science an array of mathematical calculi for studying nature. (The privileged unique position of a few traditional logical calculi is largely an artefact of textbook propaganda.) This view is no easy cultural relativism. It brings a clear task. We must chart the relevant mechanisms. Natural language presumably consists of many reasoning modules, each dealing with some aspect like temporality, quantification, plurality, etcetera.

13.5.2 Gradations

Another aspect of pluralism is that many things are matters of degree. For instance, *logicality* is a ubiquitous phenomenon (van Benthem 1986, Barwise 1989, Sher 1991), manifesting itself in the traditional logical constants, but also in very different grammatical categories (including modifiers "almost", "very" or adverbials "self"). Predicate logic has a sharp divide between logical constants and extra-logical vocabulary. But logicality lies on a spectrum. Many expressions have logical features. These may be measured by the semantic invariances of Chapter 4. Logical constants are permutation-invariant — but other expressions may show invariance for finer model equivalences. Likewise, the analysis of logical inference in Chapter 7 admits of natural gradations. Many relations between propositions perform 'inference'. This recalls *Bolzano's Program* from the early 19th century, whose main focus was the study of different styles of reasoning for the same language, studying their formal properties and connections (cf. van Benthem 1985b). The current interest in Artificial Intelligence in varieties of non-monotonic reasoning is definitely in Bolzano's spirit — and our claim has been that this form of plurality makes sense generally.

13.5.3 Architecture

Putting these points together yields the following view of natural language. The *natural logic* of van Benthem 1986 distinguishes varieties like monotonicity reasoning (predicate replacement), conservativity reasoning (domain restriction) as well as algebraic modules (e.g., for negation). Such architectures cut the cake of reasoning quite differently from traditional formal languages. In particular, monotonicy does not respect borders between first-order and second-order. In many ways, natural logic is closer to traditional logic (Sanchez Valencia 1991). Discussions in theories of meaning, heavily influenced by the agenda of modern first-order logic, seem unaware of the true logical lie of the land. Other aspects of architecture concern *interactions* between the styles of inference allowed in our pluralist universe, including their communication. We want to know what systems are involved in natural reasoning, and how they collaborate and pass information. Such architectures will also bring in *macro-structure* — in addition to fine-structure. Current logical theories operate at low levels of aggregation. We also need to grasp global structure of texts, theories or arguments, and the hierarchy of rules that govern our activities at higher levels. Bits and pieces in the literature may serve as points of departure here, such as the earlier work on the logical structure of scientific theories in the philosophy of science, or computational theories of data types. Another macro-challenge is that cognition is a *social activity* with more than one participant. Multiple agents have been studied in game theory, but hardly in logic. We need many-person theories, replacing programs by 'protocols', and allowing more interactive formats of proof.

Once again, we have not built a philosophical house in this chapter — nor even provided the foundations for building one. But we may have opened a lot of windows.

13.6 Notes

13.1. Geach 1972 warns against the 'Socratic Fallacy' of formally defining things which are reasonably clear. As for action-oriented philosophers, Burke 1995 has a dynamic logic perspective on Dewey's pragmatism. Authors on static representations through their dynamic role in cognition are Kamp 1984, Dretske 1995, Perry 1993.

13.2. An extensive survey of logical games is van Benthem 1988a. Cf. Lorenzen and Lorenz 1979, Hintikka 1973, and Lewis 1979. Functional completeness in proof theory has been investigated by Prawitz or Schroeder-Heister. Cf. the survey Sundholm 1986 — also for proof theory as a general paradigm for meaning. Dynamic logics were used in AI by Moore 1984 (epistemic logic of action), Morreau 1992 (planning). Cf. the Toronto research program of 'cognitive robotics' reported in Reiter 1994. Also relevant are the surveys of AI influences on linguistics in Thomason 1996, Asher and Pelletier 1996.

13.3. Dependence among objects is a key notion with 'arbitrary objects' in Fine 1985b, and with 'sampling' in the urn models of Rantala 1978. Proof calculi with dependence management occur in van Lambalgen 1991, and Alechina 1995a. As a bonus, they provide natural complete systems for generalized quantifiers. Sometimes, these management rules lack evident semantic counterparts (Meyer Viol 1995).

13.4. For branching temporal logic, cf. Burgess 1984 and Stirling 1989. Branching quantifiers are in Barwise 1979, plural and polyadic ones in van der Does 1992. Undecidability for anaphora is the "Any Thesis" of Hintikka 1979. The paradigm of non-monotonic reasoning is circumscription (McCarthy 1980). Many-sorted versions of higher-order logic are in Enderton 1972. (Cf. also Doets and van Benthem 1983.) For property theories, cf. Bealer 1982, Turner 1989. A definitive survey of constructive logic and mathematics is Troelstra and van Dalen 1988. Two-sorted correspondences over Beth models occur in Rodenburg 1986. For polyadic quantification, cf. van Benthem 1989b, Keenan and Westerståhl 1996. Circumscription over general models is in Morreau 1985. Various remodeling strategies (algebraic, linear, modal) occurred in Chapters 8, 9, 12. Semantic oracles are used in Cook 1978, and algorithmic ones in Buhrmann 1993.

Bibliography

Åqvist, L. 1979. A Conjectured Axiomatization of Two-Dimensional Reichenbachian Tense Logic. *Journal of Philosophical Logic* 8:1–45.

Abrusci, M. 1995. Semantics of Proofs for Non-Commutative Linear Logic. CILA preprint. Philosophical Institute, University of Bari.

Aczel, P. 1988. *Non-Well-Founded Sets.* Lecture Notes, Vol. 14. Stanford: CSLI Publications.

d'Agostino, G. 1996. *Modal Logic and Non-Well-Founded Set Theory.* Doctoral dissertation, University of Udine/University of Amsterdam. In progress.

d'Agostino, G., J. van Benthem, A. Montanari, and A. Policriti. 1995. Modal Deduction in Second-Order Logic and Set Theory. Technical Report ML-95-02. Institute for Logic, Language and Computation, University of Amsterdam.

Alechina, A. 1995a. *Modal Quantifiers.* Doctoral dissertation, Institute for Logic, Language and Computation, University of Amsterdam.

Alechina, A. 1995b. On a Decidable Generalized Quantifier Logic Corresponding to a Decidable Fragment of Predicate Logic. *Journal of Logic, Language and Information* 4:177–189.

Alechina, N., and J. van Benthem. 1996. Modal Logics of Structured Domains. In *Advances in Intensional Logic*, ed. M. de Rijke. Dordrecht: Kluwer Academic Publishers. To appear.

Aliseda-Llera, A. 1996. *Abduction.* Doctoral dissertation, Department of Philosophy, Stanford University. In preparation.

Allen, J. 1983. Maintaining Knowledge about Temporal Intervals. *Communications of the ACM* 26:832–843.

Andréka, H. 1991. *Complexity of Equations Valid in Algebras of Relations.* To appear in *Annals of Pure and Applied Logic.* D.Sc. dissertation, Hungarian Academy of Sciences, Budapest.

Andréka, H., R. Goldblatt, and I. Németi. 1996. *Relativized Quantification: Some Canonical Varieties of Sequence-Set Algebras.* Departments of Mathematics, Hungarian Academy of Sciences (Budapest) and Victoria University (Wellington).

Andréka, H., A. Kurucz, I. Németi, I. Sain, and A. Simon. 1994. Exactly Which Logics Touched by the Dynamic Trend are Decidable? In *Proceedings 9th*

Amsterdam Colloquium, ed. P. Dekker and M. Stokhof, 67–86. Institute for Logic, Language and Computation, University of Amsterdam.

Andréka, H., J. Monk, and I. Németi, (eds.). 1991. *Algebraic Logic.* Colloq. Math. Soc. J. Bolyai 54. Amsterdam: North Holland.

Andréka, H., and S. Mikulás. 1993. The Completeness of the Lambek Calculus with Respect to Relational Semantics. *Journal of Logic, Language and Information* 3:1–37.

Andréka, H., and I. Németi. 1994. Crs With, and Without Substitutions. Manuscript. Mathematical Institute, Hungarian Academy of Sciences, Budapest.

Andréka, H., I. Németi, and I. Sain. 1996. *Algebras of Relations and Algebraic Logic (an Introduction).* Budapest: Hungarian Academy of Sciences. Textbook in preparation.

Andréka, H., and R. Thompson. 1988. A Stone-Type Representation Theorem for Algebras and Relations of Higher Rank. *Transactions American Mathematical Society* 309(2):671–682.

Andréka, H., J. van Benthem, and I. Németi. 1995a. Back and Forth Between Modal Logic and Classical Logic. *Bulletin of the IGPL* 3:685–720.

Andréka, H., J. van Benthem, and I. Németi. 1995b. Submodel Preservation Theorems in Finite Variable Fragments. In *Modal Logic and Process Algebra*, ed. A. Ponse, M. de Rijke, and Y. Venema. 1–11. CSLI Publications.

Andréka, H., J. van Benthem, and I. Németi. 1996. Modal Logics and Bounded Fragments of Predicate Logic. *Journal of Philosophical Logic*, to appear.

Apt, K. 1991. Introduction to Logic Programming. In *Handbook of Theoretical Computer Science.* 494–574. Amsterdam: Elsevier Science Publishers.

Apt, K., E. Marchiori, and C. Palamidessi. 1991. A Theory of First-Order Built-Ins of Prolog. Technical report. CWI, Amsterdam.

Arsov, A., and M. Marx. 1994. Basic Arrow Logic with Relation-Algebraic Operators. In *Proceedings 9th Amsterdam Colloquium*, ed. P. Dekker and M. Stokhof, 93–112. Department of Philosophy, University of Amsterdam.

Asher, N., and J. Pelletier. 1996. Generics. In *Handbook of Logic and Language*, ed. J. van Benthem and A. ter Meulen. Amsterdam: Elsevier Science Publishers.

Backofen, R., J. Rogers, and K. Vijay-Shanker. 1995. A First-Order Axiomatization of the Theory of Finite Trees. *Journal of Logic, Language and Information* 4:5–39.

Bacon, J. 1985. Completeness of a Predicate-Functor Logic. *Journal of Symbolic Logic* 50:903–926.

Badia, A. 1996. *Query Languages with Generalized Quantifiers: Their Definition, Complexity and Expressiveness.* Doctoral Dissertation. Indiana University, Bloomington.

Baeten, J., J. Bergstra, and J.-W. Klop. 1992. Decidability of Bisimulation Equivalence for Processes Generating Context-Free Languages. Technical Report P9210. Programming Research Group, University of Amsterdam.

Baeten, J., and P. Weijland. 1990. *Process Algebra*. Tracts in Theoretical Computer Science, Vol. 18. Cambridge: Cambridge University Press.

Barendregt, H. 1992. Lambda Calculi with Types. In *Handbook of Logic in Computer Science, vol. 2*, ed. S. Abramsky, D.M. Gabbay, and T. Maibaum. 118–309. Oxford: Clarendon Press.

Barwise, J. 1975. *Admissible Sets and Structures*. Berlin: Springer Verlag.

Barwise, J. (ed.). 1977. *Handbook of Mathematical Logic*. Amsterdam: North-Holland.

Barwise, J. 1979. On Branching Quantifiers in English. *Journal of Philosophical Logic* 8:47–80.

Barwise, J. 1987. Noun Phrases, Generalized Quantifiers and Anaphora. In *Generalized Quantifiers. Logical and Linguistic Approaches*, ed. P. Gärdenfors. 1–29. Dordrecht: Reidel.

Barwise, J. 1988. Three Theories of Common Knowledge. In *Proceedings TARK-2*, ed. M. Vardi, 365–379. San Francisco: Morgan Kaufmann.

Barwise, J. 1989. *The Situation in Logic*. Stanford: CSLI Publications.

Barwise, J., and L. Moss. 1996. *Vicious Circles: On the Mathematics of Non-Wellfounded Phenomena*. Stanford: CSLI Publications. To appear.

Barwise, J., and J. Seligman. 1996. *Information Flow in Distributed Systems*. Cambridge Tracts in Theoretical Computer Science. Cambridge: Cambridge University Press. To appear.

Barwise, J., and J. van Benthem. 1996. Interpolation, Preservation, and Pebble Games. Department of Mathematics, Indiana University, Bloomington and CSLI, Stanford University.

Bealer, G. 1982. *Quality and Concept*. Oxford: Oxford University Press.

Beaver, D. 1995. *Presupposition and Assertion in Dynamic Semantics*. Doctoral dissertation, Centre for Cognitive Science, University of Edinburgh.

Beaver, D. 1996. Presupposition. In *Handbook of Logic and Language*, ed. J. van Benthem and A. ter Meulen. Amsterdam: Elsevier.

Bell, J., and M. Machover. 1977. *A Course in Mathematical Logic*. Amsterdam: North-Holland.

Belnap, N. 1977. A Useful Four-Valued Logic. In *Modern Uses of Multiple-Valued Logics*, ed. M. Dunn and G. Epstein. 8–37. Dordrecht: Reidel.

Belnap, N. and M. Perloff. 1988. Seeing To It That: A Canonical Form for Agentives. *Theoria* 54:175-199.

Ben-Shalom, D. 1994. Natural Language, Generalized Quantifiers and Modal Logic. Technical report. Department of Linguistics, UCLA and CWI Amsterdam.

van Benthem, J. 1976. *Modal Correspondence Theory*. Doctoral dissertation, Mathematical Institute, University of Amsterdam.

van Benthem, J. 1982. The Logical Study of Science. *Synthese* 51:431–472.

van Benthem, J. 1983a. *The Logic of Time*. Dordrecht: Reidel.

van Benthem, J. 1983b. The Semantics of Variety in Categorial Grammar. Technical Report 83-26. Department of Mathematics, Simon Fraser University, Burnaby (B.C.).

van Benthem, J. 1984. Correspondence Theory. In *Handbook of Philosophical Logic. Vol. 2*, ed. D.M. Gabbay and F. Guenthner. 167–247. Reidel: Dordrecht.

van Benthem, J. 1985a. *Modal Logic and Classical Logic.* Naples/Atlantic Heights: Bibliopolis/The Humanities Press.

van Benthem, J. 1985b. The Variety of Consequence, According to Bolzano. *Studia Logica* 44:389–403.

van Benthem, J. 1986. *Essays in Logical Semantics.* Dordrecht: Reidel.

van Benthem, J. 1988a. Games in Logic: A Survey. In *Representation and Reasoning*, ed. J. Hoepelman. 3–15. Tübingen: Niemeyer Verlag.

van Benthem, J. 1988b. Logical Syntax. *Theoretical Linguistics* 14:119–142.

van Benthem, J. 1988c. *A Manual of Intensional Logic.* Lecture Notes, Vol. 1. Stanford: CSLI Publications.

van Benthem, J. 1989a. Logical Constants across Varying Types. *Notre Dame Journal of Formal Logic* 30:315–342.

van Benthem, J. 1989b. Polyadic Quantifiers. *Linguistics and Philosophy* 12:437–464.

van Benthem, J. 1989c. Semantic Parallels in Natural Language and Computation. In *Logic Colloquium. Granada 1987*, ed. H.-D. Ebbinghaus et al. 331–375. Amsterdam: North-Holland.

van Benthem, J. 1991a. General Dynamics. *Theoretical Linguistics* 17:151–201.

van Benthem, J. 1991b. *Language in Action. Categories, Lambdas and Dynamic Logic.* Amsterdam: North-Holland.

van Benthem, J. 1992. Logic as Programming. *Fundamenta Informaticae* 17(4):285–317.

van Benthem, J. 1993a. Logic and the Flow of Information. In *Proceedings 9th International Congress of Logic, Methodology and Philosophy of Science. Uppsala 1991*, 693–724. Amsterdam. Elsevier Science Publishers.

van Benthem, J. 1993b. Modeling the Kinematics of Meaning. *Proceedings Aristotelean Society* 105–122.

van Benthem, J. 1994a. Dynamic Arrow Logic. In *Logic and Information Flow*, ed. J. van Eijck and A. Visser. 15–29. Cambridge (Mass.): The MIT Press.

van Benthem, J. 1994b. General Dynamic Logic. In *What is a Logical System?*, ed. D.M. Gabbay. Oxford: Oxford University Press.

van Benthem, J. 1996a. Content versus Wrapping: An Essay in Semantic Complexity. In *Arrow Logic and Multi-Modal Logic*, ed. M. Marx, M. Masuch, and L. Pólos. Studies in Logic, Language and Information. Stanford: CSLI Publications.

van Benthem, J. 1996b. Logic and Argumentation. In *Proceedings Academy Colloquium on Logic and Argumentation*, ed. J. van Benthem, F. van Eemeren, R. Grootendorst, and F. Veltman. Amsterdam: Royal Dutch Aacdemy of Sciences.

van Benthem, J. 1996c. Modal Foundations for Predicate Logic. In *Memorial Volume for Elena Rasiowa*, ed. E. Orlowska. Studia Logica Library. Dordrecht: Kluwer Academic Publishers. Extended version to appear in *Bulletin of the IGPL.*

van Benthem, J. 1996d. Modal Logic as a Theory of Information. In *Logic and Reality. Essays on the Legacy of Arthur Prior*, ed. J. Copeland. Oxford: Oxford University Press.

van Benthem, J. 1996e. Programming Operations that are Safe for Bisimulation. In *Logic Colloquium '94. Clermont-Ferrand.* Special Issue of Studia Logica.

van Benthem, J. 1996f. Proof Theory and Dynamics in Natural Language. In *Festschrift for Dov Gabbay*, ed. U. Reyle and H.-J. Ohlbach. Dordrecht: Kluwer Academic Publishers.

van Benthem, J. 1996g. Quantifiers in the World of Types. In *Quantifiers, Logic and Language*, ed. J. van der Does and J. van Eijck. Lecture Notes, Vol. 54, 47–62. Stanford: CSLI Publications.

van Benthem, J. 1996h. Shifting Contexts and Changing Assertions. In *Proceedings 4th CSLI Workshop on Logic, Language and Computation*, ed. A. Aliseda-Llera, R. van Glabbeek, and D. Westerståhl. Stanford: CSLI Publications.

van Benthem, J., and J. Bergstra. 1994. Logic of Transition Systems. *Journal of Logic, Language and Information* 3:247–283.

van Benthem, J., and G. Cepparello. 1994. Tarskian Variations: Dynamic Parameters in Classical Semantics. Technical Report CS-R9419. Centre for Mathematics and Computer Science, Amsterdam.

van Benthem, J., J. van Eijck, and A. Frolova. 1993. Changing Preferences. Technical Report CS-R9310. Centre for Mathematics and Computer Science, Amsterdam.

van Benthem, J., J. van Eijck, and V. Stebletsova. 1994. Modal Logic, Transition Systems and Processes. *Journal of Logic and Computation* 4:811–855.

van Benthem, J., and W. Meyer Viol. 1993. Logical Semantics of Programs. Lecture Notes, CWI, Amsterdam.

van Benthem, J., and D. Westerståhl. 1995. Directions in Generalized Quantifier Theory. *Studia Logica* 53:389–419.

van den Berg, M. 1996. *The Internal Structure of Discourse.* Doctoral dissertation, Institute for Logic, Language and Computation, University of Amsterdam.

Bergstra, J., I. Bethke, and A. Ponse. 1994. Process Algebra with Iteration and Nesting. *The Computer Journal* 37(4):243–258.

Bergstra, J., J. Heering, and P. Klint. 1990. Module Algebra. *Journal of the ACM* 37(2):335–372.

Bergstra, J., and J.-W. Klop. 1984. Process Algebra for Synchronous Communication. *Information and Control* 60:109–137.

Beth, E., and J. Piaget. 1966. *Mathematical Epistemology and Psychology.* Dordrecht: Reidel.

Blackburn, P. 1995. Static and Dynamic Aspects of Syntactic Structure. *Journal of Logic, Language and Information* 4:1–4.

Blackburn, P., and W. Meyer Viol. 1994. Linguistics, Logic and Finite Trees. *Bulletin of the Interest Group for Pure and Applied Logic* 2:3–29.

Blackburn, P., M. de Rijke, and Y. Venema. 1996. *A Course in Modal Logic.* Department of Computer Science, University of Warwick. To appear.

Blackburn, P., W. Meyer Viol, and M. de Rijke. 1996. A Proof System for Finite Trees. In *Proceedings of CSL'95.* Berlin: Springer Verlag.

Blackburn, P., and J. Seligman. 1995. Hybrid Languages. *Journal of Logic, Language and Information* 4(3):251–272.

Blackburn, P., and E. Spaan. 1993. A Modal Perspective on the Computational Complexity of Attribute-Value Grammar. *Journal of Logic, Language and Information* 2:129–169.

Blackburn, P., and Y. Venema. 1995. Dynamic Squares. *Journal of Philosophical Logic* 24(5):469–523.

Blok, W. 1979. The Lattice of Modal Logics. An Algebraic Investigation. *Journal of Symbolic Logic* 45:221–236.

Borghuis, T. 1994. *Coming to Terms with Modal Logic.* Doctoral dissertation, Department of Computer Science, University of Eindhoven.

Boutilier, C. 1993. Revison Sequences and Nested Conditionals. In *Proceedings 13th IJCAI*, 519–525. Washington D.C.: Morgan Kaufmann.

Boutilier, C., and M. Goldszmidt. 1993. Revision by Conditional Beliefs. In *Proceedings 11th AAAI.*

Bratko, I. 1986. *Prolog Programming for Artificial Intelligence.* Wokingham: Addison-Wesly.

Bratman, M. 1987. *Intentions, Plans, and Practical Reason.* Harvard University Press.

Bratman, M. 1991. Planning and the Stability of Intention. Research Report 159. Center for the Study of Language and Information, Stanford.

Brogi, A., P. Mancarella, D. Pedreschi, and F. Turini. 1994. Modular Logic Programming. *ACM Transactions on Programming Languages and Systems* 16:1361–1398.

Buhrmann, H. 1993. *Resource Bounded Reductions.* Doctoral dissertation, Institute for Logic, Language and Computation, University of Amsterdam.

Bull, R., and K. Segerberg. 1984. Basic Modal Logic. In *Handbook of Philosophical Logic. Vol. 2*, ed. D.M. Gabbay and F. Guenthner. 1–88. Dordrecht: Reidel.

Burgess, J. 1984. Basic Tense Logic. In *Handbook of Philosophical Logic. Vol. 2*, ed. D.M. Gabbay and F. Guenthner. 89–133. Dordrecht: Reidel.

Burke, T. 1995. A Completeness Theorem for an Action-Based Dynamic Logic. Manuscript, Center for the Study of Language and Information, Stanford.

Buszkowski, W. 1996. Mathematical Linguistics and Proof Theory. In *Handbook of Logic and Language*, ed. J. van Benthem and A. ter Meulen. Amsterdam: Elsevier Science Publishers.

Buszkowski, W., W. Marciszewski, and J. van Benthem (ed.). 1988. *Categorial Grammar.* Amsterdam. John Benjamin.

Buvac, S. 1995. Quantificational Logic of Context. In *Proceedings of the Workshop on Modeling Context in Knowledge Representation and Reasoning, Proceedings IJCAI-13.*

Buvac, S., and R. Fikes (ed.). 1995. *Formalizing Context.* Cambridge (Mass.). Working Notes for AAAI-5 Fall Symposium Series.

Buvac, S., and I. Mason. 1995. Metamathematics of Contexts. *Fundamenta Informaticae* 23(3):263–301.

Calvino, I. 1988. *Six Memos for the Next Millennium.* Cambridge (Mass.): Harvard University Press.

Carpenter, B. 1992. *The Logic of Typed Feature Structures, with Applications to Unification-Based Grammars, Logic Programming and Constraint Resolution.* Cambridge: Cambridge University Press.

Carpenter, B. 1996. *A Type-Logical Approach to Natural Language Semantics.* Cambridge (Mass.): The MIT Press.

Cepparello, G. 1995. *Studies in Dynamic Logic.* Doctoral dissertation, Scuola Normale Superiore, Pisa.

Chagrova, L. 1991. An Undecidable Problem in Correspondence Theory. *Journal of Symbolic Logic* 56:1261–1272.

Chang, C., and J. Keisler. 1973. *Model Theory.* Amsterdam: North-Holland.

Chierchia, G. 1995. *The Dynamics of Meaning.* Chicago and London: The University of Chicago Press.

Clark, H. 1996 *Using Language.* Cambridge: Cambridge University Press

Cook, S. 1978. Soundness and Completeness of an Axiom System for Program Verification. *SIAM Journal of Computing* 7:70–90.

Csirmaz, L., D.M. Gabbay, and M. de Rijke (ed.). 1995. *Logic Colloquium '92.* Studies in Logic, Language and Information. Stanford: CSLI Publications.

van Deemter, K. 1991. *On the Composition of Meaning.* Doctoral dissertation, Institute for Language, Logic and Information, University of Amsterdam.

Dekker, P. 1993. *Transsentential Meditations.* Doctoral dissertation, Institute for Logic, Language and Computation, University of Amsterdam.

De Nicola, R., and F. Vaandrager. 1990. Three Logics of Branching Bisimulation. In *Proceedings 5th LICS Conference*, 118–129. Computer Society Press.

Devlin, K. 1991. *Logic and Information.* Cambridge: Cambridge University Press.

van der Does, J. 1992. *Applied Quantifier Logics.* Doctoral dissertation, Institute for Logic, Language and Computation, University of Amsterdam.

van der Does, J. 1994. Might Be. Some Proof Theory for a Substructural Dynamic Logic. Manuscript, Institute for Logic, Language and Computation, University of Amsterdam.

van der Does, J., and J. van Eijck (ed.). 1991. *Generalized Quantifiers: Theory and Applications.* Dutch Ph.D. Network for Logic, Language and Information. Appeared as *Quantifiers, Logic and Language*, Lecturen Notes, Vol. 54, Stanford: CSLI Publications, 1995.

Doets, K. 1987. *Completeness and Definability. Applications of the Ehrenfeucht Game in Second-Order and Intensional Logic.* Doctoral dissertation, Mathematical Institute, University of Amsterdam.

Doets, K. 1994. *From Logic to Logic Programming.* Cambridge (Mass.): The MIT Press.

Doets, K. 1996. *Basic Model Theory.* Studies in Logic, Language and Information. Stanford: CSLI Publications.

Doets, K., and J. van Benthem. 1983. Higher-Order Logic. In *Handbook of Philosophical Logic. Vol. 1*, ed. D.M. Gabbay and F. Guenthner. 275–329. Dordrecht: Reidel.

Doherty, P., W. Lukasiewicz, and A. Szalas. 1994. Computing Circumscription Revisited: A Reduction Algorithm. Technical Report LiTH-IDA-R-94-42. Institutionen før Datavetenskap, University of Linköping.

Došen, K. 1985. A Completeness Theorem for the Lambek Calculus of Syntactic Categories. *Zeitschrift fuer mathematische Logik und Grundlagen der Mathematik* 31:235–241.

Došen, K. 1992. Modal Logic as Metalogic. *Journal of Logic, Language and Information* 1:173–202.

Došen, K., and P. Schröder-Heister (ed.). 1994. *Substructural Logics.* Oxford: Clarendon Press.

Dreben, B., and W. Goldfarb. 1979. *The Decision Problem: Solvable Classes of Quantificational Formulas.* Reading (Mass.): Addison-Wesley.

Dretske, F. 1995. *Naturalizing the Mind.* Cambridge (Mass.): The MIT Press.

Dummett, M. 1976. What is a Theory of Meaning? In *Truth and Meaning*, ed. G. Evans and J. McDowell. 67–137. Oxford: Oxford University Press.

Dunn, M. 1985. Relevance Logic and Entailment. In *Handbook of Philosophical Logic. Vol. 3*, ed. D.M. Gabbay and F. Guenthner. 117–224. Dordrecht: Reidel.

Dunn, M., and R. Thomason (ed.). 1988. *Philosophical Logic and Artificial Intelligence.* Special issue of *Journal of Philosophical Logic* 17(4).

Dunn, M. 1996. A Comparative Study of Various Model-Theoretic Treatments of Negation. Indiana University Logic Group, Bloomington.

Ebbinghaus, H., and J. Flum. 1995. *Finite Model Theory.* Berlin: Springer Verlag.

van Eijck, J. 1991. Presupposition Failure — A Comedy of Errors. Technical report. CWI, Amsterdam.

van Eijck, J. 1993. The Dynamics of Theory Extension. In *Proceedings 9th Amsterdam Colloquium*, ed. P. Dekker and M. Stokhof, 249–267. Department of Philosophy, University of Amsterdam.

van Eijck, J., and G. Cepparello. 1994. Dynamic Modal Predicate Logic. In *Dynamics, Polarity and Quantification*, ed. M. Kanazawa and Ch. Piñon. 251–276. Stanford: CSLI Publications.

van Eijck, J., and F.-J. de Vries. 1991. Dynamic Interpretation and Hoare Deduction. *Journal of Logic, Language and Information* 1:1–44.

van Eijck, J., and F.-J. de Vries. 1995. Reasoning About Update Logic. *Journal of Philosophical Logic* 24:19–45.

van Eijck, J., and H. Kamp. 1996. Representing Discourse in Context. In *Handbook of Logic and Language*, ed. J. van Benthem and A. ter Meulen. Amsterdam: Elsevier Science Publishers.

van Eijck, J., and A. Visser. 1994. *Dynamic Logic and Information Flow.* Cambridge (Mass.): The MIT Press.

van Emde Boas, P. 1990. Machine Models and Simulations. In *Handbook of Theoretical Computer Science*, ed. J. van Leeuwen. 3–66. Amsterdam: Elsevier Science Publishers.

Enç, M. 1981. *Tense Without Scope: An Analysis of Nouns as Indexicals.* Doctoral dissertation, Department of Linguistics, University of Wisconsin. Madison.

Enderton, H. 1972. *A Mathematical Introduction to Logic.* New York: Academic Press.

Engelfriet, J. 1995. Minimal Temporal Epistemic Logic. Technical Report IR-388. Department of Computer Science, Free University, Amsterdam.

Etchemendy, J. 1990. *The Concept of Logical Consequence.* Cambridge (Mass.): Harvard University Press.

Fagin, R., J. Halpern, Y. Moses, and M. Vardi. 1995. *Reasoning About Knowledge.* Cambridge (Mass.): The MIT Press.

Fernando, T. 1992. Transition Systems and Dynamic Semantics. Technical Report CS-R9217. Centre for Mathematics and Computer Science, Amsterdam.

Fine, K. 1985a. Logics Containing K4. Part II. *Journal of Symbolic Logic* 50:619–651.

Fine, K. 1985b. *Reasoning With Arbitrary Objects.* Oxford: Blackwell.

Finger, M., and D.M. Gabbay. 1992. Adding a Temporal Dimension to a Logic System. *Journal of Logic, Language and Information* 1:203–233.

Fischer, M., and R. Ladner. 1979. Propositional Dynamic Logic of Regular Programs. *Journal of Computer and Systems Science* 18:194–211.

Fitting, M. 1993. Basic Modal Logic. In *Handbook of Logic in Artificial Intelligence and Logic Programming, Vol. 1.* 368–448. Gabbay, D.M. and Hogger, C. and Robinson, J.: Oxford University Press.

Fokkink, W. 1994. *Clocks, Trees and Stars in Process Theory.* Doctoral dissertation, Institute for Logic, Language and Computation, University of Amsterdam.

Friedman, N. 1996. *Modeling Belief in Dynamic Systems.* Doctoral dissertation, Department of Computer Science, Stanford University.

Friedman, N. and J. Halpern. 1994. Conditional Logics of Belief Change. In *Proceedings AAAI 1994.*

Friedman, N., and J. Halpern. 1994a. A Knowledge-Based Framework for Belief Change. Part I: Foundations. In *Proceedings Fifth TARK Conference*, ed. R. Fagin, 44–64.

Friedman, N., and J. Halpern. 1994b. Modeling Belief in Dynamic Systems. Part II: Revision and Update. In *Proceedings KR '94*, ed. J. Doyle, E. Sandewall, and P. Torasso, 190–201.

Friedman, N. and J. Halpern. 1996 *Plausability Measures: A User's Guide.* Department of Computer Science, Stanford University and IBM Almaden Research Center, San Jose.

Fuhrmann, A. 1990. On the Modal Logic of Theory Change. In *The Logic of Theory Change*, ed. A. Fuhrmann and M. Morreau. LNAI, Vol. 465. Berlin: Springer Verlag.

Gabbay, D.M. 1981a. Expressive Functional Completeness in Tense Logic. In *Aspects of Philosophical Logic*, ed. U. Mönnich. 91–117. Dordrecht: Reidel.

Gabbay, D.M. 1981b. An Irreflexivity Lemma With Applications to Axiomatizations of Conditions on Tense Frames. In *Aspects of Philosophical Logic*, ed. U. Mönnich. 67–89. Dordrecht: Reidel.

Gabbay, D.M. 1992. *Labeled Deductive Systems.* Department of Computing, Imperial College, London.

Gabbay, D.M., and F. Guenthner (ed.). 1983–1989. *Handbook of Philosophical Logic.* Dordrecht: Reidel.

Gabbay, D.M., and R. Kempson. 1991. Natural-Language Content: A Proof-Theoretic Perspective. In *Proceedings 8th Amsterdam Colloquium*, ed. P. Dekker and M. Stokhof, 173–196. Department of Philosophy, Univesrity of Amsterdam.

Gärdenfors, P. 1988. *Knowledge in Flux.* Cambridge (Mass.): The MIT Press.

Gärdenfors, P., and H. Rott. 1995. Belief Revision. In *Handbook of Logic in Artificial Intelligence and Logic Programming, vol. IV*, ed. D.M. Gabbay, C. Hogger, and J. Robinson. 35–132. Oxford: Oxford University Press.

Gargov, G., S. Passy, and T. Tinchev. 1987. Modal Environment for Boolean Speculations. In *Mathematical Logic and its Applications*, ed. D. Skordev. 253–263. New York: Plenum Press.

Geach, P. 1972. *Logic Matters.* Oxford: Basil Blackwell.

Gerbrandy, J. 1994. 'Might' in Dynamic Semantics. Master's thesis, Department of Philosophy, University of Amsterdam.

Gerbrandy, J. 1995. Dynamic Epistemic Semantics. On How to Come to Know That you are Dirty When you are Dirty. Manuscript, Institute for Logic, Language and Computation, University of Amsterdam.

Ghilardi, S. 1991. Incompleteness Results in Kripke Semantics. *Journal of Symbolic Logic* 56:517–538.

Girard, J.-Y. 1987. Linear Logic. *Theoretical Computer Science* 50:1–102.

Girard, J.-Y. 1993. On the Unity of Logic. *Annals of Pure and Applied Logic* 59:201–217.

Girard, J.-Y., Y. Lafont, and P. Taylor. 1990. *Proofs and Types.* Cambridge: Cambridge University Press.

Givant, S. 1993. The Structure of Relation Algebras Generated by Relativizations. In *Contemporary Mathematics* 156. American Mathematical Society.

van Glabbeek, R. 1990. The Linear Time – Branching Time Spectrum. In *CONCUR '90*, LNCS, Vol. 458, 278–297. Berlin: Springer Verlag.

van Glabbeek, R., and G. Plotkin. 1995. Configuration Structures. Department of Computer Science, Stanford University.

Goldblatt, R. 1982. *Axiomatizing the Logic of Computer Programming.* Berlin: Springer.

Goldblatt, R. 1987. *Logics of Time and Computation.* Lecture Notes. Stanford: CSLI Publications.

Goldblatt, R. 1988. Varieties of Complex Algebras. *Annals of Pure and Applied Logic* 44:173–242.

Goldblatt, R. 1992. Concurrent Dynamic Logic with Independent Modalities. *Studia Logica* 51:551–578.

Goldblatt, R., and S. Thomason. 1974. Axiomatic Classses in Propositional Modal Logic. In *Algebra and Logic*, ed. J. Crossley, 163–173. Berlin: Springer.

Goranko, V. 1990. Modal Definability in Enriched Languages. *Notre Dame Journal of Formal Logic* 31:81–105.

Goranko, V. 1995. Hierarchies of Modal and Temporal Logics with Reference Pointers. *Journal of Logic, Language and Information*, to appear.

Grätzer, G. 1968. *Universal Algebra.* Princeton: Van Nostrand.

Groenboom, R., and G. Renardel de Lavalette. 1994. Reasoning About Dynamic Features in Specification Languages (A Modal View on Creation and Modification). In *Semantics of Specification Languages*, ed. D. Andrews, J.-F. Groote, and C. Middelburg. Workshops in Computing. Berlin: Springer Verlag.

Groenendijk, J., and M. Stokhof. 1991. Dynamic Predicate Logic. *Linguistics and Philosophy* 14:39–100.

Groenendijk, J., and M. Stokhof. 1996. Questions. In *Handbook of Logic and Language*, ed. J. van Benthem and A. ter Meulen. Amsterdam: Elsevier Science Publishers.

Groenendijk, J., M. Stokhof, and F. Veltman. 1994. Update Semantics for Modal Predicate Logic. Technical Report LP-94-14. Institute for Logic, Language and Computation, University of Amsterdam.

Groenendijk, J., M. Stokhof, and F. Veltman. 1996a. Coreference and Modality. In *Handbook of Contemporary Semantic Theory*, ed. S. Lappin. 179–214. Oxford: Blackwell.

Groenendijk, J., M. Stokhof, and F. Veltman. 1996b. This Might Be It. In *Language, Logic and Computation: The 1994 Moraga Proceedings*, ed. D. Westerståhl and J. Seligman. Stanford. CSLI Publications.

Groeneveld, W. 1995. *Logical Investigations into Dynamic Semantics.* Doctoral dissertation, Institute for Logic, Language and Computation, University of Amsterdam.

Groeneveld, W., J. van der Does, and F. Veltman. 1996. An Update on *Might.* To appear in *Journal of Logic, Language and Information.*

Grohe, M. 1996. Equivalence in Finite-Variable Logics is Complete for Polynomial-Time. To appear in *Proceedings 1996 IEEE Symposium on Foundations of Computer Science.* Department of Mathematics, Stanford University.

Grosz, B., and C. Sidner. 1986. Attention, Intention, and the Structure of Discourse. *ACL* 12:175–204.

Gurevich, Y. 1985. Logic and the Challenge of Computer Science. Technical Report CRL-TR-10-85. Computing Research Laboratory, University of Michigan, Ann Arbor.

Halpern, J., and Y. Moses. 1985. Towards a Theory of Knowledge and Ignorance. In *Logics and Models of Concurrent Systems*, ed. K. Apt. 459–476. Berlin: Springer Verlag.

Harel, D. 1984. Dynamic Logic. In *Handbook of Philosophical Logic. Vol. 2*, ed. D.M. Gabbay and F. Guenthner. 497–604. Dordrecht: Reidel.

Harel, D., D. Kozen, and J. Tiuryn. 1995. *Dynamic Logic.* The Weizmann Institute, Cornell University and University of Warsaw.

Harman, G. 1985. *Change in View: Principles of Reasoning.* Cambridge (Mass.): The MIT Press/Bradford Books.

Harrah, D. 1984. The Logic of Questions. In *Handbook of Philosophical Logic. Vol. 2*, ed. D.M. Gabbay and F. Guenthner. 715–764. Dordrecht: Reidel.

Hartonas, C. and M. Dunn. 1993. Duality Theory for Partial Orders, Semilattices, Galois Connections, and Lattices. Report IULG-93-26. Indiana University Logic Group, Bloomington.

Heim, I. 1982. *The Semantics of Definite and Indefinite Noun Phrases.* Doctoral dissertation, UMass, Amherst. Published in 1989 by Garland, New York.

Hendriks, H. 1993. *Studied Flexibility.* Doctoral dissertation, Institute for Logic, Language and Computation, University of Amsterdam.

Henkin, L. 1950. Completeness in the Theory of Types. *Journal of Symbolic Logic* 15:81–91.

Henkin, L. 1967. *Logical Systems Containing Only a Finite Number of Symbols.* Séminaire de Mathématiques Supérieures, Vol. 21. Montréal: Les Presses de l'Université de Montréal.

Henkin, L., J.-D. Monk, and A. Tarski. 1971. *Cylindric Algebras, Part I.* Amsterdam: North-Holland.

Henkin, L., J.-D. Monk, and A. Tarski. 1985. *Cylindric Algebras, Part II.* Amsterdam: North-Holland.

Henkin, L., J.-D. Monk, A. Tarski, H. Andréka and I. Németi, 1981. *Cylindric Set Algebras.* Lecture Notes in Mathematics 883 Berlin: Springer Verlag.

Hennessy, M., and R. Milner. 1985. Algebraic Laws for Nondeterminism and Concurrency. *Journal of the ACM* 32:137–161.

Hintikka, J. 1962. *Knowledge and Belief.* Ithaca (N.Y.): Cornell University Press.

Hintikka, J. 1973. *Logic, Language Games and Information.* Oxford: Clarendon Press.

Hintikka, J. 1979. Quantifiers in Natural Languages: Some Logical Problems. In *Game-Theoretical Semantics*, ed. E. Saarinen. 81–118. Dordrecht: Reidel.

Hintikka, J., and J. Kulas. 1983. *The Game of Language*. Dordrecht: Reidel.

Hirsch, R., and I. Hodkinson. 1995. Axiomatising Various Classes of Relation and Cylindric Algebras. In *Proceedings Workshop on Logic, Language, Information, and Computation, Recife, July 1995*. To appear in *Bulletin of the IGPL*.

Hirsch, R., and I. Hodkinson. 1996. Step by Step — Building Representations in Algebraic Logic. *Journal of Symbolic Logic*, to appear.

Hodges, W. 1983. Elementary Predicate Logic. In *Handbook of Philosophical Logic. Vol. 1*, ed. D.M. Gabbay and F. Guenthner. 1–131. Dordrecht: Reidel.

Hodges, W. 1993. *Model Theory*. Cambridge: Cambridge University Press.

van der Hoek, W., B. van Linder, and J.-J. Meyer. 1994. Formalising Motivational Attitudes of Agents. Technical report. Department of Computer Science, University of Utrecht.

van der Hoek, W., B. van Linder, and J.-J. Meyer. 1995. Actions That Make You Change Your Mind. In *Knowledge and Belief in Philosophy and AI*, ed. A. Laux and H. Wansing. 103–146. Leipzig: Akademie Verlag.

Hofstadter, D. 1979. *Gödel, Escher, Bach*. New York: Basic Books.

Hollenberg, M. 1995a. An Axiomatization of Strong Negation and Relational Composition. Logic Group Preprint Series 151. Institute for Philosophy, Rijksuniversiteit Utrecht.

Hollenberg, M. 1995b. Finite Safety for Bisimulation. Institute for Philosophy, Rijksuniversiteit Utrecht.

Hollenberg, M. 1996a. Bisimulation Respecting First-Order Operations. Logic Group Preprint Series 156. Institute for Philosophy, Rijksuniversiteit Utrecht.

Hollenberg, M. 1996b. General Safety for Bisimulation. In *Proceedings 10th Amsterdam Colloquium*, ed. P. Dekker and P. Stokhof. Institute for Logic, Language and Computation, University of Amsterdam.

Hollenberg, M., and K. Vermeulen. 1994. Counting Variables in a Dynamic Setting. Logic Group Preprint Series 125. Filosofisch Instituut, Rijksuniversiteit, Utrecht.

Hopcroft, J.E., and J.D. Ullman. 1979. *Introduction to Automata Theory, Languages and Computation*. Reading (Mass.): Addison-Wesley.

Horty, J., and N. Belnap. 1995. The Deliberative STIT: A study of Action, Omission, Ability, and Obligation. *Journal of Philosophical Logic* 24:583–644.

Huang, Z. 1994. *Logics for Agents with Bounded Rationality*. Doctoral dissertation, CCSOM and Institute for Logic, Language and Computation, University of Amsterdam.

Huang, Z.-S., L. Pólos, and M. Masuch. 1996. ALX, an Action Logic for Agents with Bounded Rationality. *Artificial Intelligence*, to appear.

Huizinga, J. 1938. *Homo Ludens, Proeve ener Bepaling van het Spelelement der Cultuur*. Leiden: Tjeenk Willink.

Immerman, I., and D. Kozen. 1987. Definability With Bounded Number of Bound Variables. In *Proceedings 2nd IEEE Symposium on Logic in Computer Science*, 236–244.

Immerman, N. 1981. Number of Quantifiers is Better than Number of Tape Cells. *Journal of Computer and Systems Sciences* 22:65–72.

Immerman, N. 1982. Relational Queries Computable in Polynomial Time. In *Proceedings 14th Annual ACM Symposium on the Theory of Computing*, 147–152.

Immerman, N. 1995. Descriptive Complexity: A Logician's Approach to Computation. *Notices of the American Mathematical Society* 42(10):1127–1133.

Immerman, N. 1996. *Descriptive Complexity*. Berlin: Springer Verlag. To appear.

Janin, D. 1996. *Propriétés Logiques du Non-Déterminisme et μ-Calcul Modal.* Doctoral dissertation, LaBRI, Université de Bordeaux.

Janin, D., and I. Walukiewicz. 1996. On the Expressive Completeness of the Propositional μ-Calculus with Respect to Monadic Second-Order Logic. LaBRI, Department of Mathematics and Informatics, University of Bordeaux and BRICS, Department of Computer Science, University of Aarhus.

Janssen, T. 1983. *Foundations and Applications of Montague Grammar.* Doctoral dissertation, Department of Mathematics, University of Amsterdam.

Jaspars, J. 1994. *Calculi for Constructive Communication.* Doctoral dissertation, Institute for Language and Knowledge Technology, University of Tilburg and Institute for Logic, Language and Computation, University of Amsterdam.

Jaspars, J. 1995. Partial Up and Down Logic. Technical Report CS-R9511. CWI, Amsterdam.

Jaspars, J., and E. Krahmer. 1995. Unified Dynamics. Technical report. Centre for Mathematics and Computer Science, Amsterdam.

Jaspars, J., and E. Krahmer. 1996. A Programme of Modal Unification of Dynamic Theories. In *Proceedings 10th Amsterdam Colloquium*, ed. P. Dekker and M. Stokhof. 425–444. Institute for Logic, Language and Computation, University of Amsterdam.

Jaspars, J., and E. Krahmer. 1996. Preferences in Dynamic Semantics. In *Proceedings 10th Amsterdam Colloquium*, ed. P. Dekker and M. Stokhof. 445-464. Institute for Logic, Language and Computation, University of Amsterdam.

Johnson, D. 1990. A Catalog of Complexity Classes. In *Handbook of Theoretical Computer Science*, ed. J. van Leeuwen. 69–161. Amsterdam: Elsevier Science Publishers.

de Jongh, D., and A. Troelstra. 1966. On the Connection of Partially Ordered Sets with Some Pseudo-Boolean Algebras. *Indagationes Mathematicae* 28:317–329.

Jónsson, B. 1987. The Theory of Binary Relations. Technical report. Department of Mathematics, VanderBilt University, Nashville (Tenn.).

Joshi, A., and S. Weinstein. 1981. Control of Inference: Role of Some Aspects of Discourse Structure-Centering. In *Proceedings IJCAI 1981.* Vancouver, Canada.

Kalsbeek, M. 1995. *Meta-Logics for Logic Programming.* Doctoral dissertation, Institute for Logic, Language and Information, University of Amsterdam.

Kameyama, M. 1993. The Linguistic Information in Dynamic Discourse. Technical report. CSLI Research Report, Center for the Study of Language and Information, Stanford.

Kameyama, M. 1994. Indefeasible Semantics and Defeasible Pragmatics. Technical Report CS-R9441. Center for Mathematics and Computer Science, Amsterdam.

Kamlah, W., and P. Lorenzen. 1967. *Logische Propädeutik.* Mannheim: Bibliographisches Institut.

Kamp, H. 1971. Formal Properties of "Now". *Theoria* 37:227–273.

Kamp, H. 1979. Instants, Events and Temporal Discourse. In *Semantics from Different Points of View*, ed. R. Bäuerle et al. 376–417. Berlin: Springer Verlag.

Kamp, H. 1984. A Theory of Truth and Semantic Representation. In *Truth, Interpretation and Information*, ed. J. Groenendijk et al. 1–41. Dordrecht: Foris.

Kamp, H., and U. Reyle. 1993. *From Discourse to Logic.* Dordrecht: Kluwer Academic Publishers.

Kanazawa, M. 1993a. Completeness and Decidability of the Mixed Style of Inference with Composition. In *Proceedings of the Ninth Amsterdam Colloquium*, ed. P. Dekker and M. Stokhof, 377–390. Institute for Logic, Language and Computation, University of Amsterdam.

Kanazawa, M. 1993b. Dynamic Generalized Quantifiers and Monotonicity. Technical Report LP-93-02. Institute for Logic, Language and Computation, University of Amsterdam. Appeared under the title 'Weak versus Strong Readings of Donkey Sentences and Monotonicity Inference in a Dynamic Setting' in *Linguistics and Philosophy* 17:109–158, 1994.

Kandulski, M. 1988. The Non-Associative Lambek Calculus. In *Categorial Grammar*, ed. W. Buszkowski, W. Marciszewski, and J. van Benthem. 141–151. Amsterdam: John Benjamin.

Kanellakis, P. 1990. Elements of Relational Database Theory. In *Handbook of Theoretical Computer Science*, ed. J. van Leeuwen. 1073–1156. Amsterdam: Elsevier Science Publishers.

Kanovich, M. 1993. Petri Nets, Horn Programs, Linear Logic, and Vector Games. Department of Computer Science, Russian Humanitarian University, Moscow.

Kartunen, L., and S. Peters. 1979. Conventional Implicature. In *Syntax and Semantics 11: Presupposition*, ed. C. Oh and D. Dinneen. 1–56. New York: Academic Press.

Kasper, B., and W. Rounds. 1990. The Logic of Unification in Grammar. *Linguistics and Philosophy* 13:33–58.

Keenan, E., and L. Faltz. 1985. *Boolean Semantics for Natural Language.* Dordrecht: Reidel.

Keenan, E., and D. Westerståhl. 1996. Quantifiers. In *Handbook of Logic and Language*, ed. J. van Benthem and A. ter Meulen. Amsterdam: Elsevier Science Publishers.

Keisler, H. 1971. *Model Theory for Infinitary Logic.* Amsterdam: North-Holland.

Keisler, H. 1977. Fundamentals of Model Theory. In *Handbook of Mathematical Logic*, ed. J. Barwise. 47–103. Amsterdam: North-Holland.

Kempson, R. (ed.). 1995. *Deduction and Language*. London and Saarbruecken: Special issue of *Bulletin of the Interest Group in Pure and Applied Logics* 3:2/3.

Kneale, W., and M. 1962. *The Development of Logic*. Oxford: Clarendon Press.

Kolaitis, F., and J. Väänänen. 1992. Generalized Quantifiers and Pebble Games on Finite Structures. In *Proceedings of the 7th IEEE Symposium on Logic in Computer Science*, 348–359. Full version to appear in *Annals of Pure and Applied Logic*.

Kolaitis, P. Infinitary Logic in Finite Model Theory. To appear in *Proceedings 10th International Congress of Logic, Methodology and Philosophy of Science*, eds. M. Dalla Chiara, K. Doets, D. Mundici and J. van Benthem. Kluwer: Dordrecht.

Kowalski, R. 1979. *Logic for Problem Solving*. New York: North-Holland.

Kowalski, R., and M. van Emden. 1976. The Semantics of Predicate Logic as a Programming Language. *Journal of the ACM* 23:733–742.

Kozen, D. 1994. On Action Algebras. In *Logic and Information Flow*, ed. J. van Eijck and A. Visser. 78–88. Cambridge (Mass.): The MIT Press.

Kracht, M. 1995. Syntactic Codes and Grammar Refinement. *Journal of Logic, Language and Information* 4:41–60.

Kreisel, G. 1967. Informal Rigour and Completeness Proofs. In *Problems in the Philosophy of Mathematics*, ed. I. Lakatos. 138–186. Amsterdam: North-Holland.

Kripke, S. 1972. Naming and Necessity. In *Semantics of Natural Language*, eds. D. Davidson and G. Harman. Reidel: Dordrecht.

Kripke, S. 1990. *Naming and Necessity*. Oxford: Blackwell.

Kugel, P. 1986. Thinking May be More Than Computing. *Cognition* 22:137–198.

Kühler, A. 1994. Logics of Multigraphs. Master's thesis, Institute for Logic, Language and Computation, University of Amsterdam.

Kurtonina, N. 1995. *Frames and Labels. A Modal Analysis of Categorial Deduction.* Doctoral dissertation, Onderzoeksinstituut voor Taal en Spraak, University of Utrecht and Institute for Logic, Language and Computation, University of Amsterdam.

Kurtonina, N., and M. Moortgat. 1996. Structural Control. In *Specifying Syntactic Structure*, ed. P. Blackburn and M. de Rijke. Studies in Logic, Language and Information. CSLI Publications.

Kurtonina, N., and M. de Rijke. 1996. Bisimulations for Temporal Logic. Report RR 304. Department of Computer Science, University of Warwick.

Lakatos, I. 1976. *Proofs and Refutations*. Cambridge: Cambridge University Press.

van Lambalgen, M. 1991. Natural Deduction for Generalized Quantifiers. In *Generalized Quantifiers: Theory and Applications*, ed. J. van der Does and J. van Eijck. 143–154. Dutch Ph.D. Network for Logic, Language and Information.

Lambek, J. 1958. The Mathematics of Sentence Structure. *American Mathematical Monthly* 65:154–170.

Lambek, J., and P. Scott. 1986. *Introduction to Higher-Order Categorial Logic.* Cambridge: Cambridge University Press.

Landman, F. 1986. *Towards a Theory of Information.* GRASS series, Vol. 6. Dordrecht: Foris.

Langholm, T. 1988. *Partiality, Truth and Persistence.* Lecture Notes, Vol. 15. Stanford: CSLI Publications.

Lewis, D. 1969. *Convention.* Cambridge (Mass.): Harvard University Press.

Lewis, D. 1972. General Semantics. In *Semantics of Natural Language.* 169–218. Dordrecht: Reidel.

Lewis, D. 1979. Score Keeping in a Language Game. *Journal of Philosophical Logic* 8:339–359.

Lloyd, J. 1985. *Foundations of Logic Programming.* Berlin: Springer Verlag.

Lorenzen, P., and K. Lorenz. 1979. *Dialogische Logik.* Darmstadt: Wissenschaftliche Buchgesellschaft.

MacNamara, J., and G. Reyes (ed.). 1994. *The Logical Foundations of Cognition.* New York: Oxford University Press.

Madarász, J. 1996. Craig Interpolation and Amalgamation. To appear in *Journal of Applied Non-Classical Logic.*

Maddux, R. 1978. *Topics in Relation Algebra.* Doctoral dissertation, Department of Mathematics, University of California, Berkeley.

Maddux, R. 1990. The Origin of Relation Algebras in the Development and Axiomatization of the Calculus of Relations. Technical report. Department of Mathematics, Iowa State University, Ames.

Maddux, R. 1995. Relation-Algebraic Semantics. Technical report. Department of Mathematics, Iowa State University, Ames.

Makinson, D. 1994. General Non-Monotonic Logic. In *Handbook of Logic in Artificial Intelligence and Logic Programming, vol. III,* ed. D.M. Gabbay, C. Hogger, and J. Robinson. 35–110. Oxford: Oxford University Press.

Marx, M. 1995. *Arrow Logic and Relativized Algebras of Relations.* Doctoral dissertation, CCSOM and ILLC, University of Amsterdam.

Marx, M., S. Mikulás, I. Németi, and I. Sain. 1996. Investigations in Arrow Logic. In *Arrow Logic and Multi-Modal Logic.* Studies in Logic, Language and Information. Stanford: CSLI Publications.

Marx, M., S. Mikulás, I. Németi, and A. Simon. 1994. And Now for Something Completely Different: Axiomatization of Relativized Representable Relation Algebras with the Difference Operator. Technical report. Mathematical Institute, Hungarian Academy of Sciences, Budapest.

Marx, M., L. Pólos, and M. Masuch (ed.). 1996. *Arrow Logic and Multi-Modal Logic.* Studies in Logic, Language and Information. Stanford: CSLI Publications.

McCarthy, J. 1980. Circumscription — A Form of Non-Monotonic Reasoning. *Artificial Intelligence* 13:27–39.

McCarthy, J. 1993. Notes on Formalizing Context. In *Proceedings IJCAI-93.*

Mey, D. 1992. Game-Theoretical Interpretation of a Logic Without Contraction. Technical report. Department of Computer Science, Swiss Federal Institute of Technology, Zürich.

Meyer, J.-J. 1988. A Different Approach to Deontic Logic: Deontic Logic Viewed as a Variant of Dynamic Logic. *Notre Dame Journal of Formal Logic* 29(1):109–136.

Meyer Viol, W. 1995. *Instantial Logic. An Investigation of Reasoning with Instances.* Doctoral dissertation, Onderzoeksinstituut voor Taal en Spraak, University of Utrecht and Institute for Logic, Language and Computation, University of Amsterdam.

Mikulás, S. 1995. *Taming Logics.* Doctoral dissertation, Institute for Logic, Language and Computation, University of Amsterdam.

Miller, D. 1989. A Logical Analysis of Modules in Logic Programming. *Journal of Logic Programming* 6:79–108.

Milner, R. 1980. *A Calculus of Communicating Systems.* Berlin: Springer Verlag.

Mints, G. 1990. Several Formal Systems of the Logic Programming. *Computers and Artificial Intelligence* 9:19–41.

Monk, J.D. 1969. Nonfinitizability of Classes of Representable Cylindric Algebras. *Journal of Symbolic Logic* 34:331–343.

Monk, J.D. 1976. *Mathematical Logic.* New York: Springer Verlag.

Monk, J.D. 1993. Lectures on Cylindric Set Algebras. In *Algebraic Methods in Logic and in Computer Science*, ed. C. Rauszer. 253–290. Warsaw: Banach Centre.

Montague, R. 1974. *Formal Philosophy.* New Haven: Yale University Press.

Moore, R. 1984. A Formal Theory of Knowledge and Action. Artificial Intelligence Center, SRI International, Menlo Park.

Moortgat, M. 1996. Type-Logical Grammars. In *Handbook of Logic and Language*, ed. J. van Benthem and A. ter Meulen. Amsterdam: Elsevier Science Publishers.

Morreau, M. 1985. Circumscription: A Sound and Complete Form of Non-Monotonic Reasoning. Technical report. Mathematical Institute, University of Amsterdam.

Morreau, M. 1992. *Conditionals in Philosophy and Artificial Intelligence.* Doctoral dissertation, Institute for Logic, language and Computation, University of Amsterdam.

Morrill, G. 1994. *Type-Logical Grammar.* Dordrecht: Kluwer Academic Publishers.

Moschovakis, Y. 1991. Sense and Reference as Algorithm and Value. Technical report. Department of Mathematics, University of California, Los Angeles.

Moss, L., and J. Seligman. 1996. Situation Theory. In *Handbook of Logic and Language*, ed. J. van Benthem and A. ter Meulen. Amsterdam: Elsevier Science Publishers.

Muskens, R. 1991. Anaphora and the Logic of Change. In *Logics in AI. Proceedings of JELIA'90*, ed. J. van Eijck, 414–430. Berlin. Springer Verlag.

Muskens, R. 1994. Categorial Grammar and Discourse Representation Theory. In *Proceedings COLING 94, Kyoto*, 508–514. Expanded version to appear as: 'Language, Lambdas and Logic'.

Muskens, R. 1995. *Meaning and Partiality.* Studies in Logic, Language and Information. Stanford: CSLI Publications.

Muskens, R., J. van Benthem, and A. Visser. 1996. Dynamics. In *Handbook of Logic and Language*, ed. J. van Benthem and A. ter Meulen. Amsterdam: Elsevier Science Publishers.

Naumann, R. 1995. *Aspectual Composition and Dynamic Logic.* Habilitationsschrift. Düsseldorf: Allgemeine Sprachwissenschaft, Heinrich Heine Universität.

Németi, I. 1985. The Equational Theory of Cylindric Relativized Set Algebras is Decidable. Preprint No 63/85. Mathematical Institute, Hungarian Academy of Sciences, Budapest.

Németi, I. 1986. *Free Algebras and Decidability in Algebraic Logic.* Third Doctoral Dissertation. Mathematical Institute, Hungarian Academy of Sciences, Budapest.

Németi, I. 1990. On Cylindric Algebraic Model Theory. In *Algebraic Logic and Universal Algebra in Computer Science (Proceedings Conference Ames 1988)*, ed. C. Bergman, R. Maddux, and D. Pigozzi, LNCS, Vol. 425, 37–76. Berlin. Springer Verlag.

Németi, I. 1991. Algebraizations of Quantifier Logics: An Introductory Overview. *Studia Logica* 50:485–569. See also Németi 1994.

Németi, I. 1994. Algebraizations of Quantifier Logics. In *Proceedings Summer School on Algebraic Logic, Budapest 1994.*

Németi, I. 1995. Decidability of Weakened Versions of First-Order Logic. In *Logic Colloquium '92*, ed. L. Csirmaz, D.M. Gabbay, and M. de Rijke, 177–241. Studies in Logic, Language and Information. Stanford: CSLI Publications.

Németi, I. 1996. Fine-Structure Analysis of First-Order Logic. To appear in *Arrow Logic and Multi-Modal Logic.* eds. Marx, et al. Studies in Logic, Language and Information. Stanford: CSLI Publications.

Ng, K. 1984. *Relation Algebras with Transitive Closure.* Doctoral dissertation, University of California, Berkeley.

Oehrle, R., E. Bach, and D. Wheeler (ed.). 1988. *Categorial Grammars and Natural Language Structures.* Dordrecht: Reidel.

Ohlbach, H.-J. 1991. Semantic-Based Translation Methods for Modal Logics. *Journal of Logic and Computation* 1(5):691–746.

Ohlbach, H.-J., R. Schmidt, and U. Hustadt. 1995. Translating Graded Modalities into Predicate Logic. In *Proof Theory of Modal Logic*, ed. H. Wansing. 245–285. Dordrecht: Kluwer Academic Publishers.

Ono, H. 1987. Some Problems in Intermediate Predicate Logics. *Reports on Mathematical Logic* 21:55–67.

Orlowska, E. 1991. Relational Interpretation of Modal Logics. In *Algebraic Logic*, ed. H. Andréka, J. Monk, and I. Németi. 443–471. Colloq. Math. Soc. J. Bolyai. Amsterdam: North-Holland.

Park, D. 1981. Concurrency and Automata on Infinite Sequences. In *Proceedings 5th GI Conference*, 167–183. Berlin: Springer Verlag.

Pearl, J. 1996. Causation, Action and Counterfactuals. In *Rationality and Knowledge*, ed. Y. Shoham. 51–73. San Francisco: Morgan Kaufmann Publishers.

Peleg, D. 1987. Concurrent Dynamic Logic. *Journal of the ACM* 34:450–479.

Perry, J. 1993. *The Problem of the Essential Indexical, and Other Essays.* New York: Oxford University Press.

Pigozzi, D. 1971. Amalgamation, Congruence Extension, and Interpolation Properties in Algebras. *Algebra Universalis* 1:269–439.

Plotkin, G. 1980. Lambda Definability in the Full Type Hierarchy. In *Introduction to Combinators and Lambda Calculus*, ed. J. Seldin and J. Hindley. 363–373. Cambridge: Cambridge University Press.

Polanyi, L., and R. Scha. 1984. A Syntactic Approach to Discourse Semantics. In *Proceedings COLING 1984, Stanford*, 413–419.

Ponse, A. 1994. Process Algebra and Dynamic Logic. In *Logic and Information Flow*, ed. J. van Eijck and A. Visser. 125–145. Cambridge (Mass.): The MIT Press.

Ponse, A., M. de Rijke, and Y. Venema (ed.). 1995. *Modal Logic and Process Algebra.* Lecture Notes, Vol. 53. Stanford: CSLI Publications.

Pratt, V. 1976. Semantical Considerations on Floyd-Hoare Logic. In *Proceedings 18th IEEE Symposium on the Foundations of Computer Science*, 326–337.

Pratt, V. 1994a. A Road Map of Some Two-Dimensional Logics. In *Logic and Information Flow*, ed. J. van Eijck and A. Visser. 149–162. Cambridge (Mass.): The MIT Press.

Pratt, V. 1994b. The Second Calculus of Relations. Technical report. Department of Computer Science, Stanford University.

Pratt, V. *Dual Interaction: Computation, Mathematics and Mind.* Forthcoming. Department of Computer Science, Stanford University.

Prüst, H., M. van den Berg, and R. Scha. 1994. A Formal Discourse Grammar and Verb Phrase Anaphora. *Linguistics and Philosophy* 17:261–327.

Ramanujan, R. 1996. Local Knowledge Assertions in a Changing World. In *Rationality and Knowledge*, ed. Y. Shoham. 1–14. San Francisco: Morgan Kaufmann Publishers.

Rantala, V. 1978. Urn Models: A New Kind of Non-Standard Model for First-Order Logic. In *Game-Theoretical Semantics*, ed. E. Saarinen. Dordrecht: Kluwer Academic Publishers.

Reiter, R. 1994. Cognitive Robotics. Lecture IJCAI, Chambéry. Cf. 'GOLOG: A Logic Programming Language,' `http://www.cs.toronto.edu/~cogrobo/`.

Renardel, G. 1996. A Variant of QDL. To appear in *Proceedings of the 5th CSLI Workshop on Logic, Language and Computation*, eds. C. Condoravdi and G. Renardel. Stanford: CSLI Publications.

Resek, D. and R. Thompson. 1991. Characterizing Relativized Cylindric Algebras. In *Algebraic Logic*, ed. H. Andréka, J. Monk, and I. Németi. 519–538. Colloq. Math. Soc. J. Bolyai 54. Amsterdam: North Holland.

de Rijke, M. 1992. A System of Dynamic Modal Logic. Technical Report # CSLI-92-170. Stanford University. To appear in *Journal of Philosophical Logic*.

de Rijke, M. 1993a. Correspondence Theory for Extended Modal Logic. Technical Report ML-93-16. Institute for Logic, Language and Computation, University of Amsterdam. To appear in *Mathematical Logic Quarterly*.

de Rijke, M. 1993b. *Extending Modal Logic*. Doctoral dissertation, Institute for Logic, Language and Computation, University of Amsterdam.

de Rijke, M. 1994. Meeting Some Neighbours. In *Logic and Information Flow*, ed. J. van Eijck and A. Visser. 170–195. Cambridge (Mass.): The MIT Press.

de Rijke, M. 1995. Modal Model Theory. Technical Report CS-R9517. CWI, Amsterdam. To appear in *Annals of Pure and Applied Logic*.

Rodenburg, P. 1986. *Intuitionistic Correspondence Theory*. Doctoral dissertation, Mathematical Institute, University of Amsterdam.

Rogers, J. 1996. *A Descriptive Approach to Language-Theoretic Complexity*. Studies in Logic, Language and Information. Stanford: CSLI Publications. To appear.

Roorda, D. 1991. *Resource Logics. A Proof-Theoretic Investigation*. Doctoral dissertation, Institute for Logic, Language and Computation, University of Amsterdam.

Rosen, E. 1995. Modal Logic over Finite Structures. Technical Report ML-95-08. Institute for Logic, Language and Computation, University of Amsterdam.

Rosenschein, S., and L. Pack Kaelbling. 1986. The Synthesis of Digital Machines with Provable Epistemic Properties. In *Proceedings TARK-86*, 83–98. Los Altos.

Rott, H. 1996. *Making Up One's Mind*. Habilitationsschrift. Philosophical Institute, University of Konstanz.

Rounds, W. 1996. Feature Logics. In *Handbook of Logic and Language*, ed. J. van Benthem and A. ter Meulen. Amsterdam: Elsevier Science Publishers.

Ryan, M. 1992. *Ordered Presentations of Theories*. Doctoral dissertation, Department of Computing, Imperial College.

Ryan, M., P. Y. Schobbens, and O. Rodrigues. 1996. Counterfactuals and Updates as Inverse Modalities. In *Rationality and Knowledge*, ed. Y. Shoham. 163–173. San Francisco: Morgan Kaufmann Publishers.

Rybakov, V. 1985. A Criterion for Admissibility of Rules of Inference in Modal and Intuitionistic Logic. *Soviet Mathematical Dokladi* 32(2):452–455.

Sagi, G., and I. Németi. 1996. On the Modal Logic of Substitution. Completeness Results. Mathematical Institute, Hungarian Academy of Sciences, Budapest.

Sain, I. 1990. Beth's and Craig's Properties Via Epimorphisms and Amalgamation in Algebraic Logic. In *Algebraic Logic and Universal Algebra in Computer Science*, ed. C. Bergman, R. Maddux, and D. Pigozzi, LNCS, Vol. 425, 209–226. Berlin. Springer Verlag.

Sain, I., and R. Thompson. 1990. Strictly Finite Schema Axiomatization of Quasi-Polyadic Algebras. In *Algebraic Logic*. 539–572. Amsterdam: North-Holland.

Salwicki, A. 1970. Formalised Algorithmic Languages. *Bulletin Polish Academy of Sciences* 18:227–232.

Sanchez Valencia, V. 1991. *Studies on Natural Logic and Categorial Grammar.* Doctoral dissertation, Institute for Logic, Language and Computation, University of Amsterdam.

Sandewal, E. 1994. *Features and Fluents. The Representation of Knowledge about Dynamical Systems. Volume I.* Oxford: Oxford University Press.

van der Sandt, R. 1988. *Context and Presupposition.* London: Croom Helm.

Saraswat, V. 1993. *Concurrent Constraint Programming.* Cambridge: MIT Press.

Scott, D. 1982. Domains for Denotational Semantics. In *Proceedings 9th International Colloquium on Automata, Languages and Programming*, ed. M. Nielsen and E. Schmidt, LNCS, Vol. 40, 577–613.

Segerberg, K. 1995. Hypertheories in Doxastic Logic. In *Proceedings 10th Amsterdam Colloquium*, ed. P. Dekker and M. Stokhof.

Seuren, P. 1985. *Discourse Semantics.* Oxford: Blackwell.

Sher, G. 1991. *The Bounds of Logic.* Cambridge (Mass.): The MIT Press.

Shoham, Y. 1988. *Reasoning about Change: Time and Causality from the Standpoint of Artificial Intelligence.* Cambridge (Mass.): The MIT Press.

Shoham, Y. (ed.). 1996. *Rationality and Knowledge.* Proceedings of TARK 5, Renesse (Zeeland). San Francisco: Morgan Kaufmann Publishers.

Simon, A., A. Kurucz, I. Németi, and I. Sain. 1993. Undecidability Issues of Some Boolean Algebras with Operators and Logics Related to Lambek Calculus. Workshop on Algebraization of Logic, Fifth European Summer School in Logic, Language and Information, Lisbon. Appeared as Undecidable Varieties of Semi-lattice-Ordered Semigroups, of Boolean Algebras with Operators, and Logics Extending Lambek Calculus. In *Bulletin of the IGPL* 1:1, 1993. 91–98.

Spaan, E. 1993. *Complexity of Modal Logics.* Doctoral dissertation, Institute for Logic, Language and Computation, University of Amsterdam.

Spohn, W. 1988. Ordinal Conditional Functions: A Dynamic Theory of Epistemic States. In *Causation in Decision, Belief Change and Statistics II*, ed. W.L. Harper et al. 105–134. Dordrecht: Kluwer Academic Publishers.

Stalnaker, R. 1972. Pragmatics. In *Semantics of Natural Language*, ed. D. Davidson and G. Harman. 380–397. Dordrecht: Reidel.

Stärk, R. 1994. The Declarative Semantics of the Prolog Selection Rule. In *9th Annual Symposium on Logic in Computer Science*, 252–261. Paris.

Stärk, R. 1995. A Transformation of Propositional Prolog Programs into Classical Logic. Technical report. Department of Computer Science, Stanford University.

Stirling, C. 1989. Modal and Temporal Logics. In *Handboook of Logic in Computer Science*, ed. S. Abramsky, D.M. Gabbay, and T. Maibaum. Oxford University Press.

Sturm, H. 1996. *Modale Fragmente von $L_{\omega\omega}$ und $L_{\omega_1\omega}$.* Doctoral dissertation. Department of Philosophy, Ludwig Maximilians Universität, München.

Sundholm, G. 1986. Proof Theory and Meaning. In *Handbook of Philosophical Logic. Vol. 3*, ed. D.M. Gabbay and F. Guenthner. 471–506. Dordrecht: Reidel.

Suppe, F. (ed.). 1977. *The Structure of Scientific Theories.* Urbana: University of Illinois Press.

Tarski, A. 1986. What Are Logical Notions? In *History and Philosophy of Logic.* 143–154.

Tarski, A., and S. Givant. 1987. *Formalization of Set Theory Without Variables.* Colloquium Publications, Vol. 41. Providence, Rhode Island: AMS.

Thijsse, E. 1992. *Partial Logic and Knowledge Representation.* Doctoral dissertation, Institute for Language and Knowledge Technology, University of Tilburg.

Thom, R. 1973. Langage et Catastrophes: Élements pour une Sémantique Topologique. In *Proceedings Conference on Dynamical Systems.* New York. Academic Press.

Thomason, R. 1996. Nonmonotonicity in Linguistics. In *Handbook of Logic and Language*, ed. J. van Benthem and A. ter Meulen. Amsterdam: Elsevier Science Publisher.

Thompson, R. 1981. *Transformational Structure of Algebraic Logics.* Doctoral dissertation, Department of Mathematics, University of California, Berkeley.

Thompson, R. 1990. Noncommutative Cylindric Algebras and Relativizations of Cylindric Algebras. In *Algebraic Logic and Universal Algebra in Computer Science.* LNCS, Vol. 425, 273–278. Berlin: Springer Verlag.

Toulmin, S. 1958. *The Uses of Argument.* Cambridge: Cambridge University Press.

Treur, J., and J. Engelfriet. 1994. Temporal Theories of Reasoning. In *Proceedings of JELIA 1994*, ed. C. MacNish, D. Pearce, and L. Pereira, 279–299. LNAI. Berlin. Springer Verlag.

Troelstra, A., and D. van Dalen. 1988. *Constructivism in Mathematics.* Amsterdam: North-Holland.

Troelstra, A., and H. Schwichtenberg. 1996. *Basic Proof Theory.* Cambridge: Cambridge University Press.

Turner, R. 1989. Two Issues in the Foundations of Semantics. In *Properties, Types and Meaning*, ed. G. Chierchia, B. Partee, and R. Turner. 63–84. Dordrecht: Reidel.

Turner, R. 1996. Types. A chapter in *Handbook of Logic and Language*, eds. A. ter Meulen and J. van Benthem. Amsterdam: to appear.

Vakarelov, D. 1996. Modal Arrow Logic. In *Advances in Intensional Logic*, ed. M. de Rijke. Dordrecht: Kluwer Academic Publishers.

Veltman, F. 1984. Data Semantics. In *Truth, Interpretation and Information*, ed. J. Groenendijk et al. 43–64. Dordrecht: Foris.

Veltman, F. 1985. *Logics for Conditionals.* Doctoral dissertation, Department of Philosophy, University of Amsterdam.

Veltman, F. 1991. Defaults in Update Semantics. Technical Report LP-91-02. ILLC, University of Amsterdam. To appear in *Journal of Philosophical Logic.*

Veltman, F. 1992. Three Complete Might Logics. Manuscript, Institute for Logic, Language and Computation, University of Amsterdam.

Venema, Y. 1991. *Many-Dimensional Modal Logic.* Doctoral dissertation, Department of Mathematics and Computer Science, University of Amsterdam.

Venema, Y. 1994. A Crash Course in Arrow Logic. Technical Report 107. Logic Group Preprint Series, Utrecht University. To appear in *Arrow Logic and Multi-Modal Logic*, M. Marx, M. Masuch, and L. Polós, eds.

Venema, Y. 1995a. Cylindric Modal Logic. *Journal of Symbolic Logic* 60:591–623.

Venema, Y. 1995b. A Modal Logic of Substitution and Quantification. In *Logic Colloquium '92*, ed. L. Csirmaz, D.M. Gabbay, and M. de Rijke, 293–309. CSLI Publications.

Venema, Y., and M. Marx. 1996a. *Many-Dimensional Modal Logic and Arrow Logic.* Oxford: Oxford University Press. To appear.

Venema, Y., and M. Marx. 1996b. A Modal Logic of Relations. In *Memorial Volume for Elena Rasiowa*, ed. E. Orlowska. Studia Logica Library. Dordrecht: Kluwer Academic Publishers.

Vermeulen, C. 1994. *Exploring the Dynamic Environment.* Doctoral dissertation, Philosophical Institute, Rijksuniversiteit Utrecht.

Vermeulen, C. 1995. Merging Without Mystery, or: Variables in Dynamic Semantics. *Journal of Philosophical Logic* 24:405–450.

Versmissen, J. 1996. *Grammatical Composition: Modes, Models, Modalities.* Doctoral dissertation, Onderzoeksinstituut voor Taal en Spraak, Universiteit Utrecht.

Visser, A. 1994a. Actions Under Presuppositions. In *Logic and Information Flow*, ed. J. van Eijck and A. Visser. 196–233. Cambridge (Mass.): The MIT Press.

Visser, A. 1994b. Meanings in Time. Manuscript, Philosophical Institute, Rijksuniversiteit Utrecht.

Visser, A. 1995. Relational Validity and Dynamic Predicate Logic. Logic Group Preprint Series 144. Philosophical Institute, Rijksuniversiteit Utrecht.

Visser, A., and K. Vermeulen. 1995. Dynamic Bracketing and Discourse Representation. Logic Group Preprint Series 131. Department of Philosophy, Rijksuniversiteit Utrecht.

Wansing, H. 1993. *The Logic of Information Structures.* LNAI, Vol. 681. Berlin: Springer Verlag.

Wansing, H. 1996. Predicate Logics on Display. Institute of Logic and Philosophy of Science, University of Leipzig.

Westerståhl, D. 1984. Determiners and Context Sets. In *Generalized Quantifiers in Natural Language*, ed. J. van Benthem and A. ter Meulen. 45–71. Dordrecht: Foris.

Westerståhl, D. 1989. Quantifiers in Formal and Natural Languages. In *Handbook of Philosophical Logic. Vol. 4*, ed. D.M. Gabbay and F. Guenthner. 1–131. Dordrecht: Reidel.

Westerståhl, D. 1995. On Some More or Less Modal Sublogics of Predicate Logic. Philosophical Institute, University of Stockholm.

Whitehead, A. 1929. *Process and Reality.* Cambridge: Cambridge University Press.

Zeevat, H. 1991. A Compositional Approach to DRT. *Linguistics and Philosophy* 12:95–131.

Zeinstra, L. 1990. Reasoning as Discourse. Master's thesis, Department of Philosophy, Rijksuniversiteit Utrecht.

Index